D1191782

Linear Algebra

Modular Mathematics Series

Linear Algebra

R B J T Allenby

School of Mathematics,
University of Leeds

Edward Arnold
A member of the Hodder Headline Group
LONDON SYDNEY AUCKLAND

First published in Great Britain 1995 by
Edward Arnold, a division of Hodder Headline PLC,
338 Euston Road, London NW1 3BH

British Library Cataloguing in Publication Data
A catalogue record for this book is available from the British Library

ISBN 0 340 61044 1

1 2 3 4 5 95 96 97 98 99

Typeset in 10/13 Times by
Phoenix Photosetting, Lordswood, Chatham, Kent
Printed and bound in Great Britain by
J W Arrowsmith Ltd, Bristol

In memory of my parents

R T A 1914–1990
N M A 1915–1991

Contents

Series Preface

This series is designed particularly, but not exclusively, for students reading degree programmes based on semester-long modules. Each text will cover the essential core of an area of mathematics and lay the foundation for further study in that area. Some texts may include more material than can be comfortably covered in a single module, the intention there being that the topics to be studied can be selected to meet the needs of the student. Historical contexts, real life situations, and linkages with other areas of mathematics and more advanced topics are included. Traditional worked examples and exercises are augmented by more open-ended exercises and tutorial problems suitable for group work or self-study. Where appropriate, the use of computer packages is encouraged. The first level texts assume only the A-level core curriculum.

Professor Chris D. Collinson
Dr Johnston Anderson
Mr Peter Holmes

Preface

This book is, somewhat loosely, based on a (module length) series of lectures which have been given, over the years, to the first year honours mathematics students in the University of Leeds. The 'looseness' derives from the fact that I have included here rather more material than I could have got through in the 24 lectures at my disposal. In particular I had no time for the applications nor the historical and biographical details I have included here – nor, in fact, for all parts of the main text. This is chiefly because I have expanded and rearranged my original notes substantially to take account of the fact that, as distinct from the lecturer, the writer is not available to answer questions.

The point of including the applications is to inform those genuinely interested (and, perhaps more importantly, those who, being less readily convinced, ask – or *should* ask – 'O.K., but what's all this stuff *good* for?') of why there is merit in studying linear algebra. In fact the biographies also help in this regard since they sometimes indicate from what problem(s) an idea or whole theme emerged.

Another item which found no place in my lectures is the collection of computer package problems. I had some fun making them up, as you might tell from, for example, Problems 1 and 3 of Chapter 8 and Problem 2 (and not forgetting the hand calculator problem) of Chapter 5. Most are pretty trivial in content (after all *I* had to be able to do them!) and my main hope is that the reader will invent some more interesting (or outrageous) ones for him or herself. Incidentally I used the Maple package – mainly because it is on Leeds' computer system – and, I'm told, likely to be available to widespread audiences for some time.

A word about the chapter contents. The first five chapters are fairly computational in nature and, because they deal with ideas (systems of simultaneous linear equations, matrices, their inverses and their determinants) which will be familiar to a good many readers, they should be reasonably straightforward. So, to show that one can have slightly deeper thoughts even about straightforward material – and, in particular, to encourage *YOU* to have such thoughts – the text includes what I hope are thought provoking Tutorial Problems which those readers in higher education can discuss . . . with their tutors! Such problems persist throughout the book, there being some 28 in total.

Chapter 6 marks a slight change of emphasis. It recognises that there are many examples occuring in mathematics whose underlying 'arithmetic' is similar to that of matrices and of 2- and 3-dimensional vectors. Accordingly, to get a common view of them all, we show how we can present this information abstractly, dealing less with *specific* sets of objects (matrices, vectors etc.) and more with sets of 'things', under the general concept of *vector space*. As this change is aided by 13 supporting 'Examples' (and only *one* theorem!) the transition shouldn't be too painful! Another reason for introducing a little bit of abstraction at this point is that the student is bound to have to meet some of it at some time in his or her undergraduate career and, with so many 'almost identical' examples to hand, it seems like too good an opportunity to miss.

Chapters 7, 8 and 9 then develop, with the support of many examples, further ways of organising and categorising elements of vector spaces in a coherent way. Chapter 10 discusses the fundamental mathematical concept of 'function' as it arises in a vector space setting. The special functions we consider, namely *linear transformations*, can be represented naturally by matrices. Chapter 11 (which is quite computational) then investigates how one might simplify these matrices for the purpose of aiding calculation.

At a quick glance the book might appear to move slowly from being very computational to being 'more words than numbers'. I personally think it is more a case of moving from the familiar to the newer. Nevertheless, even the more 'wordy bits' are backed up by numerous examples. Furthermore, apart from the almost 350 exercises to be found at the ends of chapters, there are over 60 'Problems' given as the subject is developed, the purpose of each being to reinforce what has just been described in the text.

Finally it is my great pleasure to thank a number of people who have helped me to produce this book. These include Dr Jeremy Gray of the Open University for his assistance, willingly given, relating to the biographies, Professor Dr Konrad Jacobs who (as with my other books) has generously supplied the photographs and my colleague Dr Eric Wallace who learnt Maple to test out my computer package problems (and sometimes found me wanting!)

Last, but not least, I should like to thank Dr Johnston Anderson both for inviting me to write this book and for reminding me that explanations which can readily be given *verbally* in lectures do need to be *written* (and sometimes at greater length) in a book. In addition, his suggestions concerning the elimination of a number of mathematical 'jokes' were, in hindsight, correct. Needless to say he takes no responsibility for the quality of those remaining.

Finally, in these days when we are required to be assessed on every aspect of our lives, I really do implore any reader (especially of the undergraduate kind) who has other than normal feelings of pleasure or annoyance in what he or she finds (or doesn't find) in this book to let me know of these feelings so that I may write a better book next time round. And if any reader should know of any really good mathematical joke(s) especially ones concerning linear algebra. . . .

RBJTA
Leeds 1994

1 • Systems of Simultaneous Linear Equations

Solutions of systems of simultaneous linear equations (also called linear systems) arise in very many real life situations; for just a couple of instances see the Applications at the end of the chapter. Here we show how to solve such systems (much as a computer would) by simplifying them in a systematic way. We shall also see how to interpret the results we obtain geometrically.

In 1849 the French mathematician Joseph Alfred Serret (30 August 1819–2 March 1885) wrote: 'Algebra is, properly speaking, the analysis of equations.' (This is no longer an accurate statement – although the motivating factor behind the theory of groups was the investigation of the solutions of polynomial equations.) Accordingly, *linear* algebra should (amongst other things) involve the study of *linear* equations – that is, equations involving the 'unknowns' (or 'indeterminates' or 'variables') $x, y, z, \ldots$ from which terms such as $\sqrt{x}$, xy, xy^3z^2, e^x, $\sin x$, $1/x$, $\log x$, etc. which are not of degree 1 in the variables, are excluded. The prefix 'linear' derives from the fact that such equations can, at least in the case of two and three unknowns, be represented geometrically by straight lines and planes in 2- and 3-dimensional space. In particular, linear algebra is much concerned with finding the solutions to a given system of simultaneous linear equations. We will begin by looking at some simple examples.

Example 1

Solve

$$2x+5y=3 \quad (1.1)$$
$$3x-2y=14. \quad (1.2)$$

Recall that, to 'solve' the pair of equations (1.1) and (1.2) simultaneously, we must find all possible (pairs of) values of x and y which make both (1.1) and (1.2) true at the same time.

To obtain the full solution to Example 1 we may eliminate x, say, from equation (1.2). We can do this by taking twice equation (1.2) from three times equation (1.1) – in brief $3\times(1.1)-2\times(1.2)$ – which gives

$$6x+15y-(6x-4y)=9-28,$$

that is,

$$0x+19y=-19. \quad (1.3)$$

Thus equations (1.1) and (1.2) are replaced by the pair

$$2x+5y=3 \tag{1.1}$$
$$0x+19y=-19. \tag{1.3}$$

Equation (1.3) tells us at once that $y=-1$ and then (1.1) can be used to deduce that $x=4$.

As an example (with somewhat more adventurous coefficients!) let us try the following.

Example 2

Solve

$$6.8x+10.2y=2.72 \tag{1.4}$$
$$7.8x+11.7y=3.11. \tag{1.5}$$

Proceeding as in Example 1, we obtain a new third equation by forming $7.8\times(1.4)-6.8\times(1.5)$.

This gives (!)

$$6.8x+10.2y=2.72 \tag{1.4}$$
$$0x+0y=0.068. \tag{1.6}$$

Since there are no numbers x and y satisfying (1.6) we infer that the given pair of equations can have no (simultaneous) solution.

As a final example consider the (almost identical) equations in the following.

Example 3

Solve

$$6.8x+10.2y=2.72 \tag{1.4}$$
$$7.8x+11.7y=3.12. \tag{1.7}$$

This time $7.8\times(1.4)-6.8\times(1.7)$ leads to

$$6.8x+10.2y=2.72 \tag{1.4}$$
$$0x+0y=0. \tag{1.8}$$

Here equation (1.8) imposes no restrictions whatsoever on x and y and so it can be ignored. Indeed the pair of equations (1.4) and (1.7) is seen to be equivalent to the single equation (1.4). If we let y take any (real) number value α, say, (1.4) shows that x must take the value $(2.72-10.2\alpha)/6.8=0.4-1.5\alpha$.

Thus, even in the very simple case of a pair of (simultaneous) equations in just two unknowns, we see that there are (at least) three outcomes: in Example 1 the equations led to a unique solution; in Example 2 they led to no solution; in Example 3 they led to infinitely many solutions, one for each real number α.

We can see why there are exactly three possible outcomes if we study Examples 1, 2 and 3 geometrically. In Figs 1.1, 1.2 and 1.3 we draw the pairs of lines represented by the pairs of equations in Examples 1, 2 and 3 respectively. In Fig. 1.1 the lines meet in a single point whose coordinates give the unique solution of Example 1. In Fig. 1.2 the lines are parallel; their having no point in common corresponds to the given equations

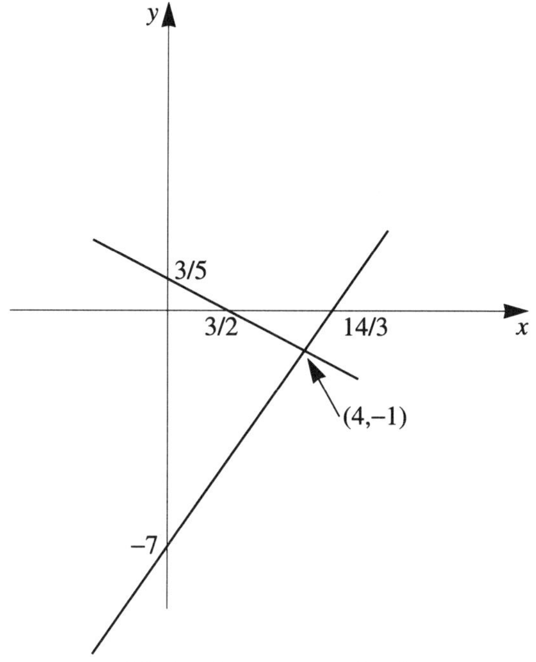

Fig 1.1 Lines meet at (4,–1)

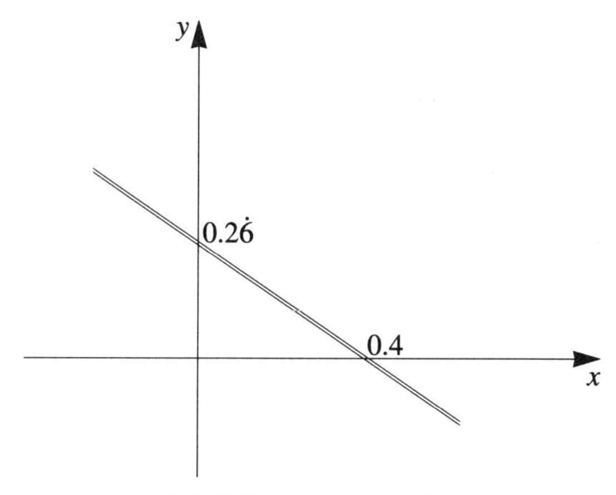

Fig 1.2 Lines are parallel

having no (common) solution. In Fig. 1.3 the two lines coincide along their entire lengths. Thus every point on this line gives rise to a solution of the given pair of equations.

The same kind of analysis can be performed on systems of three equations in three unknowns.

PROBLEM 1.1

Solve each of the following systems of equations by using the first equation to eliminate x from the second and third equations and then the 'new' second equation to eliminate y from the 'new' third equation.

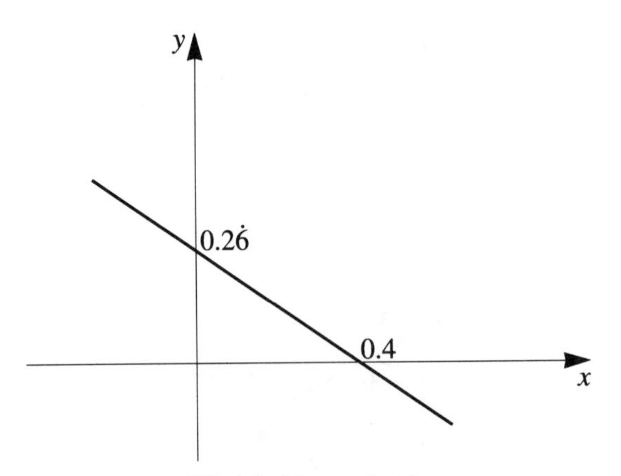

Fig 1.3 Lines coincide

(a) $x+y+z=2$, $x+2y+3z=5$, $4x+7y+56z=28$

(b) $x+y+z=2$, $x+2y+3z=5$, $3x+y-z=2$

(c) $x+y+z=2$, $x+2y+3z=5$, $2x+y=1$.

You should find: for (a) $x=-\frac{35}{46}$, $y=\frac{58}{23}$, $z=\frac{11}{46}$; for (b) there is no solution; for (c) that $x=-1+\alpha$, $y=3-2\alpha$, $z=\alpha$ is a solution no matter what value is given to α. Thus again we see that we may obtain, for a given system of simultaneous equations, (a) a unique solution, (b) no solution and (c) infinitely many solutions. Corresponding geometrical interpretations are illustrated in Figs 1.4–1.6[a].

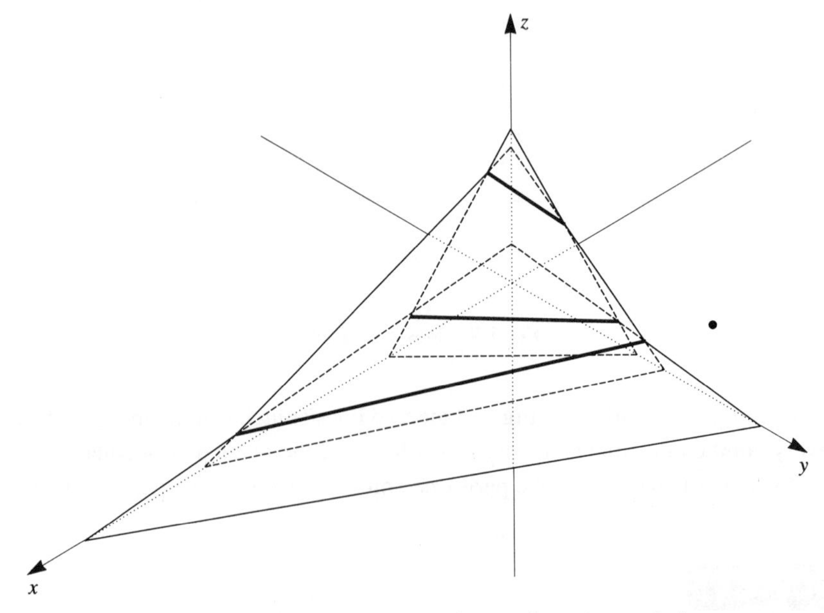

Fig 1.4 Bold lines meet at point shown: the unique solution

[a]We leave the drawing of other possible pictures to you.

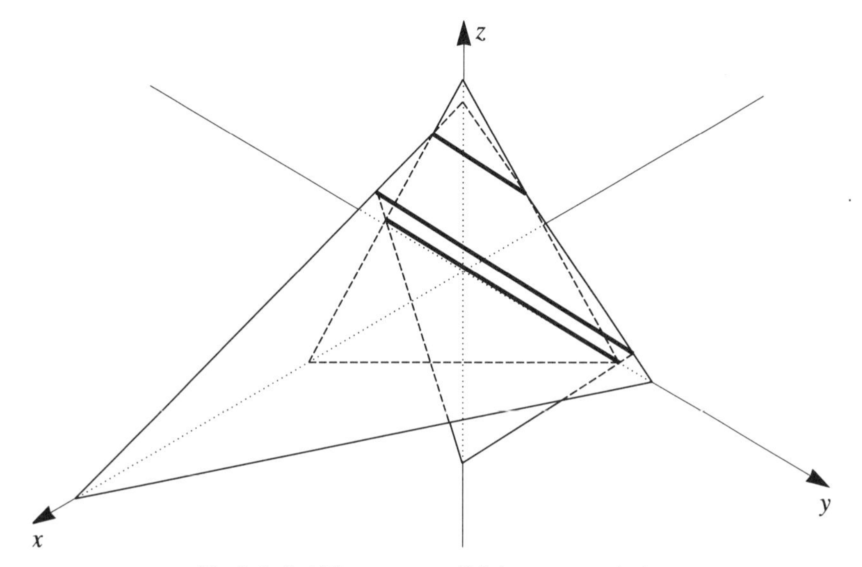

Fig 1.5 Bold lines are parallel: hence no solution

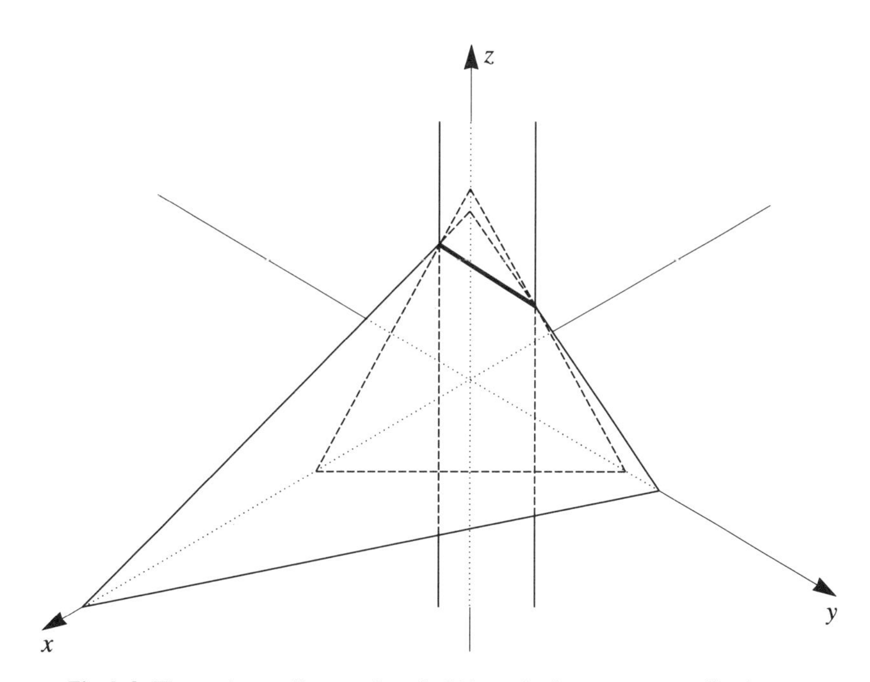

Fig 1.6 Three planes all meet along bold line: the line comprises all solutions

Of course, the obvious question arises. Are these three outcomes the only possibilities for systems of three equations in three unknowns? What about four equations in four unknowns? We shall answer these and similar questions later. But, first, some even more general examples in which the number of equations may be different from the number of unknowns.

Example 4

Solve (i.e. find all solutions – if any – of) the linear system

$$\left.\begin{aligned} x-3y+2z&=-4 &\quad(1.9)\\ 2x-5y+z&=-1 &\quad(1.10)\\ 5x-14y+8z&=-13 &\quad(1.11)\\ 5x-10y-2z&=8. &\quad(1.12)\end{aligned}\right\}\text{(E1)}$$

Here we are asking that the *three* unknowns x, y, z are subject to *four* constraints – one imposed by each of the four equations. Consequently our intuition may tell us that equations (E1) should have no solution. Let us see if such intuition is correct.

As in our earlier examples we start by using the first equation to eliminate the unknown x from all the remaining equations. We obtain

$$\left.\begin{aligned} x-3y+2z&=-4 &\quad(1.9)\\ y-3z&=7 &\quad(1.13)\\ y-2z&=7 &\quad(1.14)\\ 5y-12z&=28 &\quad(1.15)\end{aligned}\right\}\text{(E2)}$$

where $(1.13)=(1.10)-2\times(1.9)$, $(1.14)=(1.11)-5\times(1.9)$ and $(1.15)=(1.12)-5\times(1.9)$.

Noting that equations (1.13)–(1.15) form a system of three equations in two unknowns, we now use equation (1.13) to eliminate y from equations (1.14) and (1.15) giving

$$\left.\begin{aligned} x-3y+2z&=-4 &\quad(1.9)\\ y-3z&=7 &\quad(1.13)\\ z&=0 &\quad(1.16)\\ 3z&=-7 &\quad(1.17)\end{aligned}\right\}\text{(E3)}$$

where $(1.16)=(1.14)-1\times(1.13)$ and $(1.17)=(1.15)-5\times(1.13)$.

Now it is obvious that the system of equations (E3) has no (simultaneous) solution. For no triple of numbers for x, y, z can be found which can satisfy just the final two equations in (E3), never mind all four equations simultaneously! Thus our intuition concerning the lack of solution to the given system of equations (E1) is correct. Because it has no solution we say that the system of equations (E1) is **inconsistent**.

Whilst intuition is often a valuable commodity, its indiscriminate use (especially in mathematics) should be avoided. The following example shows our intuition in Example 4 was a bit premature.

Example 5

Solve, if possible, the linear system

$$\left.\begin{aligned} x-3y+2z&=-4\\ 2x-5y+z&=-1\\ 5x-14y+8z&=-13\\ 5x-11y-2z&=8\end{aligned}\right\}\qquad\text{(F1)}$$

[Note that the system (F1) is identical to the system (E1) except in one coefficient.]

Applying the same elimination technique to (F1) as we did to (E1) we obtain

$$\left.\begin{aligned} x-3y+2z&=-4\\ y-3z&=7\\ y-2z&=7\\ 4y-12z&=28 \end{aligned}\right\} \qquad \text{(F2)}$$

$$\left.\begin{aligned} x-3y+2z&=-4 && (1.9)\\ y-3z&=7 && (1.13)\\ z&=0 && (1.16)\\ (0z&=0). && (1.18) \end{aligned}\right\} \text{(F3)}$$

We can now solve (F3) by first noting that (1.18) imposes no restriction on the value of z. Next (1.16) tells us that z has to be equal to 0. Then, putting this information into equation (1.13), we find that y must be equal to 7. Finally, substituting $y=7$ and $z=0$ in equation (1.9), we see that x must be equal to 17. (This method of procedure is called **back substitution**.)

We therefore conclude that (F1) has a unique solution. Because there is a solution in this case, we say that the linear system (F1) is **consistent**.

Before tackling another example let us quickly investigate the linear system

$$\begin{aligned} u-3v+2w&=-4\\ 2u-5v+w&=-1\\ 5u-14v+8w&=-13\\ 5u-11v-2w&=8. \end{aligned}$$

From Example 5 it is immediate that this linear system has (unique) solution $u=17$, $v=7$, $w=0$. This indicates that the names we give to the unknowns are essentially irrelevant: all that are important are the **coefficients** 1, −3, 2; 2, −5, 1; 5, −14, 8; 5, −11, −2 on the left-hand sides of the equations and the **constant terms** −4, −1, −13, 8 on the right-hand sides.

Example 6

Consequently, to solve

$$\left.\begin{aligned} x-3y+2z&=-4\\ 2x-5y+z&=-1\\ 5x-14y+7z&=-13\\ 5x-11y-2z&=8 \end{aligned}\right\} \qquad \text{(G1)}$$

we only need write, successively,

$$\begin{array}{rrr|r} 1 & -3 & 2 & -4\\ 2 & -5 & 1 & -1\\ 5 & -14 & 7 & -13\\ 5 & -11 & -2 & 8 \end{array} \rightarrow \begin{array}{rrr|r} 1 & -3 & 2 & -4\\ 0 & 1 & -3 & 7\\ 0 & 1 & -3 & 7\\ 0 & 4 & -12 & 28 \end{array} \rightarrow \begin{array}{rrr|r} 1 & -3 & 2 & -4\\ 0 & 1 & -3 & 7\\ 0 & 0 & 0 & 0\\ 0 & 0 & 0 & 0 \end{array},$$

the vertical bar separating the coefficients from the constant terms being just for convenience. (We leave you to check that each of these tables has been correctly calculated from its predecessor.)

The final display of numbers shows that to solve (G1) we only have to solve the (simpler) linear system

$$\left.\begin{aligned} x-3y+2z&=-4\\ y-3z&=7.\end{aligned}\right\} \qquad \text{(G2)}$$

As in Example 3 and Problem 1.1(c) there is insufficient information to determine the values of x, y and z uniquely. Indeed, if we choose z quite arbitrarily, $z=\alpha$, say, then the second equation of (G2) tells us that $y=7+3\alpha$ and the first equation of (G2) reveals that

$$x=-4+3(7+3\alpha)-2\alpha=17+7\alpha.$$

Thus the system (G1) is certainly consistent and we say that $x = 17 + 7\alpha$, $y = 7 + 3\alpha$ and $z = \alpha$ is the **general solution** of the system (G2). If we choose a particular value for α, for example $\alpha = 3\pi/7$, we obtain $x = 17 + 3\pi$, $y = 7 + 9\pi/7$, $z = 3\pi/7$, which is just one **particular solution** (of the infinitely many solutions) of the given linear system. Since we freely chose $z = \alpha$ we call z a **free variable** and α an **arbitrary constant**. There is, of course, no good reason why we should have chosen z as the free variable in preference to y. Indeed, choosing $y(= \beta)$ as the free variable (and arbitrary constant respectively) we find that (G2) forces z to be $(\beta - 7)/3$ and $x = (7\beta + 2)/3$. The particular solution obtained above can now be obtained by choosing $\beta = 7 + 9\pi/7$.

PROBLEM 1.2

Find the formulae for y and z if, in equations (G2) you choose x as the free variable and then put $x=\gamma$.

One final example, this time of a linear system with more unknowns than equations, will highlight important points to be made shortly.

Example 7

Solve the system

$$\left.\begin{aligned} 5x-11y+6z+7u+12v&=31 && (1.19)\\ 2x+5y-z-9u+3v&=3 && (1.20)\\ 4x-12y+8z-8v&=-16 && (1.21)\\ 5x-y-6z+2u+13v&=26. && (1.22)\end{aligned}\right\}\text{(H1)}$$

Here, to get rid of the terms $2x$, $4x$, $5x$ respectively from equations (1.20)–(1.22), it seems that we have to take, systematically, $\frac{2}{5}$, $\frac{4}{5}$ and 1 times equation (1.19) from equations (1.20), (1.21) and (1.22). This will introduce quite a lot of fractional coefficients. In general this will be unavoidable but it can be avoided here by first dividing equation (1.21) by 4. [Alternatively we might have replaced equation (1.19) by (1.19) − (1.21).] Then, to make subsequent calculations easier, we can interchange this 'new' equation with equation (1.19). Proceeding in this way we first obtain:

$$\left.\begin{aligned} x-3y+2z-2v&=-4 && (1.23)\\ 2x-5y-z-9u+3v&=3 && (1.20)\\ 5x-11y+6z+7u+12v&=31 && (1.24)\\ 5x-y-6z+2u+13v&=26. && (1.22)\end{aligned}\right\}(\text{H2})$$

We then continue in the usual manner, using (1.23) to eliminate x from equations (1.20), (1.24) and (1.22). . . .

This method of systematically removing $x, y, z, \ldots$ from a linear system is called **Gaussian elimination** (or, sometimes, reduction to echelon form.) It is very straightforward but there are one or two questions one can raise, the answers to which are not immediately apparent, as we shall see below.

Before posing the questions referred to above we note that all the linear systems so far considered are particular examples of the **general system of m linear equations in the n unknowns $x_1, x_2, \ldots, x_n$**, namely

$$\left.\begin{aligned} a_{11}x_1+a_{12}x_2+\ldots+a_{1n}x_n&=b_1\\ a_{21}x_1+a_{22}x_2+\ldots+a_{2n}x_n&=b_2\\ &\cdots\\ a_{i1}x_1+a_{i2}x_2+\ldots+a_{in}x_n&=b_i\\ &\cdots\\ a_{m1}x_1+a_{m2}x_2+\ldots+a_{mn}x_n&=b_m.\end{aligned}\right\}\qquad (\text{L})$$

$\leftarrow$ called the ith (or general) equation and often written $\sum_{j=1}^{n} a_{ij}x_j = b_i$.

Despite Gauss's name being attached to the above method of solving linear systems, an identical approach to solving similar problems can be found in Babylonian and Chinese writings which date back to about 200–300 BC.

As an example, consider the following problem to be found in the Chinese text, *Nine Chapters of the Mathematical Art*. Bundles of top (T), medium (M) and low (L) grade rice are such that three bundles of T plus two of M and one of L make 39 dǒu. Writing this $3T + 2M + 1L = 39$ we can write the two other bits of information given in the problem as $2T + 3M + 1L = 34$ and $1T + 2M + 3L = 26$. The question asks: how many dǒu are there in a bundle of each grade of rice?

Of course one simply has to solve the three simultaneous equations for T, M and L. How did the Chinese do it? They placed counting rods (we have replaced them with arabic numerals) on a board as in (a) below and then gave the instruction: 'Use the 3 in the final column to multiply the second column. Then modify the (new) second column by subtracting twice the corresponding entries from the third column.' This gives (b). The third column is now used similarly to modify the first column, yielding (c). Finally, the 'new' second column is used, likewise, to modify the first column, giving (d).

	(a)			(b)			(c)			(d)		
(T)	1	2	3	1	0	3	0	0	3	0	0	3
(M)	2	3	2	2	5	2	4	5	2	0	5	2
(L)	3	1	1	3	1	1	8	1	1	36	1	1
(dǒu)	26	34	39	26	24	39	39	24	39	99	24	39

Carl Friedrich Gauss, 30 April 1777–23 February 1855, was one of the best mathematicians that the world has ever seen. Precocious as an infant, it is said that, aged three, he corrected an error in his father's accounts and, aged eight, wrote down immediately on being asked the sum of the numbers from 1 to 100, presumably through reorganising

$1 + 2 + \ldots + 99 + 100$ as $(1 + 100) + (2 + 99) + \ldots + (50 + 51) = 50 \times 101$.

Aged 18 he showed it possible to construct, by straight edge and compass alone, a regular 17-gon, the first 'new' such *n*-gon for 2000 years. And so it went on. Following observations of the asteroid Pallas, Gauss showed how the above elimination techniques could be used to solve a system of six simultaneous linear equations in six unknowns and hence determine the asteroid's orbit. Although this method of solution had been known for 2000 years, Gauss's name was attached to it – one of his more trivial contributions!

(The Greek letter Σ indicates 'summation' over all terms of the form $a_{ij}x_j$ as j ranges over all the whole numbers from 1 up to n. [Likewise $\sum_{i=1}^{5} t^i$ indicates $t + t^2 + t^3 + t^4 + t^5$.])

As with the examples we have given, the mn numbers a_{ij} (where $1 \le i \le m$ and $1 \le j \le n$) and the m numbers b_i are called the **coefficients** and the **constant** terms, respectively, of (L). Notice that the **double suffix notation** immediately identifies a_{ij} as the coefficient of x_j in the ith equation of (*L*). By a **solution** of (L) we mean an n-tuple $(r_1, r_2, \ldots, r_n)$ of numbers which satisfy all the equations of (L) simultaneously when each r_i is substituted for the corresponding x_i. To **solve** (L) is to find all solutions of (L). We can now pose the questions.

QUESTION 1

The examples above showed that a linear system such as (L) can have: (i) no solution or (ii) a unique solution or (iii) infinitely many solutions. Is it possible for (L) to have, say, exactly two solutions? We shall answer this fully without relying on geometrical interpretations (which, for systems involving more than three variables, might be difficult to interpret and so regarded with suspicion?) in the Exercises at the end of Chapter 6.

QUESTION 2

(A somewhat more subtle point!) In solving equations (G1) we 'reduced' (G1) to the system of just two equations (G2) and then solved (G2). But, how do we know that the solutions of (G2) are in any way related to, never mind being THE SAME AS, the solutions of the given system (G1)? You may choose to take my word for it that it is so, or you may feel (intuitively?) that it must be the case. But these are clearly unreliable positions to adopt.

No! Presented with the assertion that the solutions of (G1) are precisely those of (G2) the mathematician who had not come across this (very provocative?) statement before would call for some justification in the form of a **proof**. (Proofs are described briefly in the appendix.)

Let us construct our first general proof. Since most of the more important assertions in mathematics are given the title **theorem** we are about to offer a proof of our first theorem.

In order to state our theorem succinctly, we need to list the different types of 'moves' we allowed ourselves when solving the various systems of equations above. Specifically we allowed (especially in Example 7):

e(1) Multiplication of any equation by a non-zero constant (in (H1) we used $\frac{1}{4}$ as a multiplying factor).
e(2) The interchange of any two equations.
e(3) [The most used move.] The addition of some multiple of one equation to any other equation. [We used three such moves successively to change (F1) into (F2) in Example 5.]

Moves of types e(1), e(2), e(3) are called **elementary operations**. Using them we can change a given set of equations into another whose solution (or lack of!) is more readily spotted. We are now ready to state (and prove) our first theorem!

• *Theorem 1*

Let (L) be the system of equations given above and let (L′) be any system obtained from (L) by the use of elementary operations. Then the systems (L) and (L′) have precisely the same solutions. ●

COMMENT We shall first prove the theorem in the case of a single application of a move of type e(3) – leaving you to give the proof for the case of a single move of each of the types e(1) and e(2) – and then consider the case of a succession of such moves.

PROOF

(Case e(3)) Suppose $(r_1, r_2, \ldots, r_n)$ is a (common) solution of all the equations of (L). Assume (L′) is obtained from (L) by replacing equation (*i*) of (L) by {equation (*i*) plus k times equation (*j*) of (L)} i.e. by

$$(a_{i1} + ka_{j1})x_1 + (a_{i2} + ka_{j2})x_2 + \ldots + (a_{in} + ka_{jn})x_n = b_i + kb_j. \tag{1.25}$$

Then the n-tuple $(r_1, r_2, \ldots, r_n)$ satisfies this equation too since

$$(a_{i1} + ka_{j1})r_1 + (a_{i2} + ka_{j2})r_2 + \ldots + (a_{in} + ka_{jn})r_n$$

$$= (a_{i1}r_1 + a_{i2}r_2 + \ldots + a_{in}r_n) + k(a_{j1}r_1 + a_{j2}r_2 + \ldots + a_{jn}r_n)$$

which is indeed equal to $\qquad b_i \qquad + kb_j$

Conversely, any (common) solution of all the equations of (L′) satisfies, in particular, its ith equation (i.e. equation (1.25)) and its jth equation and hence (equation $(i) - k$ times equation (*j*)). But this latter is just the ith equation of (L).

This completes the proof of Theorem 1 in the case of a single elementary operation of type e(3). After checking (i.e. proving) that the same result holds for a single elementary operation of each of the types e(1) and e(2) we may deduce immediately that, if repeated use of e(1), e(2) and e(3) yields a system (L″), then each solution of (L) is a solution of (L″) and vice versa. ●

[Notice how, just as (L″) is obtained from (L) by a sequence of elementary operations so can (L), likewise, be obtained from (L″) by 'undoing' each of these operations in the reverse order. Each 'undoing' is an elementary operation: see Exercise 16 at the end of Chapter 2.]

TUTORIAL PROBLEM 1.1

How do you respond if asked to solve

$$\begin{aligned} 3x - 0y + 4z &= -1 \\ 5x + 0y - z &= 6 \\ -x + 0y + 6z &= -7? \end{aligned}$$

(Does the solution allow you to choose y arbitrarily? If so, are you also allowed to choose, arbitrarily, other unknowns u, v, w, . . . which have not been expressly written down? Systems with columns of zero coefficients can sometimes arise naturally – see the note in Exercise 11 of Chapter 11, where the answers to these questions will, fortunately, be clear.)

TUTORIAL PROBLEM 1.2

Why are we not allowed to multiply by 0 in (elementary) operation e(1)? [We *are* allowed to take $k = 0$ in operation e(3). Is this fair?]

Theorem 1 is important because it clarifies which operations on a linear system are allowable – if we are to obtain a simpler system of equations with exactly the same solutions.

Applications

There are many real-life problems whose solution involves systems of linear equations. Here, we offer just two in order to give a flavour of what is possible.

1. In the electric circuit below we find the currents I_k. We apply Ohm's law (voltage V = current $I \times$ resistance R) and Kirchhoff's laws: (i) at each node, current in = current out; (ii) the voltage drop round a closed loop is zero.

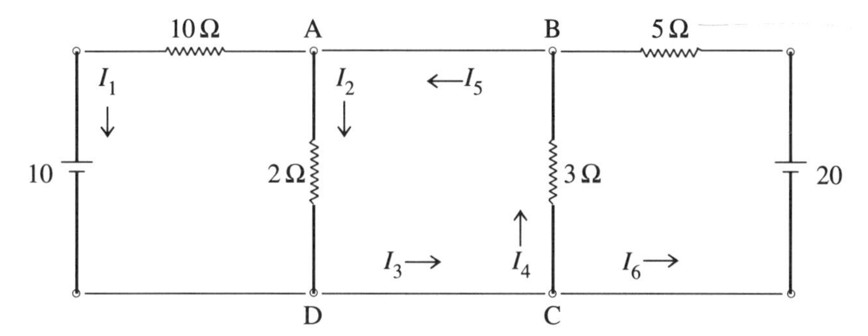

Using Kirchhoff's law (i) we obtain the simultaneous equations

$$\begin{array}{llr} \text{(node A)} & -I_1 - I_2 \qquad\qquad + I_5 \qquad = 0 & (1.26) \\ \text{(node B)} & \qquad\qquad\qquad I_4 - I_5 + I_6 = 0 & (1.27) \\ \text{(node C)} & \qquad\qquad I_3 - I_4 \qquad - I_6 = 0 & (1.28) \\ \text{(node D)} & I_1 + I_2 - I_3 \qquad\qquad\qquad = 0. & (1.29) \end{array}$$

The solution of this system involves three (not two) free variables. [Note that equations $(1.26)+(1.29)$ and $(1.27)+(1.28)$ both yield $I_3 - I_5 = 0$.] Thus we need three further equations to determine the currents I_k uniquely. Fortunately we can get these by applying Kirchhoff's law (ii) and Ohm's law to the three small rectangles in the diagram. Taking them in turn we get

$$\begin{aligned} 10I_1 - 2I_2 \qquad\qquad\qquad &= -10 \\ 2I_2 + 0I_3 + 3I_4 + 0I_5 \qquad &= \ \ 0 \\ -3I_4 \qquad + 5I_6 &= \ 20. \end{aligned}$$

These three equations are enough, together with those above, to determine the currents uniquely. We leave this to you.

2. Consider the (differential) equation $\mathrm{d}^2y/\mathrm{d}x^2 = f(x)$ which we assume to be satisfied on the range $x = 0$ to $x = 1$. Here $f(x)$ is a given function of x and we are to determine the function $y(x)$ which solves this equation, given the boundary conditions $y(0) = a$, $y(1) = b$. This *continuous* problem can be replaced by a *discrete* one – indeed by a system of linear equations, perhaps more readily solved on a computer – as follows.

For small values of h we can approximate $\mathrm{d}y/\mathrm{d}x$ by

$$\frac{y(x+h) - y(x)}{h}.$$

In a similar manner d^2y/dx^2 can be approximated by

$$\frac{y(x+h)-2y(x)+y(x-h)}{h^2} \tag{1.30}$$

If we now split the range [0, 1] into n equal parts of length $h(=1/n)$, if we write y_j for the value of $y(x)$ at $x=jh$ and if we use the above approximation (1.30) for d^2y/dx^2, we can replace our original equation by

$$y_{j+1}-2y_j+y_{j-1}=h^2 f(jh).$$

where j can take each value from 1 to $n-1$ which becomes a system of $n-1$ equations in $n-1$ unknowns $y_1, y_2, \ldots y_{n-1}$ (if we incorporate the given y_0, y_n into the right-hand sides.)

Of course, we would expect our approximations to give a more accurate description of $y(x)$ the smaller we take h. Unfortunately this leads to a larger set of equations in a greater number of unknowns. Clearly, if n is large, we shall have to use a computer to solve the system for us.

This can be done in various ways, one of which, the **Gauss–Seidel** method, requires that we rewrite each equation in (L) (when $m=n$) as:

$$x_i=\frac{1}{a_{ii}}(b_i-a_{i1}x_1-\ldots-a_{i,i-1}x_{i-1}-a_{i,i+1}x_{i+1}-\ldots-a_{in}x_n) \quad (i=1,\ldots n). \tag{1.31}$$

We then guess at a solution for (x_1), $x_2, \ldots, x_n$ and substitute $x_2, \ldots, x_n$ in the first equation of (1.31), namely

$$x_1=\frac{1}{a_{11}}(b_1-a_{12}x_2-a_{13}x_3-\ldots-a_{1n}x_n)$$

to find a new value for x_1. Using this new value for x_1 and the assumed values of $x_3, \ldots, x_n$ we use the second equation of (1.31) to find a new value for x_2. Continuing in this way we find new values for $x_1, x_2, \ldots, x_n$ and then start all over again using these new values of the x_i in the first equation of (1.31) to find a second new value for x_1, etc.

Incidentally Strang reports that the Gauss–Seidel method was unknown to Gauss and was not recommended by Seidel! (Philipp Ludwig von Seidel, 24 October 1821–13 August 1896.)

Very interesting questions arise in connection with this procedure. For example:

(i) When will such a procedure 'converge' to numbers which are near the true values of the x_i?

(ii) What effect will round-off errors, caused by the computer's inability to work with total accuracy, introduce?

(iii) What methods are best as regards the 'cost' (i.e. computer time) of solving the linear system?

These questions are raised here to impress upon you that deep thoughts can be had concerning such an apparently simple object as a set of simultaneous linear equations. Answers may be found in books on numerical linear algebra, examples of which are given in the Bibliography.

Summary

A given system (L) of m linear equations in n unknowns can always be fully solved by **Gaussian elimination**, that is, by replacing (L) by a system (M) which has the same set of solutions as (L) and in which each equation involves fewer unknowns than its predecessors. This is achieved by performing a succession of **elementary operations**. The **consistency** or **inconsistency** of (M) and hence of (L) can be read off immediately. Further, if the equations are consistent, the full solution, possibly involving **arbitrary constants** or **free variables**, can be determined by **back substitution**.

The full set of solutions to (L) may be (i) empty or (ii) infinite or (iii) contain just one element. These are the only possibilities. This can be demonstrated in the cases $n = 2, 3$ by geometric means. Numerical (computer) methods may also be used to solve systems of equations but, then, convergence problems and round-off errors must be considered.

EXERCISES ON CHAPTER 1

1. Solve

$$\begin{aligned} 2x+3y&=6\\ x-4y&=21. \end{aligned}$$

2. Solve

$$\begin{aligned} 5x+8y&=8\\ x+5y&=3. \end{aligned}$$

3. Use your ($\leq$ 8 place) hand calculator to solve all four linear systems:

(a) $6.073x+2.693y=3.921$, $4.772x+2.116y=3.082$;

(b) $6.073x+2.692y=3.921$, $4.771x+2.116y=3.082$;

(c) $6.073x+2.693y=3.922$, $4.772x+2.116y=3.081$;

(d) $6.07x+2.69y=3.92$, $4.77x+2.12y=3.08$.

Can you explain the rather remarkable solutions you get?

4. Solve, quite formally, without worrying about the specific values of a, b, c and d, the system

$$\begin{aligned} ax+by&=u\\ cx+dy&=v. \end{aligned}$$

(That is, find formulae for x and y in terms of a, b, c, d, u and v. See Chapter 5 if you are not sure what to do.)

(i) Under what conditions on a, b, c, d will the solution be unique?

(ii) Can you now explain the curious results of Exercise 3 where a very small variation in the coefficients or constants leads to an enormous change in the answers? [Linear systems with this property are called **ill-conditioned**.]

5. (i) Retaining the fractions throughout, solve the linear system

$$\frac{1}{29}x+\frac{1}{42}y=1$$
$$\frac{1}{67}x+\frac{1}{97}y=1.$$

(ii) Solve the same system on your hand calculator after:
(a) replacing the coefficients by decimals, correct to three decimal places;
(b) replacing the coefficients by decimals, correct to five decimal places.

In Exercises 6–10 find the complete set of solutions of the systems of equations given.

6. (a)
$$\begin{aligned} x-2y+t&=7\\ 3x+3z-4t&=1\\ 7x-2y+6z-7t&=9; \end{aligned}$$
(b)
$$\begin{aligned} x-2y+4t&=1\\ 3x-6y+z+5t&=21\\ 2x-2y-2z-3t&=8\\ 3x-y+2z-t&=9. \end{aligned}$$

7. (a)
$$\begin{aligned} y-2z&=-1\\ x-4y-4z&=-1\\ 5x+y+4z&=40; \end{aligned}$$
(b)
$$\begin{aligned} 18x-2y+5z&=-4\\ 29x+14y-5z&=-7\\ 13x+2y+z&=-3. \end{aligned}$$

[Hint: do you *have* to eliminate x first?]

8.
$$\begin{aligned} y-2z&=-1\\ x+3y+4z&=14\\ x-4y-4z&=-1\\ 5x+y+4z&=40. \end{aligned}$$

9.
$$\begin{aligned} 2x-y&=13\\ x+3y+4z&=14\\ x-4y-4z&=-1\\ 5x+y+4z&=40. \end{aligned}$$

10.
$$\begin{aligned} x+3y+4z&=13\\ x-4y-4z&=-1\\ 5x+y+4z&=40. \end{aligned}$$

11. One can prove (by induction, see the Appendix) that

$$1+2+\ldots+n\left(=\sum_{k=1}^{n}k\right)=\frac{n(n+1)}{2}$$

and that

$$1^2+2^2+\ldots+n^2\left(=\sum_{k=1}^{n}k^2\right)=\frac{n(n+1)(2n+1)}{6}.$$

What answer should we try for $1^3+2^3+\ldots+n^3\left(=\sum_{k=1}^{n}k^3\right)$? Looking at the two given cases we might try $\sum_{k=1}^{n}k^3=an^4+bn^3+cn^2+dn+e$ and then determine a, b, c, d and e from the five simultaneous equation obtained by setting $n=1, 2, 3, 4, 5$ successively. Determine potential formulae for $\sum_{k=1}^{n}k^3$ and $\sum_{k=1}^{n}k^4$ by these means (and then confirm your 'answers' by mathematical induction.)

12. Find the value(s) of k, if any, for which the following linear system is consistent:

$$\begin{aligned} x+2y-z&=7\\ 3x-y+2z&=9\\ 7x+7y-2z&=k. \end{aligned}$$

13. Suppose that

$$\begin{aligned} x+2y+3z&=a\\ 2x+3y+4z&=b\\ 3x+4y+kz&=c \end{aligned}$$

where k, a, b, c are constants. Show that these equations are consistent for all values of k except, perhaps, $k=5$ and find the relation between a, b and c if the equations are consistent when $k=5$. If $a=1$, $b=2$ and $c=4$, find those values of k, if any, for which x, y, z are all integers.

14. Find the value(s) of k (if any) for which the linear system

$$\begin{aligned} x+y+kz&=1\\ x+ky+z&=1\\ kx+y+z&=-2. \end{aligned}$$

has (i) a unique solution; (ii) no solution; (iii) infinitely many solutions. Find the complete solution in cases (i) and (iii). [If you do any divisions, take care that what you are dividing by is not zero!]

15. Find the values of a, b, c in order that the parabola $y=ax^2+bx+c$ passes through the points $(-4,56)$, $(1,11)$, $(3,7)$. [Hint: since $(-4,56)$ lies on the curve we have $56=(-4)^2a+(-4)b+c$.]

16. Check that the system

$$\begin{aligned} y-2z&=-1 \quad (1.32)\\ x-4y-4z&=-1 \quad (1.33)\\ 5x+y+4z&=40 \quad (1.34)\\ x+y+z&=8 \quad (1.35) \end{aligned}$$

has no solution. On the other hand, the system with complex number coefficients

$$ix + (1-4i)y - (2+4i)z = -1 - i \quad (1.32) + i(1.33)$$
$$(5+i)x + (1+i)y + (4+i)z = 40 + 8i \quad (1.34) + i(1.35)$$

where $i = \sqrt{-1}$, can be solved for x, y, z. How can you explain this apparent anomaly?

17. Explain, geometrically, why a system of three equations in two unknowns x, y can be inconsistent whilst each pair of equations has a unique solution. Can you extend your argument to the case of four equations in three unknowns x, y, z?

18. Explain, geometrically, why it is possible to have a system of, say, 20 equations in 3 unknowns x, y, z such that both the entire system as well as each trio of equations has a unique solution.

19. Let (L) be a linear system in which all the coefficients are rational numbers. Given that (L) has a solution in real numbers, show that (L) has a solution in which all the values of the unknowns are rational numbers.

COMPUTER PROBLEMS

1. Program your computer to solve, if possible, by the Gauss–Seidel method:

$$\begin{aligned} 4.1x + 2.5y + 1.6z &= 7.3 \\ 2.4x - 5.2y + 1.8z &= 10.2 \\ 1.1x + 4.3y - 8.3z &= -3.3. \end{aligned}$$

2. (a) Solve, by hand, using Gaussian elimination, the system of equations

$$\begin{aligned} x + 2y + 2z &= 1 \\ 2x + y + 2z &= 2 \\ 2x + 2y + z &= 3. \end{aligned}$$

Now use the Gauss–Seidel method on the computer by first 'solving' these equations for x, y, z (in terms of y and z; z and x; x and y respectively as in Application 2) and then trying various initial guesses for x, y, z (including values near to the true values of x, y, z). Can you explain your (peculiar) findings?

(b) Try Gauss–Seidel on the system

$$\begin{aligned} 2z + x + 2y &= 1 \\ 2z + 2x + y &= 2 \\ z + 2x + 2y &= 3. \end{aligned}$$

Compare this system with that in part (a).

COMPUTER PACKAGE PROBLEMS

1. Let (A) be: $2.6x + 2.2y + 1.9z = 11.08$
Let (B) be: $2.7x + 3.5y + 1.7z = 14.20$
Let (C) be: $1.4x + 1.9y + 2.4z = 8.50$
Let (D) be: $2.3x + 2.5y + 2.0z = 11.28.$

Solve (if possible) the linear systems comprising: (i) equations (A) and (B): (ii) equations (A), (B), (C); (iii) equations (A), (B), (D); (iv) equations (A), (B), (C), (D) – after replacing 11.28 by 11.29 in equation (D).

2. Let (A) be: $1.61x + 1.80y - 1.77z - 2.52w + 1.41t = -0.28527$
Let (B) be: $3.72x + 1.61y + 1.30z + 1.80w - 8.21t = 1.83106$
Let (C) be: $1.11x + 3.79y - 6.61z - 9.36w + 12.44t = -2.68686$.
Solve (if possible) the linear systems comprising (i) equations (A) and (B); (ii) equations (A), (B), (C). [If, in solving (ii), you obtain, as I did, an answer involving two free variables (i.e. arbitrary constants) first compare equation 3 times equation (A) minus equation (B) with equation (C). Then try entering all the coefficients of these equations in rational form. That is, replace 1.61 by 161/100, etc.]

3. Consider the system

$$\begin{aligned}
x + \frac{1}{2}y + \frac{1}{3}z + \frac{1}{4}w + \frac{1}{5}t &= a \\
\frac{1}{2}x + \frac{1}{3}y + \frac{1}{4}z + \frac{1}{5}w + \frac{1}{6}t &= b \\
\frac{1}{3}x + \frac{1}{4}y + \frac{1}{5}z + \frac{1}{6}w + \frac{1}{7}t &= c \\
\frac{1}{4}x + \frac{1}{5}y + \frac{1}{6}z + \frac{1}{7}w + \frac{1}{8}t &= d \\
\frac{1}{5}x + \frac{1}{6}y + \frac{1}{7}z + \frac{1}{8}w + \frac{1}{9}t &= e.
\end{aligned}$$

(i) Solve this system of equations when $a = b = c = d = e = 1$.
(ii) Solve the system when $a = c = e = 1.1$ and $b = d = 0.9$.
(iii) Solve the system when $a = c = e = \frac{11}{10}$ and $b = d = \frac{9}{10}$.

4. Solve the system

$$\begin{aligned}
3x + 5y + 7z &= a \\
2x + 4y + 6z &= b \\
x + 2y + 4z &= c.
\end{aligned}$$

5. Solve the system

$$\begin{aligned}
kx + 2y + 3z &= 1 \\
x + ky + 3z &= 2 \\
x + 2y + kz &= 3
\end{aligned}$$

in terms of k. Now determine, using your computer package, those values of k for which the equations are inconsistent.

2 · Matrices

This chapter recognises that to solve a linear system one only needs to study its (rectangular) array of coefficients and constants. Elementary operations can be applied to the rows of such arrays (called matrices) just as they were to the equations of linear systems in Chapter 1. It is quite convenient to translate statements about equations into matrix form but, until Chapter 3, these arrays themselves are merely notational – useful but lifeless. Elementary operations on matrices will prove to be very useful, for example in finding inverses of matrices (Chapter 4) and for evaluating determinants (Chapter 5). Furthermore computers like to perform them!

As we saw in Example 6 of Chapter 1, we can, in solving systems of equations, ignore the symbols for the unknowns and deal only with their arrays of coefficients. For this, and many other reasons, it is useful to introduce a term to refer to such arrays. We do this in the following definition.

• *Definition 1*

A rectangular array of mn numbers set out in m rows and n columns is called an **m by n** (written **$m \times n$**) **matrix**. The numbers are called the **entries** (or **elements** or **components**) of the matrix. •

We usually write brackets (of one shape or another) around the array to aid us in thinking of it as a single entity. Here are some simple examples.

• *Examples 1*

(i) $\begin{bmatrix} e^{\pi} & \pi^{e} & \frac{3}{8} \\ 2 & -2.3 & 7.11 \end{bmatrix}$ is a 2×3 matrix;

(ii) $[5]$ is a 1×1 matrix;

(iii) the equations (L) of Chapter 1 generate two important matrices:

$$A = \begin{bmatrix} a_{11} & a_{12} & \cdots & a_{1n} \\ a_{21} & a_{22} & \cdots & a_{2n} \\ \\ a_{m1} & a_{m2} & \cdots & a_{mn} \end{bmatrix}; \qquad A_1 = \left[\begin{array}{cccc|c} a_{11} & a_{12} & \cdots & a_{1n} & b_1 \\ a_{21} & a_{22} & \cdots & a_{2n} & b_2 \\ \\ a_{m1} & a_{m2} & \cdots & a_{mn} & b_m \end{array}\right].$$

The matrix of coefficients of (L) **The augmented matrix of (L)**

[The vertical line in the second matrix is merely one of convenience.] These matrices are visibly of types $m \times n$ and $m \times (n + 1)$ respectively. In the first matrix the $a_{i1}, a_{i2}, \ldots, a_{in}$ constitute the ***i*th row**; the $a_{1j}, a_{2j}, \ldots, a_{mj}$ constitute the ***j*th column**. The element common to the ith row and the jth column is the ***ij*th element** (or **entry**) a_{ij}. •

For later use we note that (usually only in the case when A is **square**, that is, when $m = n$) the elements a_{ii} of the matrix A are called the **diagonal elements** of A, the set

$\{a_{11}, a_{22}, \ldots, a_{nn}\}$ then comprising the **(main) diagonal** of A. Finally, if the only entries which are non-zero lie on this main diagonal then A is called a **diagonal** matrix. In symbols: $a_{ij} = 0$ whenever i and j are different.

The term matrix was introduced into mathematics by James Joseph Sylvester in 1850. He saw a matrix as 'a rectangular array of terms out of which different systems of determinants may be engendered as from the womb of a common parent . . .'. One can, of course, argue that the Chinese were (usefully) using the matrix concept 200 years BC (see Chapter 1). Some two millennia later the concept of square array of numbers emerged again, being used, in particular, by Gauss as an abbreviation in the study of linear substitutions (≡ transformations – see Chapter 10) on ternary quadratic forms. Later, in 1829, Cauchy put the coefficients themselves of such forms into an array, calling such an array a 'tableau'.

Sylvester was born into a Jewish family on 3 September 1814. On finishing his studies at Cambridge his faith prevented his being awarded the degree since he could not accept the 39 Articles of the Church of England. He was made an FRS in 1839, but he did not spend all his life in academia. In fact, after leaving the University of Virginia in 1841, he spent some time in both the financial and legal worlds (whilst still researching and tutoring in mathematics, one of his tutees being Florence Nightingale).

Sylvester was very proud of the great number of words he introduced into mathematics – such as 'allotrocious factors', 'catalecticant'. (Most are no longer used!) He was also proud of his poetry amongst which was the 'Rosalind' poem (400 lines all ending in words rhyming with Rosalind) and he took singing lessons from Gounod. In mathematics he was best known for his work on invariants; in this book his name is connected with his law of inertia and his law of nullity. He died on 15 March 1897.

CHALLENGE

Those readers who, having seen matrices before, feel they are having too easy a ride may, for fun, care to contemplate which is larger, e^{π} or π^{e}? (Use of a computer is forbidden!)

NOTE I

(i) So far matrices are completely sterile, being only a notational device. We shall later (beginning in Chapter 3) breathe life into them!

(ii) We have already referred to the first matrix in Example 1(iii) by the symbol A. For greater clarity we might have written $A_{m \times n}$ or $A_{m,n}$ or $[a_{ij}]_{m \times n}$ etc. instead. Likewise the second matrix can be denoted briefly by $[A \mid b]$ where b denotes the $m \times 1$ column matrix with entries $b_1, \ldots, b_m$.

When solving linear systems we found it useful to try and 'fiddle' some zero coefficients into strategic places at the fronts of equations. In several applications it will be convenient to do the same kind of thing to matrices. So we introduce an appropriate name.

• *Definition 2*

Any $m \times n$ matrix (respectively system of equations) in which each non-zero row (respectively equation) begins with more zeros (respectively zero coefficients) than the previous row (respectively equation) is said to be in **(row) echelon form**. •

Example 2

$$\begin{bmatrix} 1 & 2 & 3 & 4 \\ 0 & 5 & 6 & 7 \\ 0 & 0 & 0 & 0 \end{bmatrix} \quad \text{and} \quad \begin{bmatrix} 0 & 1 & 0 & 0 \\ 0 & 0 & 0 & 1 \\ 0 & 0 & 0 & 0 \end{bmatrix}$$

are in echelon form whilst

$$\begin{bmatrix} 1 & 2 & 3 & 4 \\ 0 & 0 & 5 & 6 \\ 0 & 0 & 7 & 8 \end{bmatrix} \quad \text{and} \quad \begin{bmatrix} 1 & 2 & 3 & 4 \\ 0 & 0 & 0 & 5 \\ 0 & 6 & 0 & 0 \end{bmatrix}$$

are not.

Notice the 'staircase' effect, with long but not high steps allowed. Note, too, that a given matrix, and, hence, any echelon matrix derived from it, may have columns of zeros.

TUTORIAL PROBLEM 2.1

Definition 2 seems to me to admit

$$\begin{bmatrix} 0 & 0 & 0 & 0 & 0 \\ 0 & 0 & 0 & 0 & 0 \end{bmatrix}$$

as an echelon matrix. Do your agree?

The following extension of Definition 2 is sometimes useful.

Definition 3

If, in addition to the requirements of Definition 2, the 'leading' (non-zero) coefficient in each (non-zero) row, the so-called **pivot** of the row, is 1 and if, in each 'leading' column (i.e. a column containing a pivot), all other entries are 0, we say that the matrix is in **reduced (row) echelon form**.

Example 3

$$\text{(i)} \begin{bmatrix} 1 & 3 & 0 & 0 \\ 0 & 1 & -2 & 0 \\ 0 & 0 & 0 & 0 \end{bmatrix} \text{ is not, (ii)} \begin{bmatrix} 1 & -2 & 0 & 0 & 3 \\ 0 & 0 & 1 & 0 & -7 \\ 0 & 0 & 0 & 1 & 4 \\ 0 & 0 & 0 & 0 & 0 \end{bmatrix} \text{ is and (iii)} \begin{bmatrix} 1 & 0 & 0 & 0 \\ 0 & -1 & 0 & 0 \\ 0 & 0 & 0 & 1 \end{bmatrix}$$

is not in reduced echelon form.

It is pretty clear, but we have not formally proved it, that each system of equations can be changed into echelon (or, even, reduced echelon) form by using only the three types of elementary operation introduced in Chapter 1. Similarly, matrices can be changed into (reduced) echelon form by means of corresponding operations called **elementary row operations** performed on their rows.

Rather than give the (rather messy) formal general proof we offer the following example.

Example 4

Let us reduce the matrix

$$A := \begin{bmatrix} 0 & 2 & -7 & 1 & -1 \\ 0 & 1 & -3 & 2 & -4 \\ 0 & 5 & -16 & 7 & -13 \\ 0 & 5 & -11 & 22 & 9 \end{bmatrix}$$

to echelon form. Using elementary row operations of the type indicated we get, successively,

$$A \overset{\text{(i)}}{\rightarrow} \begin{bmatrix} 0 & 1 & -3 & 2 & -4 \\ 0 & 2 & -7 & 1 & -1 \\ 0 & 5 & -16 & 7 & -13 \\ 0 & 5 & -11 & 22 & 9 \end{bmatrix} \overset{\text{(ii)}}{\rightarrow} \begin{bmatrix} 0 & 1 & -3 & 2 & -4 \\ 0 & 0 & -1 & -3 & 7 \\ 0 & 0 & -1 & -3 & 7 \\ 0 & 0 & 4 & 12 & 29 \end{bmatrix} \overset{\text{(iii)}}{\rightarrow} \begin{bmatrix} 0 & 1 & -3 & 2 & -4 \\ 0 & 0 & -1 & -3 & 7 \\ 0 & 0 & 0 & 0 & 0 \\ 0 & 0 & 0 & 0 & 57 \end{bmatrix}$$

$$\overset{\text{(iv)}}{\rightarrow} \begin{bmatrix} 0 & 1 & -3 & 2 & -4 \\ 0 & 0 & -1 & -3 & 7 \\ 0 & 0 & 0 & 0 & 57 \\ 0 & 0 & 0 & 0 & 0 \end{bmatrix}.$$

(i) interchanges rows 1 and 2; (ii) subtracts multiples of the new row 1 from rows 2, 3 and 4; (iii) subtracts multiples of the latest row 2 from rows 3 and 4 and (iv) interchanges the latest rows 3 and 4. This gives an echelon form for A. Pivots are 1, −1 and 57.

The last matrix above then reduces to the form

$$\begin{bmatrix} 0 & 1 & 0 & 11 & 0 \\ 0 & 0 & 1 & 3 & 0 \\ 0 & 0 & 0 & 0 & 1 \\ 0 & 0 & 0 & 0 & 0 \end{bmatrix}$$

as you should be able to see fairly readily.

[Can you identify exactly which elementary row operations we have just used? For the convenience of the reader it is usual to list the operations used in some succinct manner. See Example 6 below.]

PROBLEM 2.1

Change to (row) echelon form and then reduced (row) echelon form

$$\text{(i)} \begin{bmatrix} 1 & 1 & 3 & 1 & 2 \\ 2 & 1 & 5 & 1 & 4 \\ 4 & 2 & -3 & 2 & 2 \\ 1 & -1 & 1 & -1 & -1 \end{bmatrix}; \quad \text{(ii)} \begin{bmatrix} 1 & 2 & 4 & 2 & 5 \\ 3 & 2 & 8 & 2 & 6 \\ 7 & 4 & 5 & 4 & 8 \\ 5 & 1 & -2 & 1 & 1 \end{bmatrix}.$$

PROBLEM 2.2

Try to write out the formal general proof referred to above. [The main nuisance derives from the number of subcases e.g. (i) $a_{11} \neq 0$; (ii) $a_{11} = 0$.]

TUTORIAL PROBLEM 2.2

(i) If you and I convert a given matrix to echelon form do we necessarily end up with the same echelon matrix? (See Exercise 3.)

(ii) What if we are both asked to change the matrix to reduced echelon form? (See Exercise 4 at the end of the chapter.)

(iii) Is it possible to convert a matrix, not all of whose entries are 0, to an (echelon) matrix containing only zeros? [This may strike you as silly. If so, try to spell out precisely (i.e. prove) why you know it cannot be done – and then compare this with the 'proof' that it can(!) in Exercise 9.]

(iv) Choose for yourself, quite randomly, a rectangular matrix, say of size 4×7. Change it to row echelon form R, say, and also to *column echelon form* C, say, (using elementary column operations – the natural analogues of our row operations). Do you think R will have more (or fewer or the same) number of non-zero rows as C has non-zero columns? Choose yourself several such matrices and see what happens.

• *Definition 4*

If two matrices A and B are obtainable from one another by a sequence of elementary row operations we say that A and B are **row equivalent**. (See Exercise 16.)

Example 5

Let

$$A=\begin{bmatrix}1&0&6&6\\1&9&4&2\\2&0&0&1\end{bmatrix},\quad B=\begin{bmatrix}2&0&12&12\\0&9&-2&-4\\9&0&6&10\end{bmatrix}\quad\text{and}\quad C=\begin{bmatrix}0&-9&2&4\\7&0&-6&-2\\2&27&6&0\end{bmatrix}.$$

Then A and B are row equivalent as are B and C. This necessarily implies (without further work) that A and C are row equivalent.

PROBLEM 2.3

Confirm this last assertion by giving a suitable argument (i.e. proof).

We can now (a) restate Theorem 1 of Chapter 1 and (b) encapsulate some of our previous findings in terms of row equivalence. We have the following theorem.

Theorem I

Suppose

$$\left[\begin{array}{cccc|c}a_{11}&a_{12}&\dots&a_{1n}&b_1\\a_{21}&a_{22}&\dots&a_{2n}&b_2\\\dots&\dots&\dots&\dots&..\\a_{m1}&a_{m2}&\dots&a_{mn}&b_m\end{array}\right]\quad(\text{briefly }[A\mid b])$$

and

$$\left[\begin{array}{cccc|c}c_{11}&c_{12}&\dots&c_{1n}&d_1\\c_{21}&c_{22}&\dots&c_{2n}&d_2\\\dots&\dots&\dots&\dots&..\\c_{m1}&c_{m2}&\dots&c_{mn}&d_m\end{array}\right]\quad(\text{briefly }[C\mid d])$$

are row equivalent matrices. Then:

(a) the systems

$$\begin{aligned}a_{11}x_1+a_{12}x_2+\dots+a_{1n}x_n&=b_1\\a_{21}x_1+a_{22}x_2+\dots+a_{2n}x_n&=b_2\\\dots\quad.\quad\dots\quad.\dots.\quad\dots\quad&\quad..\\a_{m1}x_1+a_{m2}x_2+\dots+a_{mn}x_n&=b_m\end{aligned}\tag{L}$$

and

$$\begin{aligned}c_{11}x_1+c_{12}x_2+\dots+c_{1n}x_n&=d_1\\c_{21}x_1+c_{22}x_2+\dots+c_{2n}x_n&=d_2\\\dots\quad.\quad\dots\quad.\dots.\quad\dots\quad&\quad..\\c_{m1}x_1+c_{m2}x_2+\dots+c_{mn}x_n&=d_m\end{aligned}\tag{M}$$

have exactly the same set of solutions. (They may each have none!)

(b) Suppose $\lfloor C \mid d \rfloor$ is in echelon form. Then the equations (M) are consistent (i.e. have a solution) iff[a] $\lfloor C \mid d \rfloor$ does not have a pivot in its final column. ●

PROBLEM 2.4

Find the one value of k which makes the following system of equations consistent:

$$\begin{aligned} 3x+2y-z+5u+6v&=10\\ 5x-y-z+7u+9v&=-1\\ 4x+7y-2z+8u+9v&=k. \end{aligned}$$ ●

Although a system of two or more equations may have no solution (that is, it may be inconsistent) any system in which all the numbers b_i are zero (called a **homogeneous linear system**) always has at least one solution – namely the **trivial solution** in which each unknown takes the value 0.

In fact, the method of changing a system of equations to reduced echelon form allows us to say much more.

The following example shows us the way.

Example 6

Solve

$$\left.\begin{aligned} -x-y-2z+7t&=0\\ 2x+2y+7z&=0\\ x+y+3z-2t&=0 \end{aligned}\right\} \qquad \text{(J1)}$$

In matrix terms this becomes (on noting that we may omit the final columns from the augmented and subsequent matrices since they would all be full of 0s).

$$\begin{bmatrix} -1 & -1 & -2 & 7\\ 2 & 2 & 7 & 0\\ 1 & 1 & 3 & -2 \end{bmatrix} \rho_1 \leftrightarrow \rho_3 \begin{bmatrix} 1 & 1 & 3 & -2\\ 2 & 2 & 7 & 0\\ -1 & -1 & -2 & 7 \end{bmatrix}$$

The Greek letters ρ (rho) [how fortunate!] here indicate that this new matrix has been obtained from the original by interchanging rows 1 and 3. [The use of rho symbols is helpful to the reader (and at revision times to the student!) in that it readily indicates exactly how each 'new' matrix is obtained from an 'old' one. Cf. Example 4.]

Continuing in this way we get

$$\begin{bmatrix} 1 & 1 & 3 & -2\\ 2 & 2 & 7 & 0\\ -1 & -1 & -2 & 7 \end{bmatrix} \begin{matrix} \rho_2 \to \rho_2 - 2\rho_1\\ \rho_3 \to \rho_3 + \rho_1 \end{matrix} \begin{bmatrix} 1 & 1 & 3 & -2\\ 0 & 0 & 1 & 4\\ 0 & 0 & 1 & 5 \end{bmatrix}. \qquad \text{(J2)}$$

This time the information between the matrices indicates that the new matrix was obtained from the previous one by replacing the (old) second row by {the old row 2

[a]Meaning 'if and only if' – see the Appendix.

minus twice the old row 1} and by replacing the (old) third row by {the old row 3 plus the old row 1}.

Of course, no one who is more than half awake should need to pass to reduced echelon form before being able to solve equations (J1) by using equations (J2). But, for completeness, we continue with our calculation. We get, successively:

$$\begin{bmatrix} 1 & 1 & 3 & -2 \\ 0 & 0 & 1 & 4 \\ 0 & 0 & 1 & 5 \end{bmatrix} \rho_3 \to \rho_3 - \rho_2 \begin{bmatrix} 1 & 1 & 3 & -2 \\ 0 & 0 & 1 & 4 \\ 0 & 0 & 0 & 1 \end{bmatrix}$$

$$\rho_1 \to \rho_1 - 3\rho_2 \begin{bmatrix} 1 & 1 & 0 & -14 \\ 0 & 0 & 1 & 4 \\ 0 & 0 & 0 & 1 \end{bmatrix} \begin{matrix} \rho_1 \to \rho_1 + 14\rho_3 \\ \rho_2 \to \rho_2 - 4\rho_3 \end{matrix} \begin{bmatrix} 1 & 1 & 0 & 0 \\ 0 & 0 & 1 & 0 \\ 0 & 0 & 0 & 1 \end{bmatrix}. \qquad \text{(J3)}$$

The last row of (J3) forces us to take $t = 0$ and the preceding row forces z to be 0. However, the top row allows us complete freedom of choice in the value of y (or, if you like, x!). Clearly we can choose y (or x) arbitrarily and then use the consequence of line 1 (namely that $x = -y$) to determine the other. Hence, a full solution of (J1) is given by $x = -c$, $y = c$, $z = 0$, $t = 0$ [also written briefly as $(x, y, z, t) = (-c, c, 0, 0)$] where c is an arbitrary constant.

We can now state the general theorem and sketch a proof.

• Theorem 2

The homogeneous system $\sum_{j=1}^{n} a_{ij} x_j = 0 \; (i = 1, 2, \ldots, m)$ of m equations in n unknowns has infinitely many solutions, if $m < n$.

PROOF

Let A be the $(m \times n)$ matrix of coefficients of the given system and let B be the reduced echelon form of A. Suppose that B has exactly r non-zero rows [in Example 6 we have $r = 3$] and hence exactly r pivots in columns $c_1, c_2, \ldots, c_r$, say. [In the example $c_1 = 1$, $c_2 = 3$ and $c_3 = 4$.] Then each of the unknowns $x_{c_1}, x_{c_2}, \ldots, x_{c_r}$ corresponding to these columns can be expressed in terms of the remaining $n - r$ unknowns (sometimes, as with t and z in Example 6, requiring none of these unknowns, sometimes, as with x in Example 6, requiring at least one). Since $r \leq m$ and since $m < n$ we see that $n - r > 0$. Hence there is certainly at least one unknown whose value may be chosen to be any real number. Thus the given set of homogeneous equations has infinitely many solutions, as claimed. ●

TUTORIAL PROBLEM 2.3

The full solution to (J1) comprises (because of the c) a sort of 'single infinity' of particular solutions. On the other hand, the general solution to the 'homogeneous version' of Exercise 6(a) of Chapter 1 appears to involve two free variables. We would not really expect the number of arbitrary constants to depend upon which sequence of elementary operations we use to solve (i.e. convert to echelon form) a given set of equations. But how can we be sure it does not? Also: can we identify, from the matrix of coefficients alone, whether a homogeneous system has a single, double, etc. infinity of solutions? We have already seen that the number of free variables is not simply {number of unknowns – number of equations}, i.e. the number of columns minus the number of rows of the matrix. So, what is the correct formula? We shall answer this question properly in Chapter 9.

Application

There are less trivial applications of the matrix concept than those relating to solving equations. Here is one.

Suppose a town council requires four jobs to be undertaken simultaneously. Four firms tender for each of the jobs as follows.

Firm ↓ / Job →	1	2	3	4
X	20	22	25	20
Y	32	29	35	33
Z	73	73	76	71
T	42	38	45	40

← Tenders in thousands of pounds (or dollars, etc.)

Each firm is to be allotted one job. How should the council allocate the jobs in order to minimise the total cost?

The following *algorithm* (which is called the Hungarian algorithm and is equally applicable to any $n \times n$ array) solves the problem for us. [Roughly speaking, an algorithm is a set of instructions to be followed slavishly without the need for any thought. Of course the fact that the algorithm does what is claimed needs proof – and to find a proof may require a very great deal of thought!]

Let $[a_{ij}]$ be the 4×4 matrix above.

1. For $i = 1, \ldots, 4$ let ρ_i be the smallest entry in row i of $[a_{ij}]$. Form the matrix $[b_{ij}]$ where $b_{ij} = a_{ij} - \rho_i$. In our case this means

$[b_{ij}] =$

0	2	5	0	(i.e. subtract 20 from all row 1 entries)
3	0	6	4	(i.e. subtract 29 from all row 2 entries)
2	2	5	0	(i.e. subtract 71 from all row 3 entries)
4	0	7	2	(i.e. subtract 38 from all row 4 entries).

2. For $j = 1, \ldots, 4$ let κ_j be the smallest entry in column j of $[b_{ij}]$. Form the matrix $[c_{ij}]$ where $c_{ij} = b_{ij} - \kappa_j$. In our case this means

$[c_{ij}] =$

0	2	0	0
3	0	1	4
2	2	0	0
4	0	2	2

(i.e. subtract 0,0,5,0 from each element in columns 1,2,3,4 respectively).

3. Find the least number λ of horizontal and vertical lines which cover all zero entries in $[c_{ij}]$. If $\lambda < 4$ go to step 4. Otherwise go to step 5. [In our case $\lambda = 3$ so we proceed to step 4.]

4. Let μ be the least entry not covered by a line in step 3. Form the matrix $[d_{ij}]$ where

$$d_{ij} = \begin{cases} c_{ij} - \mu & \text{if } c \text{ is uncovered} \\ c_{ij} & \text{if } c \text{ is covered once} \\ c_{ij} + \mu & \text{if } c \text{ is covered twice} \end{cases}$$

and then return to step 3. [In our case $\mu = 1$ and so we get

$[d_{ij}] =$

0	3	0	0
2	0	0	3
2	3	0	0
3	0	1	1

and return to step 3. This time we need four (horizontal and vertical) lines to cover all the zeros so we pass to step 5.]

5. Find a set $d_{i_1,j_1}, \ldots, d_{i_4 j_4}$ of zeros such that no two lie in the same column and no two lie in the same row of $[d_{ij}]$. Then the tenders $a_{i_1,j_1}, \ldots, a_{i_4 j_4}$ give the required solution. (It may not be unique.)
[In our example these d can (must?) be chosen to be the starred zeros in

*0	3	0	0
2	0	*0	3
2	3	0	*0
3	*0	1	1

.

Thus the council minimises cost by allocating jobs 1, 2, 3, 4 respectively to firms X, T, Y, Z.]

PROBLEM 2.5

Find a book in your library which proves that the above algorithm is valid.

Summary

Matrices are (until Chapter 3) merely rectangular arrays of real (or perhaps complex) numbers. Their introduction here, solely on notational grounds, arose from suppressing unnecessary symbols (the unknowns) when solving systems of linear equations. The above application then showed their usefulness in other settings. As with sets of equations, matrices can be transformed into **(row) echelon form** – even **reduced (row) echelon form** – using **elementary (row) operations** which are the exact analogues of those previously applied to systems of equations. Changing to echelon form allows us to prove easily that a homogeneous system of equations with fewer equations than unknowns has infinitely many solutions. At this point matrices are still quite lifeless. We inject life into them in the next chapter.

EXERCISES ON CHAPTER 2

1. Reduce the following matrices to (row) echelon form using sequences of elementary row operations:

$$\text{(a)}\begin{bmatrix} 3 & 2 & 9 \\ -2 & 1 & -6 \\ 7 & -3 & 3 \end{bmatrix};\quad \text{(b)}\begin{bmatrix} -1 & 3 & 1 & 4 & 0 \\ 3 & -9 & 2 & 4 & -1 \\ 2 & -6 & 2 & 4 & 0 \\ 5 & -15 & 5 & 11 & 3 \end{bmatrix};\quad \text{(c)}\begin{bmatrix} 0 & 1 & 2 \\ 0 & 3 & 4 \\ 5 & 6 & 7 \end{bmatrix};$$

$$\text{(d)}\begin{bmatrix} -1 & 1 & 3 & 1 \\ 1 & 0 & 5 & 6 \\ -3 & 2 & 1 & -4 \end{bmatrix};\quad \text{(e)}\begin{bmatrix} 1 & 2 & 3 & 4 \\ 4 & 1 & 2 & 3 \\ 3 & 4 & 1 & 2 \\ 2 & 3 & 4 & 1 \end{bmatrix}.$$

2. Now reduce the matrices of Exercise 1 to reduced (row) echelon form.

3. For each matrix in Exercise 1 find an echelon form different from that you gave in Exercise 1.

4. Now reduce the matrices you have obtained in Exercise 3 to reduce (row) echelon form.

5. Compare the matrices you obtained in (i) Exercises 2(a) and 4(a); (ii) 2(b) and 4(b); (iii) 2(d) and 4(d). What do you notice? Do you think this sort of equality always holds? [Can you see why, if the equality did hold, the first and last problems posed in Tutorial Problem 2.3 above would be answered immediately?]

6. Obtain (or just write down if you can!) the full solution (if any) to the linear systems whose augmented matrix can be changed into the given echelon matrix.

$$\text{(a)}\left[\begin{array}{ccc|c} 1 & 2 & 3 & 4 \\ 0 & 3 & 4 & 5 \end{array}\right];\quad \text{(b)}\left[\begin{array}{cccc|c} 1 & 2 & 3 & 4 & 4 \\ 0 & 3 & 4 & 5 & 5 \\ 0 & 0 & 0 & 0 & 0 \end{array}\right];\quad \text{(c)}\left[\begin{array}{cccc|c} 1 & 2 & 3 & 4 & 4 \\ 0 & 3 & 4 & 5 & 5 \\ 0 & 0 & 0 & 0 & 239 \end{array}\right].$$

7. Give the complete solution – if any – of the linear system corresponding to the augmented matrix

$$\left[\begin{array}{ccccc|c} 1 & 2 & -1 & 3 & -1 & a \\ 3 & 4 & 2 & 5 & -2 & b \\ 1 & 0 & 4 & -1 & 0 & c \\ 3 & 2 & 7 & 1 & -1 & d \end{array}\right]$$

in the cases where (a, b, c, d) is equal to (i) $(0, 0, 0, 0)$; (ii) $(1, 3, 1, 3)$; (iii) $(1, 3, 1, 4)$.

8. Solve, if possible, the linear system given economically by the matrix

$$\begin{array}{cccccc|c} a_{11} & a_{12} & a_{13} & a_{14} & a_{15} & a_{16} & 417 \\ a_{21} & a_{22} & a_{23} & a_{24} & a_{25} & a_{26} & 624 \\ a_{31} & a_{32} & a_{33} & a_{34} & a_{35} & a_{36} & 519 \\ a_{41} & a_{42} & a_{43} & a_{44} & a_{45} & a_{46} & 773 \\ 461 & 357 & 603 & 228 & 315 & 359 & \end{array}$$

meaning, for example: $\sum_{j=1}^{6} a_{3j} = 519$ and $\sum_{i=1}^{4} a_{i5} = 315$.

(A little thought might save a lot of effort!)

9. What is wrong with the following 'proof' that each $m \times n$ matrix can be changed into an echelon matrix in which each element is 0?
'Proof' Use the elementary row operations $\rho_1 \to \rho_1 + \rho_2$, $\rho_2 \to \rho_2 + \rho_1$. This produces new first and second rows which are equal. Now on these rows use the pair of elementary operations $\rho_1 \to \rho_1 - \rho_2$, $\rho_2 \to \rho_2 - \rho_1$. This replaces each of the first two (new) rows by a complete row of zeros. Continue this way with the other rows. QED.

10. Let A be the matrix

$$\begin{bmatrix} a & b & c \\ c & a & b \\ b & c & a \end{bmatrix}.$$

Show that, if $a + b + c = 0$, then A can be changed to an echelon matrix with at most two non-zero rows.

11. Let A be the matrix

$$\begin{bmatrix} 1 & a & a^2 \\ 1 & b & b^2 \\ 1 & c & c^2 \end{bmatrix}.$$

Show that, if $a < b < c$, then A can be changed to an echelon matrix with three non-zero rows.

12. Let A be the matrix

$$\begin{bmatrix} a & b & c \\ b & d & e \\ c & e & f \end{bmatrix} \text{ and let } B \text{ be } \begin{bmatrix} a & b & c \\ -b & d & e \\ -c & -e & f \end{bmatrix}.$$

If A can be changed to an echelon matrix with three non-zero rows, must the same be true of B? [A proof that it can or a counterexample (see the Appendix) showing that it cannot is required.]

13. Show that, if a square (i.e. $n \times n$) matrix A is in reduced echelon form then either (i) A has at least one row full of zeros or (ii) $a_{ii} = 1$ for $1 \le i \le n$ and $a_{ij} = 0$ if $i \ne j$. [(ii) defines the (important) $\boldsymbol{n \times n}$ **identity matrix** I_n.]

14. Let A be an $m \times n$ matrix ($m \ge 2$) and let (I) $\rho_1 \to \rho_1 + 3\rho_2$, (II) $\rho_2 \to 2\rho_2$ be the two row operations indicated. Let B_1 and B_2 be the matrices obtained from A by applying these operations (respectively) in the orders I, II and II, I. Show by means of a specific example that, in general B_1 and B_2 will be different. Can you, nevertheless, give an example of a 2×2 matrix A (other than $\begin{bmatrix} 0 & 0 \\ 0 & 0 \end{bmatrix}$) in which B_1 and B_2 are the same?

15. Let A be the matrix

$$\begin{bmatrix} 1 & 3 & 4 & -2 & 0 \\ -2 & 1 & -2 & 0 & 3 \\ -1 & 4 & 2 & 0 & 0 \end{bmatrix}$$

and let B be the matrix obtained from A by applying, successively, the operations (I) $\rho_2 \to 3\rho_2$; (II) $\rho_3 \to \rho_3 + 5\rho_2$; (III) $\rho_1 \to \rho_1 - \rho_3$; (IV) $\rho_2 \leftrightarrow \rho_1$. Write down the sequence of operations which 'undoes' this sequence and which, consequently, changes B back to A.

16. Prove that the 'undoing' of each elementary operation (technically called its **inverse**) is again an elementary operation. Deduce what is implied in Definition 4: if A is any $m \times n$ matrix which is row equivalent to B then B is row equivalent to A. (We may then simply say that A and B are **row equivalent**).

17. Prove that the two (reduced echelon) matrices

$$\begin{bmatrix} 1 & 0 & 1 \\ 0 & 1 & 1 \end{bmatrix} \text{ and } \begin{bmatrix} 1 & 0 & 2 \\ 0 & 1 & 1 \end{bmatrix}$$

are not row equivalent.

18. Let (L), as in Chapter 1, be given. Suppose that $(r_1, r_2, \ldots, r_n)$ and $(2r_1, 2r_2, \ldots, 2r_n)$ are solutions of (L). Prove that (L) is a system of homogeneous equations. [Hint: what system of equations is satisfied by the difference of these two solutions?]

19. Show that a system of m homogeneous equations in n unknowns has either one solution or infinitely many, no matter whether $m < n$ or not. Explain this for the case $n = 3$, by geometrical means.

20. Show that each system (L) is equivalent to a homogeneous system of $m - 1$ equations together with at most one additional equation.

21. A matrimonial agency has on its books five men and four women seeking marriage partners. The agency's director is able, for each pair, to determine a happiness factor as in the table below. Apply the Hungarian method to determine all pairings of four men and four women whose total happiness factor is as great as possible.

	Allan	Barry	Clint	David	Errol
Alice	6	7	7	6	8
Betty	7	9	5	7	8
Carol	4	8	5	5	6
Debra	5	8	6	7	8

(Have you noticed how people in mathematical problems always seem to have alphabetically associated names – and sometimes even have names of the same length?) [*Hints*: (1) As our array is not square, introduce a fifth 'fictitious' lady, Ecila, say, whose happiness factor with all the men is zero! (2) To maximise a given array is to minimise its negative!]

COMPUTER PROBLEM (THEORY)

1. Reduce the matrix

$$\begin{bmatrix} 1 & 2 & 3 & 4 \\ 0 & 5 & 6 & 7 \\ 0 & 0 & 0 & 8 \end{bmatrix}$$

by (i) first going to the form

$$\begin{bmatrix} 1 & 0 & X & X \\ 0 & 1 & X & X \\ 0 & 0 & 0 & 8 \end{bmatrix} \text{ and then to } \begin{bmatrix} 1 & 0 & X & 0 \\ 0 & 1 & X & 0 \\ 0 & 0 & 0 & 1 \end{bmatrix};$$

(ii) by going first to

$$\begin{bmatrix} 1 & 2 & 3 & 0 \\ 0 & 5 & 6 & 0 \\ 0 & 0 & 0 & 1 \end{bmatrix} \text{ and then to } \begin{bmatrix} 1 & 0 & X & 0 \\ 0 & 1 & X & 0 \\ 0 & 0 & 0 & 1 \end{bmatrix}.$$

Count the number of additions/subtractions and multiplications/divisions used in getting to

$$\begin{bmatrix} 1 & 0 & X & 0 \\ 0 & 1 & X & 0 \\ 0 & 0 & 0 & 1 \end{bmatrix}.$$

What lesson about changing a matrix to reduced echelon form do you learn?

COMPUTER PACKAGE PROBLEMS

1. For $n = 2, 3, 4, 5, 6$, find the row echelon and reduced row echelon forms of the $n \times n$ Hilbert matrix $H_n = [h_{ij}]$ defined by $h_{ij} = 1/(i + j - 1)$.
2. Find reduced echelon forms of the matrices:

$$\begin{bmatrix} 0 & 1 & 0 \\ 4 & 0 & 2 \\ 0 & 3 & 0 \end{bmatrix}; \quad \begin{bmatrix} 0 & 1 & 0 & 0 \\ 6 & 0 & 2 & 0 \\ 0 & 5 & 0 & 3 \\ 0 & 0 & 4 & 0 \end{bmatrix}; \quad \begin{bmatrix} 0 & 1 & 0 & 0 & 0 \\ 8 & 0 & 2 & 0 & 0 \\ 0 & 7 & 0 & 3 & 0 \\ 0 & 0 & 6 & 0 & 4 \\ 0 & 0 & 0 & 5 & 0 \end{bmatrix}; \quad \begin{bmatrix} 0 & 1 & 0 & 0 & 0 & 0 \\ 10 & 0 & 2 & 0 & 0 & 0 \\ 0 & 9 & 0 & 3 & 0 & 0 \\ 0 & 0 & 8 & 0 & 4 & 0 \\ 0 & 0 & 0 & 7 & 0 & 5 \\ 0 & 0 & 0 & 0 & 6 & 0 \end{bmatrix}.$$

Do you spot any pattern? Do you think it persists?

3. Are any two of the following matrices row equivalent?

$$\begin{bmatrix} 7 & -13 & 17 & 11 & -11 \\ 6 & -11 & 14 & 9 & -8 \\ -13 & 21 & -25 & -16 & 12 \\ 13 & -23 & 27 & 17 & -9 \end{bmatrix}; \quad \begin{bmatrix} 27 & -49 & 61 & 39 & -31 \\ 32 & -53 & 64 & 41 & -32 \\ 14 & -28 & 36 & 23 & -19 \\ 20 & -36 & 44 & 28 & -20 \end{bmatrix}; \quad \begin{bmatrix} -20 & 28 & -30 & 31 & 9 \\ 13 & -23 & 27 & -28 & -9 \\ -21 & 30 & -33 & 34 & 12 \\ 14 & -22 & 24 & -25 & -5 \end{bmatrix}.$$

3 · The Arithmetic of Matrices

In this chapter we bring matrices to life by showing how they can be added, subtracted and multiplied. In particular the addition of matrices reminds us of vector addition and suggests a natural extension of 2- and 3-dimensional vectors to greater dimensions. Multiplication of matrices, motivated by the substitution of one linear system in another, leads to a method of representing linear systems economically as well as to the possibility of 'mapping' from one dimension into another. We identify the main arithmetical rules satisfied by matrix addition and multiplication and, as two applications from what is, essentially, an infinite list(!) we describe (i) a method (the Markov process) of evaluating long-term trends and (ii) a method for sending secret messages.

In Chapter 2 we introduced the concept of matrix: an $m \times n$ matrix is merely a rectangular array of mn numbers arranged into m rows and n columns and, as such, is a fairly lifeless object. We will now see how matrices develop a life of their own by showing how they may (sensibly) be added and multiplied together.

Historically, it was matrix multiplication which appeared first but, as the definition is not especially easy to assimilate, we prefer to begin with a couple of simpler notions.

The most fundamental question one can ask if one is wishing to develop an arithmetic of matrices is: when should two matrices be regarded as equal? The only (?) sensible answer seems to be given by the following definition.

• Definition 1

Matrices $A = [a_{ij}]_{m\times n}$ and $B = [b_{kl}]_{r\times s}$ are **equal** iff $m = r$ and $n = s$ and $a_{uv} = b_{uv}$ for all u, v $(1 \le u \le m\ \{= r\}, 1 \le v \le n\ \{= s\})$. •

That is, two matrices are equal when and only when they 'have the same shape' and elements in corresponding positions are equal.

• Example 1

Given matrices

$$A = \begin{bmatrix} a & b \\ c & d \end{bmatrix}, \quad B = \begin{bmatrix} 5 & 13 \\ 3 & 8 \end{bmatrix} \quad \text{and} \quad C = \begin{bmatrix} r & s & t \\ u & v & w \end{bmatrix}$$

we see that neither A nor B can be equal to C (because C is the 'wrong shape') and that A and B are equal if, and only if, $a = 5$, $b = 13$, $c = 3$ and $d = 8$. •

How should we define the sum of two matrices? We really ought to admit that, as with all definitions in mathematics, the precise requirements of a definition are the prerogative of the person making it![a] However, the following has always seemed most appropriate.

[a]'When *I* use a word,' Humpty Dumpty said, 'it means just what I choose it to mean – neither more nor less.' (From: *Through the Looking Glass.*)

• Definition 2

Let

$$A=\begin{bmatrix} a_{11} & \dots & a_{1n} \\ \vdots & & \vdots \\ a_{m1} & \dots & a_{mn} \end{bmatrix} \quad \text{and} \quad B=\begin{bmatrix} b_{11} & \dots & b_{1n} \\ \vdots & & \vdots \\ b_{m1} & \dots & b_{mn} \end{bmatrix}$$

both be $m \times n$ matrices (so that they have the same shape). Their **sum** $A \oplus B$ is the $m \times n$ matrix

$$\begin{bmatrix} a_{11}+b_{11} & \dots & a_{1n}+b_{1n} \\ \vdots & & \vdots \\ a_{m1}+b_{m1} & \dots & a_{mn}+b_{mn} \end{bmatrix}.$$

That is, addition is *componentwise*. •

We use the symbol $\oplus$ (rather than +) to remind us that, whilst we are not actually adding numbers, we are doing something very similar – namely, adding arrays of numbers.

Example 2

$$\begin{bmatrix} 2 & 4 & 7 & -1 \\ 0 & -3 & 1 & 0 \\ 1 & 2 & 3 & 1 \end{bmatrix} \oplus \begin{bmatrix} 4 & 1 & 0 & 5 \\ -1 & 2 & -5 & 6 \\ 5 & 6 & -7 & 8 \end{bmatrix} = \begin{bmatrix} 6 & 5 & 7 & 4 \\ -1 & -1 & -4 & 6 \\ 6 & 8 & -4 & 9 \end{bmatrix}$$

whereas

$$\begin{bmatrix} 2 & 1 & 6 \\ 1 & -3 & 2 \\ 0 & 31 & -7 \end{bmatrix} \oplus \begin{bmatrix} 3 & 5 \\ 4 & 9 \\ -1 & 6 \end{bmatrix}$$

is not defined.

PROBLEM 3.1

Find numbers a, b, c and d such that

$$\begin{bmatrix} a & b \\ c & d \end{bmatrix} \oplus \begin{bmatrix} b & c \\ d & 2a \end{bmatrix} = \begin{bmatrix} 3 & -2 \\ 4 & 9 \end{bmatrix}.$$

•

In particular, the sum of the two 1×2 matrices[b] $[5 \quad 2]$ and $[2 \quad -3]$ is the 1×2 matrix $[7 \quad -1]$. The reader who is familiar with the idea of vectors in the plane will see from Fig. 3.1 that, in this case, matrix addition coincides with the usual parallelogram law for vector addition of vectors in the plane. A similar correspondence likewise exists between 1×3 matrices and vectors in 3-dimensional space. It then becomes natural to speak of the $1 \times n$ matrix $\begin{bmatrix} a_1 & a_2 & \dots & a_n \end{bmatrix}$ as being a *vector in n-dimensional space* – even though

[b] To avoid possible confusion, it is usual to insert commas between the components of a vector or a $1 \times n$ matrix, thus: [5, 2], [1, 11, 1, 111].

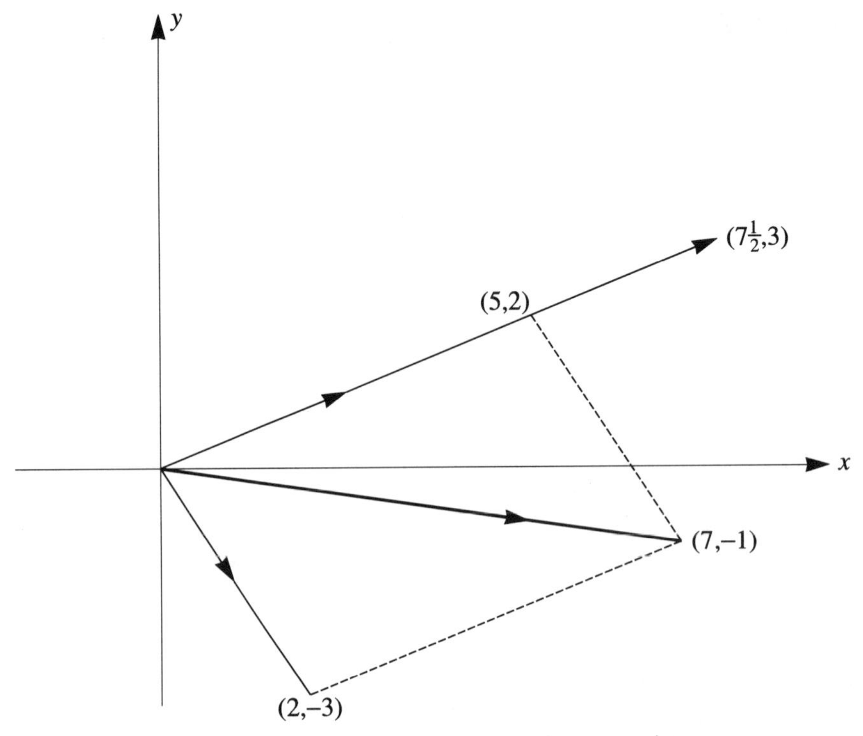

Fig 3.1 (5,2) + (2,–3) = (7,–1) and $1\frac{1}{2}$ (5,2) = ($7\frac{1}{2}$,3)

few of us can 'picture' n-dimensional space geometrically for $n \geq 4$. [Thus, for $n \geq 4$, the geometry of n-dimensional space seems hard but its corresponding algebraic version is equally easy for all n.]

Since it is the order in which the components of an n-dimensional vector occur which is important, we could equally represent such an n-vector by an $n \times 1$ matrix

$$\begin{bmatrix} a_1 \\ a_2 \\ \vdots \\ a_n \end{bmatrix}$$

rather than $\begin{bmatrix} a_1 & a_2 & \dots & a_n \end{bmatrix}$, and on many occasions we shall do just that. Later, we shall readily swap between the vector notation $\mathbf{v} = (a_1, a_2, \dots, a_n)$ and either of the above matrix forms, as we see fit and, in particular, usually use bold letters to represent $n \times 1$ and $1 \times n$ matrices.

Next we introduce multiplication into matrices. There are two types. To motivate the first consider the matrix sums $A \oplus A$ and $A \oplus A \oplus A$ where A is the matrix

$$\begin{bmatrix} a & b & c & d \\ p & q & r & s \\ x & y & z & t \end{bmatrix}.$$

Clearly

$$A \oplus A = \begin{bmatrix} 2a & 2b & 2c & 2d \\ 2p & 2q & 2r & 2s \\ 2x & 2y & 2z & 2t \end{bmatrix} \quad \text{whilst} \quad A \oplus A \oplus A = \begin{bmatrix} 3a & 3b & 3c & 3d \\ 3p & 3q & 3r & 3s \\ 3x & 3y & 3z & 3t \end{bmatrix}.$$

If, as is natural, we write the sum $A \oplus A \oplus \ldots \oplus A$ of k copies of A briefly as kA, we see that kA is the matrix each of whose elements is k times the corresponding element of A. There seems no reason why we should not extend this to any rational or even real value of k, as in the following definition.

• *Definition 3(i) Scalar Multiplication*

If α is a number (in this context often called a **scalar**) and if A is the $m \times n$ matrix above then αA is defined to be the $m \times n$ matrix

$$\begin{bmatrix} \alpha a_{11} & \ldots & \alpha a_{1n} \\ \vdots & & \vdots \\ \alpha a_{m1} & \ldots & \alpha a_{mn} \end{bmatrix} \quad (\text{briefly } [\alpha a_{ij}]).$$

•

(Thus, multiplying a $1 \times n$ (or $n \times 1$) matrix by a scalar corresponds, for $n = 2$ and 3, to the usual multiplication of a vector by a scalar. See Fig. 3.1.)

PROBLEM 3.2

Find, if possible, scalars α and β such that

$$\alpha \begin{bmatrix} 2 & 4 \\ 3 & -2 \end{bmatrix} \oplus \beta \begin{bmatrix} -3 & 5 \\ 2 & 8 \end{bmatrix} = \begin{bmatrix} -5 & 133 \\ 77 & 71 \end{bmatrix}.$$

•

To motivate the definition of the multiplication of two matrices we follow the historical path. Indeed, suppose that we have two systems of equations:

$$\begin{aligned} z_1 &= a_{11}y_1 + a_{12}y_2 + a_{13}y_3 \\ z_2 &= a_{21}y_1 + a_{22}y_2 + a_{23}y_3 \end{aligned} \quad \text{and} \quad \begin{aligned} y_1 &= b_{11}x_1 + b_{12}x_2 \\ y_2 &= b_{21}x_1 + b_{22}x_2 \\ y_3 &= b_{31}x_1 + b_{32}x_2. \end{aligned}$$

We associate, with these systems, the matrices of coefficients, namely

$$A = \begin{bmatrix} a_{11} & a_{12} & a_{13} \\ a_{21} & a_{22} & a_{23} \end{bmatrix} \quad \text{and} \quad B = \begin{bmatrix} b_{11} & b_{12} \\ b_{21} & b_{22} \\ b_{31} & b_{32} \end{bmatrix}.$$

Clearly we may substitute the y from the second system of equations into the first system and obtain the zs in terms of the xs. If we do this what matrix of coefficients do we get? It is fairly easy to check that the resulting matrix is the 2×2 matrix

$$C = \begin{bmatrix} c_{11} & c_{12} \\ c_{21} & c_{22} \end{bmatrix}$$

where, for example, $c_{21} = a_{21}b_{11} + a_{22}b_{21} + a_{23}b_{31}$ – and, generally,

$c_{ij} = a_{i1}b_{1j} + a_{i2}b_{2j} + a_{i3}b_{3j}$ where i and j are either of the integers 1 and 2. We call C the *product* of the matrices A and B. Notice how, for each i, j, the element c_{ij} of C is determined by the elements a_{i1}, a_{i2}, a_{i3} of the ith row of A and those, b_{1j}, b_{2j}, b_{3j}, of the jth column of B. Notice, too, how this definition requires that the number of columns of A must be equal to the number of rows of B – and that the number of rows (columns) of C is the same as the number of rows of A (columns of B).

PROBLEM 3.3

Given

$$\begin{aligned} z_1 &= y_1 + 3y_2 - y_3 \\ z_2 &= 2y_1 - y_2 - y_3 \end{aligned} \quad \text{and} \quad \begin{aligned} y_1 &= 2x_1 + x_2 \\ y_2 &= 3x_1 - x_2 \\ y_3 &= 4x_1 + 3x_2 \end{aligned}$$

find expressions for z_1 and z_2 in terms of x_1 and x_2. Write down the associated matrices of coefficients. ●

We adopt the above definition for the product of two general matrices in the following definition.

• Definition 3(ii) Multiplication of Two Matrices

Let A be an $m \times n$ matrix and B be an $n \times p$ matrix. (Note the positions of the two ns.) Then the **product** $A \odot B$ is the $m \times p$ matrix $[c_{ij}]_{m \times p}$ where for each i, j, we set $c_{ij} = a_{i1}b_{1j} + a_{i2}b_{2j} + \ldots + a_{in}b_{nj}$. ●

It might be worth while noting that each 'outside' pair of numbers in each product $a_{ik}b_{kj}$ is i, j whereas each 'inside' pair is a pair of equal integers ranging from 1 to n. We remind the reader that the very definition of c_{ij} explains why we insist that the number of columns of A must be equal to the number of rows of B.

Some examples should make things even clearer.

Example 3

Let

$$A = \begin{bmatrix} 3 & 1 & 7 \\ 2 & -5 & 4 \end{bmatrix}; \quad B = \begin{bmatrix} x & \alpha \\ y & \beta \\ z & \gamma \end{bmatrix}; \quad C = \begin{bmatrix} \frac{1}{2} & \frac{1}{2} \\ \frac{1}{2} & \frac{1}{2} \end{bmatrix}; \quad D = \begin{bmatrix} 1 & 2 \\ 3 & 4 \end{bmatrix}.$$

Then $A \odot B$ exists and the result

$$\begin{bmatrix} 3x+y+7z & 3\alpha+\beta+7\gamma \\ 2x-5y+4z & 2\alpha-5\beta+4\gamma \end{bmatrix} \quad \text{is } 2\times 2.$$

Also $B \odot A$ exists, the result,

$$\begin{bmatrix} x3+\alpha 2 & x1+\alpha(-5) & x7+\alpha 4 \\ y3+\beta 2 & y1+\beta(-5) & y7+\beta 4 \\ z3+\gamma 2 & z1+\gamma(-5) & z7+\gamma 4 \end{bmatrix} \quad \text{being } 3\times 3.$$

Finally observe that $D \odot A$ is the 2×3 matrix

$$\begin{bmatrix} 1.3+2.2 & 1.1+2.(-5) & 1.7+2.4 \\ 3.3+4.2 & 3.1+4.(-5) & 3.7+4.4 \end{bmatrix} = \begin{bmatrix} 7 & -9 & 15 \\ 17 & -17 & 37 \end{bmatrix}$$

and yet $A \odot D$ does not exist (since A is $2 \times \underline{3}$ and D is $\underline{2} \times 2$).

We have seen that $D \odot A$ exists and yet $A \odot D$ does not, and that $A \odot B$ and $B \odot A$ both exist and yet are (very much!) unequal (since they are not even the same shape!).

PROBLEM 3.4

Show that:

(i) $C \odot A$, $B \odot C$, $B \odot D$ all exist: $A \odot C$, $C \odot B$, $D \odot B$ do not.

(ii) $C \odot D$ and $D \odot C$ both exist and are both 2×2 but they are unequal.

(iii) Neither $A \odot A$ nor $B \odot B$ exists! And yet for each positive integer n, $C \odot C \odot \ldots \odot C$ and $D \odot D \odot \ldots \odot D$, where there are n terms in each product, both exist. [Just as for numbers, we then write $C \odot C \odot \ldots \odot C$ and $D \odot D \odot \ldots \odot D$ respectively as C^n and D^n. It can then be shown that the usual 'laws of exponents' hold, namely: if $r, s \geq 0$, we have $C^r \odot C^s = C^{r+s}$ and $(C^r)^s = C^{rs}$ where, by C^0, we mean the (appropriate sized) identity matrix. (See Exercise 13 of Chapter 2.) Of course, similar remarks apply to D.] You will find that $D^m = D^n$ only when $m = n$. We leave the joy of finding C^n, for each n, to you!

(iv) For 3×3 diagonal matrices

$$A = \begin{bmatrix} \alpha & 0 & 0 \\ 0 & \beta & 0 \\ 0 & 0 & \gamma \end{bmatrix} \quad \text{and} \quad B = \begin{bmatrix} x & 0 & 0 \\ 0 & y & 0 \\ 0 & 0 & z \end{bmatrix}$$

show that

$$AB = \begin{bmatrix} \alpha x & 0 & 0 \\ 0 & \beta y & 0 \\ 0 & 0 & \gamma z \end{bmatrix} = BA$$

with a similar result for two $n \times n$ diagonal matrices.

We break off to make a connection with the world of geometric vectors.

Suppose that $\mathbf{v}_1 = (a_1, b_1, c_1)$ and $\mathbf{v}_2 = (a_2, b_2, c_2)$ are two vectors in 3-dimensional space. You might recall that their *scalar product* (also known as their *dot product*) $\mathbf{v}_1.\mathbf{v}_2$ is defined to be the number $|\mathbf{v}_1||\mathbf{v}_2|\cos\vartheta$ where $|\mathbf{v}_1|$ is the *length* of the vector $\mathbf{v}_1$ (namely $\sqrt{(a_1^2 + b_1^2 + c_1^2)}$ — by Pythagoras' Theorem) and ϑ is the acute angle between the vectors $\mathbf{v}_1$ and $\mathbf{v}_2$. It turns out that $\mathbf{v}_1.\mathbf{v}_2$ is also equal to $a_1a_2 + b_1b_2 + c_1c_2$.

Here is a proof (see Fig. 3.2). By the cosine rule we have

$$|\mathbf{v}_1 - \mathbf{v}_2|^2 = |\mathbf{v}_1|^2 + |\mathbf{v}_2|^2 - 2|\mathbf{v}_1||\mathbf{v}_2|\cos\vartheta$$

Consequently

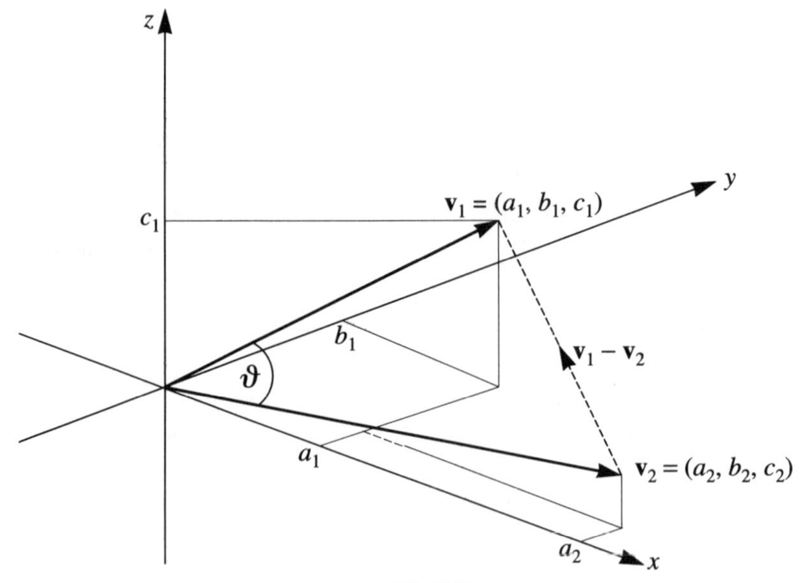

Fig 3.2

$$(a_1 - a_2)^2 + (b_1 - b_2)^2 + (c_1 - c_2)^2 = (a_1{}^2 + b_1{}^2 + c_1{}^2) + (a_2{}^2 + b_2{}^2 + c_2{}^2) - 2|\mathbf{v}_1||\mathbf{v}_2|\cos\vartheta.$$

It follows that $|\mathbf{v}_1||\mathbf{v}_2|\cos\vartheta = a_1a_2 + b_1b_2 + c_1c_2$, as claimed. In particular $\mathbf{v}_1$ and $\mathbf{v}_2$ are **orthogonal** (i.e. they are at right angles) if and only if $a_1a_2 + b_1b_2 + c_1c_2 = 0$.

Note that $\mathbf{v}_1.\mathbf{v}_2$ would be the matrix product

$$[a_1, b_1, c_1] \odot \begin{bmatrix} a_2 \\ b_2 \\ c_2 \end{bmatrix}$$

if we were prepared to identify each 1×1 matrix with its sole element.

There are yet more splendid consequences of Definition 3(ii). First note that it allows the system of equations (L) of Chapter 1 to be written succinctly as $A \odot \mathbf{x} = \mathbf{b}$ where A is the coefficient matrix – see Example 1(iii) of Chapter 2 – where

$$\mathbf{x} = \begin{bmatrix} x_1 \\ \vdots \\ x_n \end{bmatrix} \text{ and where } \mathbf{b} = \begin{bmatrix} b_1 \\ \vdots \\ b_2 \end{bmatrix}.$$

We shall build on this shortly.

In fact, merely writing the system of equations (L) in matrix form produces a burst of ideas – almost faster than one can write them down! For example:

(i) $A \odot \mathbf{x} = \mathbf{b}$ formally suggests the (unique?) solution $\mathbf{x} = \mathbf{b}/A = A^{-1} \odot \mathbf{b}$ (or should it be $\mathbf{b} \odot A^{-1}$?) But what, if anything, does this mean?

(ii) Regarding the $n \times 1$ matrix $\mathbf{x}$ as a point (or, vector) in n-dimensional space, A 'operates' on $\mathbf{x}$ to produce an $m \times 1$ matrix $A \odot \mathbf{x}$ which we may regard as a point (or, vector) in m-dimensional space. Thus the matrix A is capable of acting like a function – but giving a vector output for each (suitable) vector input.

As examples of this latter use of matrix multiplication we give the following, where we use $\mathbb{R}^n$ to denote the collection of all points (or vectors) in n-dimensional space or the corresponding collection of matrices. (See Example 6 of Chapter 6 and p. 217.)

Example 4

(i) The matrix

$$\begin{bmatrix} 0 & -1 \\ 1 & 0 \end{bmatrix}$$

'operates' on the vectors

$$\begin{bmatrix} x \\ y \end{bmatrix} \text{ of } \mathbb{R}^2 \text{ by sending } \begin{bmatrix} x \\ y \end{bmatrix} \text{ to } \begin{bmatrix} 0 & -1 \\ 1 & 0 \end{bmatrix} \odot \begin{bmatrix} x \\ y \end{bmatrix} = \begin{bmatrix} -y \\ x \end{bmatrix}.$$

Geometrically, this operation represents an anticlockwise rotation through 90 degrees about the origin.

(ii) More generally, the matrix

$$\begin{bmatrix} \cos\vartheta & -\sin\vartheta \\ \sin\vartheta & \cos\vartheta \end{bmatrix}$$

represents an anticlockwise rotation about the origin through the angle ϑ. (See Exercise 15 at the end of the chapter.)

PROBLEM 3.5

Let

A be the matrix $\begin{bmatrix} 1 & 0 \\ 0 & -1 \end{bmatrix}$, B the matrix $\begin{bmatrix} 0 & -1 \\ 1 & 0 \end{bmatrix}$ and $\mathbf{x}$ the vector $\begin{bmatrix} 1 \\ 4 \end{bmatrix}$.

Describe, geometrically, the effect on $\mathbb{R}^2$ of operating by the matrices A and B. Further determine the effect of letting the matrices $A \odot B$ and $B \odot A$ act, as functions (see above), on the vector $\mathbf{x}$.

Surely we can now say that matrices are beginning to come alive!

Just like ordinary numbers, matrices, when combined under $\oplus$, $\odot$ and scalar multiplication, satisfy some obvious rules (but *fail* to satisfy others!) We summarise some of these properties in the following theorem.

Theorem 1

Let A, B and C be any three $m \times n$ matrices and let α and β be scalars (i.e. numbers). Then:

We have	And we say that
(i) $A \oplus B$ is an $m \times n$ matrix	the set of all $m \times n$ matrices is *closed under addition*
(ii) $(A \oplus B) \oplus C = A \oplus (B \oplus C)$	the *associative law of addition* holds

(iii) there exists an $m \times n$ matrix denoted by $0_{m \times n}$ such that $A \oplus 0_{m \times n} = 0_{m \times n} \oplus A = A$	$0_{m \times n} = \begin{bmatrix} 0 & \dots & 0 \\ \vdots & & \vdots \\ 0 & \dots & 0 \end{bmatrix}$, the so-called **zero** $\boldsymbol{m \times n}$ **matrix**, is the additive identity for the set of $m \times n$ matrices.

(Thus $0_{m \times n}$ is the equivalent, for $m \times n$ matrices, of the number 0.)

(iv) there exists an $m \times n$ matrix denoted by $\ominus A$ such that $A \oplus (\ominus A) = \ominus A \oplus A = 0_{m \times n}$	$\ominus A = (-1)A$ is the additive inverse or negative for A

(Thus $\ominus A$ corresponds to A just as the number $-a$ corresponds to the number a.)

(v) $A \oplus B = B \oplus A$	the commutative law of addition holds
(vi) αA is an $m \times n$ matrix	the set of all $m \times n$ matrices is closed under scalar multiplication
(vii) $\alpha(A \oplus B) = \alpha A \oplus \alpha B$	the first distributive law for scalar multiplication holds
(viii) $(\alpha + \beta)A = \alpha A \oplus \beta A$	the second distributive law for scalar multiplication holds
(ix) $\alpha(\beta A) = (\alpha\beta)A$	the associative law for scalar multiplication holds
(x) $1A = A$	an identity for scalar multiplication exists ●

There is a special reason, revealed in Chapter 6, as to why these 10 rules have been isolated.

(i) to (v) above deal entirely with properties of matrix addition. It is natural to check whether or not their analogues hold for matrix multiplication. We shall see that some do – and some do not! However, before summarising the facts, we invite the reader to try the following exercise.

PROBLEM 3.6

Let A be the 2×3 matrix

$$\begin{bmatrix} a & b & c \\ d & e & f \end{bmatrix} \quad \text{and set} \quad B = \begin{bmatrix} 1 & 0 \\ 0 & 1 \end{bmatrix}.$$

Show that $B \odot A = A$. Is $A \odot B = A$? If not, can you find a matrix, C say, such that $A \odot C = A$? [Such a C, if it exists at all, must be of size – what?] ●

Now we examine the multiplicative analogues (labelled (I) to (V)) of properties (i) to (v).

● Theorem I (cont.)

Let A be $m \times n$, B be $n \times p$ and C be $p \times r$ matrices. Then:

(I) $A \odot B$ is $m \times p$, $B \odot C$ is $n \times r$	
(II) $(A \odot B) \odot C = A \odot (B \odot C)$	the associate law of multiplication holds

(III) there exists an $m \times m$ matrix, namely[c] I_m and an $n \times n$ matrix, namely I_n, such that $I_m \odot A = A \odot I_n = A$

$I_s = \begin{bmatrix} 1 & & 0 \\ & 1 \cdot \cdot & \\ 0 & & \cdot 1 \end{bmatrix}$ is a left (respectively right) identity for multiplication of $s \times n$ (respectively $m \times s$) matrices

(IV) There can be no multiplicative analogue to (iv) if A is not square. [Multiplicative inverses for matrices are considered in the next chapter.]

(V) There is no multiplicative analogue to (v) even if A and B are square and of the same size (as Problem 3.4(ii) shows).

Finally, three observations. First: you will recall that, for ordinary real numbers a, b and c the distributive laws always hold: that is, (i) $a(b + c) = ab + ac$ and (ii) $(a + b)c = ac + bc$. There are similar rules for matrices (when their shapes are compatible!). We have

(D) If B_1 is $n \times p$ and A_1 is $m \times n$ then $A \odot (B \oplus B_1) = (A \odot B) \oplus (A \odot B_1)$; $(A \oplus A_1) \odot B = (A \odot B) \oplus (A_1 \odot B)$

the distributive laws for matrix multiplication ●

PROBLEM 3.7

Given

$$A = \begin{bmatrix} 11 & -13 \\ -2 & 3 \end{bmatrix}, \quad B = \begin{bmatrix} -112 & 74 & 313 \\ 104 & -68 & 53 \end{bmatrix} \quad \text{and} \quad C = \begin{bmatrix} 113 & -76 & --313 \\ -103 & 68 & -52 \end{bmatrix}$$

evaluate $(A \odot B) \oplus (A \odot C)$. ●

Second: for each scalar α, each $m \times n$ matrix A and each $n \times p$ matrix B, we have

$$A \odot (\alpha B) = \alpha(A \odot B) = (\alpha A) \odot B.$$

Third: suppose real numbers a and b are such that $a.b = 0$. One may deduce immediately that either a is 0 or $b = 0$ (or both!). There is no analogous result for matrices! – as the following exercise will show.

PROBLEM 3.8

Find a (2×2) matrix

$$\begin{bmatrix} a & b \\ c & d \end{bmatrix} \text{ (other than } \begin{bmatrix} 0 & 0 \\ 0 & 0 \end{bmatrix}!)$$

such that

$$\begin{bmatrix} 27 & 63 \\ 21 & 49 \end{bmatrix} \odot \begin{bmatrix} a & b \\ c & d \end{bmatrix} = \begin{bmatrix} 0 & 0 \\ 0 & 0 \end{bmatrix}.$$

Can you find more than 1000 such matrices? ●

[c]See Exercise 13 at the end of Chapter 2.

Most of the assertions in Theorem 1 are embarrassingly easy (and also amazingly tedious) to check, i.e. prove! One which is neither is (II). The proof can be obtained in several ways. The following is probably the most naïve, and hence the clumsiest (from the point of view of elegance). The proof of this theorem for matrices of specific sizes – say $A_{2\times 4}$, $B_{4\times 3}$, $C_{3\times 5}$ – is essentially no easier to write down than the general proof. Nevertheless, any reader who is not yet at ease multiplying matrices of general sizes $m \times n$, $n \times r$, $r \times s$ might care to write out in full the proof of Theorem 2 for the special case of 2×4, 4×3 and 3×5 matrices and then try to follow the more general proof of the same theorem.

• *Theorem 2*

Matrix multiplication is associative. (From now on we shall, where convenient, replace $A \odot B$ by AB and $A \oplus B$ by $A + B$, etc.)

PROOF

Let matrices $A_{m\times n}$, $B_{n\times r}$ and $C_{r\times s}$ be given. Let the pqth elements of AB and BC be denoted by d_{pq} and e_{pq} respectively. Then the ijth element of $(AB)C$ is, by definition,

$$\begin{aligned}
d_{i1}c_{1j} + d_{i2}c_{2j} + \ldots + d_{ir}c_{rj} &= (a_{i1}b_{11} + a_{i2}b_{21} + \ldots + a_{in}b_{n1})c_{1j} \\
&\quad + (a_{i1}b_{12} + a_{i2}b_{22} + \ldots + a_{in}b_{n2})c_{2j} \\
&\quad + \ldots + (a_{i1}b_{1r} + a_{i2}b_{2r} + \ldots + a_{in}b_{nr})c_{rj} \\
&= a_{i1}(b_{11}c_{1j} + b_{12}c_{2j} + \ldots + b_{1r}c_{rj}) \\
&\quad + a_{i2}(b_{21}c_{1j} + b_{22}c_{2j} + \ldots + b_{2r}c_{rj}) \\
&\quad + \ldots + a_{in}(b_{n1}c_{1j} + b_{n2}c_{2j} + \ldots + b_{nr}c_{rj}) \\
&= a_{i1}e_{1j} + a_{i2}e_{2j} + \ldots + a_{in}e_{nj}
\end{aligned}$$

which is the ijth element of $A(BC)$.

[For those who are happy using the summation notation Σ, the above calculations and the associativity of matrix multiplication are readily expressed by the following statement. For each i, j $(1 \le i \le m,\ 1 \le j \le s)$ we have

$$\sum_{\beta=1}^{r}\left\{\sum_{\alpha=1}^{n} a_{i\alpha}b_{\alpha\beta}\right\}c_{\beta j} = \sum_{\alpha=1}^{n}\left\{a_{i\alpha}\sum_{\beta=1}^{r} b_{\alpha\beta}c_{\beta j}\right\}$$

which is often written as

$$\sum_{\beta=1}^{r}\sum_{\alpha=1}^{n} a_{i\alpha}b_{\alpha\beta}c_{\beta j}\left(=\sum_{\alpha=1}^{r}\sum_{\beta=1}^{n} a_{i\alpha}b_{\alpha\beta}c_{\beta j}\right).]$$

Since elements in all corresponding places in $(AB)C$ and $A(BC)$ are equal we deduce that $(AB)C = A(BC)$ as claimed. ●

As a result of Theorem 2 we can denote the common value of $(AB)C$ and $A(BC)$ quite unambiguously by ABC. (But see the remark following Theorem 4.)

A related consequence of Theorem 2 which is too nice to ignore is the following theorem.

• *Theorem 3*

Let A, B, C and D be matrices of shapes $k \times l$, $l \times m$, $m \times n$, $n \times p$ respectively. Then all five fourfold products $(AB)(CD)$, $((AB)C)D$, $(A(BC))D$, $A((BC)D)$ and $A(B(CD))$ exist and are equal (so that the result may be denoted unambiguously by $ABCD$).

PROOF

We shall prove that the first two are equal and leave the rest to you. Accordingly, set $X = AB$. Then, by Theorem 2, we have $X(CD) = (XC)D$. Thus, immediately, $\{AB\}(CD) = (\{AB\}C)D$ – where we have used { } to highlight the replacement of X by the product AB. ●

[In a similar way one can show that Theorem 1 leads to (expected) equalities such as $A(B + C + D) = AB + AC + AD$ and $(A + 2B)(A + 4B + 3C) = A^2 + 4AB + 3AC + 2BA + 8B^2 + 6BC$. (Note that the AB and BA terms cannot, in general, be amalgamated.)]

PROBLEM 3.9

Complete the proof of Theorem 3. ●

TUTORIAL PROBLEM 3.1

An intriguing question (asked only for fun) is: in how many ways can the product of 5, of 6, . . . etc. matrices be evaluated (there being five possible products of four matrices as in Theorem 3)?

In fact, whatever the number, one can prove , using only (II) of Theorem 1, the following theorem.

An often repeated assertion in the history of matrices is that Cayley founded the theory of matrices in a paper published in 1858. Hawkins has explained why this assertion is not acceptable. Certainly Cayley observed that matrices can be added and multiplied (as described above) but, for one thing, his efforts were almost totally ignored for 25 years. In any case, others, Gauss, for instance, had already multiplied pairs of linear substitutions after representing them as square arrays. Later, Eisenstein, following Gauss in describing such arrays by single letters such as S and T, had warned that $S \times T$ and $T \times S$ will, in general, be different. Furthermore, Eisenstein showed that arrays could be added (though he did nothing with this idea). In fact Cayley's 1858 paper contains only one result of note – the so-called Cayley–Hamilton theorem. (See Exercise 11 and computer package problem 9 in Chapter 11.) Cayley proves the result for 2×2 and says he has checked it for 3×3 matrices but, quite remarkably, sees no need to offer a proof for matrices of larger size. Hamilton's contribution was to establish the 4×4 case. (A proof for $n \times n$ matrices will appear in the follow-up volume to this.)

Ferdinand Gotthold Max Eisenstein was born in Berlin on 16 April 1823. Five brothers and sisters all died in childhood and he, himself, was continually plagued by

ill health and depression. His autobiography reveals that, aged six, he could more easily understand mathematical proof than that meat should be cut with a knife rather than with a fork! Returning to Berlin in 1843 after meeting Hamilton in Dublin, Eisenstein produced a mountain of high-class research, mainly in aspects of number theory. Gauss, who, it seems, did not suffer fools gladly, welcomed him to Göttingen for an extended visit. Despite his great fame, Eisenstein died in misery of tuberculosis on 11 October 1852 at the tragically early age of 29.

• *Theorem 4 (The Generalised Associative Law)*

Given matrices $A_1, A_2, \ldots, A_t$ (such that, for $1 \leq i \leq t-1$, the number of columns of A_i is equal to the number of rows of A_{i+1}), then all ways of working out their product, in the given order, give the same result (which may, therefore, be denoted unambiguously by $A_1A_2 \ldots A_t$). •

REMARK Note that Theorem 2 must not be regarded as 'obviously true'. Since, for example, $7-(2-4) \neq (7-2)-4$, we see that subtraction is not an associative 'operation' on the set of integers. Nor is the vector product operation, on the set of all 3-dimensional vectors. (See Exercise 13(iii).)

TUTORIAL PROBLEM 3.2

Can you dream up a new rule, $\diamond$ say, (shall we call it 'sedition'?) on the set of all integers such that, for all pairs a, b of integers, (i) $a \diamond b$ is again an integer, (ii) $a \diamond b = b \diamond a$ but (iii) $(a \diamond b) \diamond c \neq a \diamond (b \diamond c)$ for certain integers a, b, c?

Applications

(i) Markov Chains (or processes)

These were named after Andrei Andreevich Markov, 14 June 1856–20 May 1922. Rather than give a formal definition of a Markov process we give an example.

In a certain city there are three large supermarkets A, B and C. Suppose each family in the city buys its groceries from just one of these stores but the probability that each year a family changes its allegiance from one store to another is summarised by the matrix

$$\begin{array}{c} \text{from} \\ \begin{array}{ccc} \text{A} & \text{B} & \text{C} \end{array} \end{array}$$

$$T = \begin{bmatrix} 0.8 & 0.15 & 0.05 \\ 0.2 & 0.7 & 0.35 \\ 0 & 0.15 & 0.6 \end{bmatrix} \quad \text{to} \quad \begin{array}{c} \text{A} \\ \text{B} \\ \text{C} \end{array}$$

where, for example, 80% of all families patronising A remain with store A whilst 20% change to store B and (most unlikely?) none changes to 'C'. [Clearly the entries in each column must add up to 1. T is called a **transition matrix**.]

Suppose we make the (rash) assumption that T remains fixed over a long time span, and that the vector

$$\mathbf{p} = \begin{bmatrix} x \\ y \\ z \end{bmatrix}$$

describes the proportions of the population in classes A, B and C at the beginning of the survey. Then the proportions in each class after k years is given by

$$T^k \begin{bmatrix} x \\ y \\ z \end{bmatrix},$$

the transition matrix for this period being T^k. One may then ask: 'What proportions of the population will shop at each store in the long term?'

It turns out that, as k increases, T^k approaches a matrix of the form

$$\begin{bmatrix} \alpha & \alpha & \alpha \\ \beta & \beta & \beta \\ \gamma & \gamma & \gamma \end{bmatrix} \quad \text{with three identical columns} \quad s = \begin{bmatrix} \alpha \\ \beta \\ \gamma \end{bmatrix},$$

where α, β, γ give the appropriate proportions. (For a proof of this see Chapter 11.)

Here are some sample values of T^k:

$$T^5 = \begin{bmatrix} 0.48\ldots & 0.33\ldots & 0.28\ldots \\ 0.40\ldots & 0.47\ldots & 0.48\ldots \\ 0.11\ldots & 0.19\ldots & 0.23\ldots \end{bmatrix}, \quad T^{10} = \begin{bmatrix} 0.398\ldots & 0.372\ldots & 0.362\ldots \\ 0.443\ldots & 0.454\ldots & 0.458\ldots \\ 0.158\ldots & 0.173\ldots & 0.179\ldots \end{bmatrix},$$

$$T^{20} = \begin{bmatrix} 0.3808... & 0.3800... & 0.3797... \\ 0.4504... & 0.4508... & 0.4509... \\ 0.1691... & 0.1691... & 0.1693... \end{bmatrix}, \quad T^{30} = \begin{bmatrix} 0.38029... & 0.38027... & 0.38026... \\ 0.45069... & 0.45070... & 0.45071... \\ 0.16900... & 0.16901... & 0.16902... \end{bmatrix}$$

Therefore, in the long run, store B will gain almost half of all the city's custom.

(ii) Constructing Secret Codes

There are many ways of constructing codes, the object of which is to pass secret messages to desired recipients without their being decoded by unauthorised persons. One of the simplest (and easily breakable) codes replaces each letter *a*, *b*, *c* . . ., *u*, *v*, *w*, *x*, *y*, *z* of the alphabet by a 'shifted' version, for example: *f*, *g*, *h*, . . ., *z*, *a*, *b*, *c*, *d*, *e*, etc. Somewhat harder to crack appear to be the so-called public-key encryption codes which rely, for security, on the general difficulty of factoring numbers which are products of large primes. Much nearer to the former in terms of the ease with which they can be broken are codes formed as follows.

Let A be (say) a 2×2 matrix with integer coefficients, for example

$$A = \begin{bmatrix} 4 & 7 \\ 1 & 2 \end{bmatrix}$$

and let the letters *a*, *b*, *c*, . . . be replaced by the numbers 1, 2, 3, ... To send the message BL|AC|KP|OO|LF|OR|TH|EC|UP write the message in number form $2,12|1,3|11,16|\ldots$ etc. The coded message then comprises the number pairs

$$A\begin{bmatrix} 2 \\ 12 \end{bmatrix}, A\begin{bmatrix} 1 \\ 3 \end{bmatrix}, A\begin{bmatrix} 11 \\ 16 \end{bmatrix}, \ldots \text{etc.},$$

that is $92,26|25,7|156,43|\ldots$ etc.

To unscramble the coded message the recipient must find a matrix B, say, which changes $92,26|25,7|156,43|\ldots$ etc. back to $2,12|1,3|11,16|\ldots$ etc. again. Thus (s)he must find a 2×2 matrix B such that

$$BA\begin{bmatrix} 2 \\ 12 \end{bmatrix} = \begin{bmatrix} 2 \\ 12 \end{bmatrix}, \quad BA\begin{bmatrix} 1 \\ 3 \end{bmatrix} = \begin{bmatrix} 1 \\ 3 \end{bmatrix}, \text{ etc.}$$

Certainly any B for which $BA = I_2$, the 2×2 identity matrix, would fit the bill. That is, B should be a (and, as it turns out, the, unique) multiplicative inverse for A. [In particular A must have such an inverse.] Inverses of matrices are the subject of the next chapter.

TUTORIAL PROBLEM 3.3

It might be nice to encode letter pairs into other letter pairs (rather than into number pairs). Can you think of a way in which this might be done? (Is there, then, any problem with the decoding?)

Summary

Let $A_{m\times n}$ and $B_{r\times s}$ be matrices (which we denote briefly by A, B.) A and B are **equal** if (and only if) they are of the 'same shape' and elements in corresponding positions are identical. The **sum** $A \oplus B$ is defined when (and only when) $m = r$ and $n = s$ and then $A \oplus B$ is given by componentwise addition. The sum of two $n \times 1$ (or $1 \times n$) matrices is analogous to their sum when they are regarded as vectors in n-dimensional space.

For each scalar (i.e. number) α the **scalar multiple** αA of A (= $[a_{ij}]$) is the matrix whose ijth entry is αa_{ij}. The (matrix) **product** $A \odot B$ is defined when (and only when!) $n = r$. The result is the $m \times s$ matrix whose ijth entry is 'the product of the ith row of A with the jth row of B'. Under $\oplus$, $\odot$ and scalar multiplication, matrices (of appropriate shapes) satisfy many of the ordinary laws of arithmetic (see Theorem 1). In particular, if matrices A, B and C are the correct shape, we have $(A \odot B) \odot C = A \odot (B \odot C)$. Two notable exceptions are that, in general, it is not true that $A \odot B = B \odot A$, nor is it universally true that, if $A \odot B = 0_{m\times s}$ (a zero matrix) then at least one of A and B must be a zero matrix.

The definition of matrix multiplication, motivated by substitution of variables, has the useful consequence that systems of equations can be expressed succinctly in the form $A\mathbf{x} = \mathbf{b}$.

EXERCISES ON CHAPTER 3

(Here we will frequently replace $\oplus$ by + and omit $\odot$.)

1. Solve for a, b, c and d: $\begin{bmatrix} b & c \\ d & 1 \end{bmatrix} + \begin{bmatrix} a & b \\ c & d \end{bmatrix} = \begin{bmatrix} 4 & 7 \\ 0 & -2 \end{bmatrix}$.

2. Solve, if possible, for α: $\alpha\begin{bmatrix} 615 & -43 \\ 0 & 412 \end{bmatrix} = \begin{bmatrix} 47355 & -3311 \\ 0 & 31274 \end{bmatrix}$.

3. Let

$$A = \begin{bmatrix} 1 & 7 & 2 \\ 0 & -5 & 3 \end{bmatrix} \quad B = \begin{bmatrix} a & X \\ b & Y \\ c & Z \end{bmatrix} \quad C = \begin{bmatrix} \frac{1}{3} & \frac{1}{3} \\ \frac{2}{3} & \frac{2}{3} \end{bmatrix} \quad D = \begin{bmatrix} 2 & 3 \\ 4 & 5 \end{bmatrix}.$$

Find, if possible, AB, $(AB)C$, BC, CB, C^{100}, $A(BC)$, $(AB)(CD)$, $A(B(CD))$, $(AB)^2$ and A^2B^2.

4. (a) Give, if possible, examples of 2×2 matrices, A, B and C such that

$$A \neq \begin{bmatrix} 0 & 0 \\ 0 & 0 \end{bmatrix}$$

and $AB = AC$ and yet $B \neq C$.

(b) Find 3×3 matrices A and B such that $A \neq 0_{3\times 3}$ and $B \neq 0_{3\times 3}$ and yet $AB = 0_{3\times 3}$.

(c) Find, if possible, a (square) matrix such that $A^3 \neq 0$ but $A^4 = 0$.

5. Find a 2×2 matrix A, other than

$$\begin{bmatrix} \pm 1 & 0 \\ 0 & \pm 1 \end{bmatrix}, \quad \begin{bmatrix} \pm 1 & 0 \\ 0 & 0 \end{bmatrix}, \quad \begin{bmatrix} 0 & 0 \\ 0 & \pm 1 \end{bmatrix}, \quad \begin{bmatrix} \frac{1}{2} & \frac{1}{2} \\ \frac{1}{2} & \frac{1}{2} \end{bmatrix}$$

for which $A^2 = A\ (= AI_2)$. Can you find infinitely many different such A? Use this result to solve Exercise 4(a) again.

6. (a) Find 2×2 matrices A and B such that $(AB)^2 \neq A^2B^2$.
(b) Find 2×2 matrices C and D such that $CD \neq DC$ and yet $(CD)^2$ is equal to C^2D^2.

7. Let

$$X = \begin{bmatrix} -1 & 0 \\ 0 & 1 \end{bmatrix}, \quad Y = \begin{bmatrix} 0 & 1 \\ 1 & 0 \end{bmatrix} \quad \text{and} \quad Z = \begin{bmatrix} 0 & \mathrm{i} \\ -\mathrm{i} & 0 \end{bmatrix},$$

where $\mathrm{i} = \sqrt{-1}$. Prove that $XY = -YX = \mathrm{i}Z$, $YZ = -ZY = \mathrm{i}X$, $ZX = -XZ = \mathrm{i}Y$ and $X^2 = Y^2 = Z^2 = I_2$.

8. (a) Let

$$A = \begin{bmatrix} 3 & -1 \\ 2 & 5 \end{bmatrix}.$$

Find (using only pencil and paper) A^{16} and A^{13}. [Hint: $A^{16} = (((A^2)^2)^2)^2$ and $A^{13} = A^8.A^4.A$. It is nice to get A^{16} with only four multiplications!]

(b) Given that

$$\begin{bmatrix} a & b \\ c & d \end{bmatrix} = \begin{bmatrix} 1 & 1 \\ 0 & 1 \end{bmatrix}^{5039}$$

find a, b, c and d. Prove your claim is correct by mathematical induction. [Hint: to see what the induction hypothesis should be, evaluate

$$\begin{bmatrix} 1 & 1 \\ 0 & 1 \end{bmatrix}^{n}$$

for $n = 2, 3, 4$.]

9. Let A and B be matrices, B having n columns, such that AB exists. Let the matrix B be thought of as a matrix $[B_1\, B_2 \ldots B_n]$ comprising n column vectors B_i.
(a) Show that the jth column of AB is the matrix AB_j. Identify, similarly, the ith row of AB.
(b) If A has a row full of zeros prove that AB also has such a row. What, if anything, may be said about BA?
(c) If $\mathbf{x}$ is the vector

$$\begin{bmatrix} x_1 \\ \vdots \\ x_n \end{bmatrix}$$

show that $B\mathbf{x} = B_1x_1 + B_2x_2 + \ldots + B_nx_n$.

10. Let A be a 2×2 matrix such that A commutes (multiplicatively) with B (that is, $AB = BA$) for every 2×2 matrix B. Show that A has the form

$$\begin{bmatrix} \lambda & 0 \\ 0 & \lambda \end{bmatrix} \text{ for some scalar } \lambda. \text{ [Hint: begin by taking } B = \begin{bmatrix} 0 & 1 \\ 0 & 0 \end{bmatrix}.]$$

11. Let A be the matrix

$$\begin{bmatrix} a & b \\ c & d \end{bmatrix}.$$

Noting that $(a - x)(d - x) - bc = x^2 - (a + d)x + ad - bc$, calculate the matrix $A^2 - (a + d)A + (ad - bc)I_2$. (This is an instance of the celebrated Cayley–Hamilton Theorem.)

12. (a) Show $\mathbf{v}.\mathbf{v}$ is the square of the length of $\mathbf{v}$, that $\mathbf{v}.\mathbf{v} \geq 0$ and that $\mathbf{v}.\mathbf{v} = 0$ when and only when $\mathbf{v} = \mathbf{0}$.

(b) Show that $\mathbf{u}.\mathbf{v} = \mathbf{v}.\mathbf{u}$ and that $\mathbf{u}.(\mathbf{v} + \mathbf{w}) = \mathbf{u}.\mathbf{v} + \mathbf{u}.\mathbf{w}$.

(c) If $\mathbf{v} = (a_1 + ib_1, \ldots, a_n + ib_n)$ is a vector with complex number entries, and if $\overline{\mathbf{v}} = (a_1 - ib_1, \ldots, a_n - ib_n)$, show that the dot product $\mathbf{v}.\overline{\mathbf{v}}$ is a real number and that $\mathbf{v}.\overline{\mathbf{v}} > 0$ unless $\mathbf{v} = \mathbf{0}$, the zero vector.

13. For vectors $\mathbf{v}_1 = (a_1, b_1, c_1)$, $\mathbf{v}_2 = (a_2, b_2, c_2)$ in $\mathbb{R}^3$ we define their **vector product** to be the vector $\mathbf{v}_1 \wedge \mathbf{v}_2 = (b_1c_2 - b_2c_1, c_1a_2 - c_2a_1, a_1b_2 - a_2b_1)$. Show that (i) $\mathbf{v}_1 \wedge \mathbf{v}_2 = -(\mathbf{v}_2 \wedge \mathbf{v}_1)$, (ii) $\mathbf{v}_1 \wedge (\mathbf{v}_1 \wedge \mathbf{v}_2) = \mathbf{v}_2 \wedge (\mathbf{v}_1 \wedge \mathbf{v}_2) = (0, 0, 0)$ {so that $\mathbf{v}_1$ and $\mathbf{v}_2$ are each orthogonal to $\mathbf{v}_1 \wedge \mathbf{v}_2$} and find particular $\mathbf{v}_1$ and $\mathbf{v}_2$ such that (iii) $(\mathbf{v}_1 \wedge \mathbf{v}_2) \wedge \mathbf{v}_2 \neq \mathbf{v}_1 \wedge (\mathbf{v}_2 \wedge \mathbf{v}_2)$.

14. Show that

$$\begin{bmatrix} -3 & -8 & 2 \\ 1 & 3 & 4 \\ 1 & 3 & 1 \end{bmatrix} \begin{bmatrix} 1 \\ 2 \\ 5 \end{bmatrix} = \begin{bmatrix} 1 & 5 & -4 \\ 0 & 1 & 5 \\ -1 & -6 & 5 \end{bmatrix} \begin{bmatrix} 1 \\ 2 \\ 5 \end{bmatrix}.$$

(Can you think of a way to halve the work?)

15. (a) By applying the matrix

$$\begin{bmatrix} \cos\vartheta & -\sin\vartheta \\ \sin\vartheta & \cos\vartheta \end{bmatrix} \text{ to the point (or vector) } \begin{bmatrix} r\cos\alpha \\ r\sin\alpha \end{bmatrix},$$

show that the given matrix describes an anticlockwise rotation in the plane.

(b) Show that the product of

$$\begin{bmatrix} \cos\vartheta & -\sin\vartheta \\ \sin\vartheta & \cos\vartheta \end{bmatrix} \text{ and } \begin{bmatrix} \cos\psi & -\sin\psi \\ \sin\psi & \cos\psi \end{bmatrix} \text{ is } \begin{bmatrix} \cos(\vartheta+\psi) & -\sin(\vartheta+\psi) \\ \sin(\vartheta+\psi) & \cos(\vartheta+\psi) \end{bmatrix}.$$

Describe the transformations of the plane which these three matrices represent.

(c) Show, by induction on t, that for each positive integer t and for each angle ϑ we have

$$\begin{bmatrix} \cos\vartheta & -\sin\vartheta \\ \sin\vartheta & \cos\vartheta \end{bmatrix}^t = \begin{bmatrix} \cos t\vartheta & -\sin t\vartheta \\ \sin t\vartheta & \cos t\vartheta \end{bmatrix}.$$

(d) What is the geometric effect of transformation by the matrix $\begin{bmatrix} \cos\phi & \sin\phi \\ \sin\phi & -\cos\phi \end{bmatrix}$?

16. (1,1), (–1,1), (–1,–1) (1,–1) are the four corners of a square. Find where they are sent under the action of the matrix

$$\begin{bmatrix} 1 & \frac{1}{2} \\ 0 & 1 \end{bmatrix}.$$

Plot the given points (in blue) and the new points (in red) in the x–y plane to see how the matrix shears the square. Do the same for the matrix

$$\begin{bmatrix} -1 & 3 \\ 2 & -1 \end{bmatrix}.$$

17. Find all $\begin{bmatrix} x & y \\ z & t \end{bmatrix}$, if any, for which:

(i) $\begin{bmatrix} 2 & 3 \\ 4 & 5 \end{bmatrix}\begin{bmatrix} x & y \\ z & t \end{bmatrix} = \begin{bmatrix} 2 & 3 \\ 4 & 6 \end{bmatrix}$; (ii) $\begin{bmatrix} x & y \\ z & t \end{bmatrix}\begin{bmatrix} 2 & 3 \\ 4 & 5 \end{bmatrix} = \begin{bmatrix} 2 & 3 \\ 4 & 6 \end{bmatrix}$;

(iii) $\begin{bmatrix} 9 & 6 \\ 6 & 4 \end{bmatrix}\begin{bmatrix} x & y \\ z & t \end{bmatrix} = \begin{bmatrix} 1 & 2 \\ 3 & 4 \end{bmatrix}$; (iv) $\begin{bmatrix} 7 & 10 \\ 9 & 13 \end{bmatrix}\begin{bmatrix} x & y \\ z & t \end{bmatrix} = \begin{bmatrix} 0 & 0 \\ 0 & 0 \end{bmatrix}$.

18. If $\begin{bmatrix} 6 & -9 \\ -8 & 12 \end{bmatrix}\begin{bmatrix} x & y \\ z & t \end{bmatrix} = \begin{bmatrix} 0 & 0 \\ 0 & 0 \end{bmatrix}$ must $\begin{bmatrix} x & y \\ z & t \end{bmatrix}\begin{bmatrix} 6 & -9 \\ -8 & 12 \end{bmatrix} = \begin{bmatrix} 0 & 0 \\ 0 & 0 \end{bmatrix}$?

19. Find a, b, c and d such that

$$\left(\begin{bmatrix} 3 & -1 \\ 5 & 2 \end{bmatrix} + \begin{bmatrix} a & b \\ c & d \end{bmatrix}\right)\begin{bmatrix} 9 & 4 \\ 6 & -2 \end{bmatrix} = \begin{bmatrix} 7 & 1 \\ 7 & 4 \end{bmatrix}.$$

Are a, b, c and d uniquely determined?

20. Prove that if A is $m \times n$ and if α is a scalar, then $(\alpha I_m)A = \alpha A$.

21. [We temporarily return to the $\oplus$, $\odot$ notation for clarity.]

(a) Let A, B and C be matrices such that $A \odot (B \oplus C)$ exists. Prove that $(A \odot B) \oplus (A \odot C)$ exists and is equal to $A \odot (B \oplus C)$. [If you cannot deal with general matrices take A, B and C to be of sizes 2×4, 4×3 and 4×3 respectively.]

(b) Find three different 2×2 matrices A, B and C, all of whose entries are non-zero, such that $A \oplus (B \odot C) = (A \oplus B) \odot (A \oplus C)$. [This example is meant to convince you that this 'dual' of the distributive law rarely occurs.]

22. Describe the transformations of 3-dimensional space represented by the matrices

$$\begin{bmatrix} \cos\vartheta & -\sin\vartheta & 0 \\ \sin\vartheta & \cos\vartheta & 0 \\ 0 & 0 & 1 \end{bmatrix} \quad \text{and} \quad \begin{bmatrix} \cos\psi & 0 & -\sin\psi \\ 0 & 1 & 0 \\ \sin\psi & 0 & \cos\psi \end{bmatrix}.$$

Find their product. This should represent a rotation in 3-dimensional space about the origin. What is the axis of rotation if $\vartheta = \psi = \pi/2$? (Cf. Exercise 29 at the end of Chapter 11.)

23. The **trace** $\mathrm{Tr}(A)$ of the (square) $n \times n$ matrix A is defined to be the sum of its diagnonal elements. Prove that, for $n \times n$ matrices A and B, we have

$$\mathrm{Tr}(A+B) = \mathrm{Tr}(A) + \mathrm{Tr}(B) \text{ and } \mathrm{Tr}(AB) = \mathrm{Tr}(BA).$$

Use these results to deduce that there are no $n \times n$ matrices A and B such that $AB - BA = I_n$.

24. (a) The **transpose** of the $m \times n$ matrix A is the $n \times m$ matrix A^T whose ith row is the ith column of A. Thus, for example,

$$\begin{bmatrix} 1 & 2 & 3 \\ 4 & 5 & 6 \end{bmatrix}^T = \begin{bmatrix} 1 & 4 \\ 2 & 5 \\ 3 & 6 \end{bmatrix}.$$

Prove that if A and B are $m \times n$ matrices and if C is an $n \times r$ matrix, then $(A^T)^T = A$, $(A + B)^T = A^T + B^T$ and $(AC)^T = C^T A^T$. If $m = n$, deduce, for all positive integers r, that $(A^r)^T = (A^T)^r$.

(b) The matrix A is **symmetric** if and only if $A^T = A$. Show that, for each square matrix B, both BB^T and $B + B^T$ are symmetric. Suppose that C and D are symmetric $n \times n$ matrices. Show that CD is symmetric if and only if $CD = DC$.

(c) The $n \times n$ matrix A is said to be **orthogonal** if and only if $AA^T = A^T A = I_n$. Regarding the rows of A as (row) vectors $\mathbf{r}_1, \mathbf{r}_2, \ldots, \mathbf{r}_n$ and the columns of A as (column) vectors $\mathbf{c}_1, \mathbf{c}_2, \ldots, \mathbf{c}_n$ show (i) that A is orthogonal if and only if $\mathbf{r}_i.\mathbf{r}_j = 1$ if $i = j$ and 0 if $i \neq j$, a similar result being true for columns. Show, also, (ii) that A is orthogonal if and only if $A\mathbf{v}.A\mathbf{v} = \mathbf{v}.\mathbf{v}$ for each[d] $\mathbf{v} \in \mathbb{R}^n$, that is, if and only if multiplication by A preserves 'lengths' of vectors in $\mathbb{R}^n$. [Hint: $\mathbf{v}.\mathbf{v} = \mathbf{v}^T I_n \mathbf{v}$.] What is the geometric effect of the mapping $\mathbf{v} \to A\mathbf{v}$ if $n = 2$ or if $n = 3$?

25. Prove that, if the entries in each column of each of the $n \times n$ matrices A and B add up to 1, then so do the entries in each column of AB.

COMPUTER PROBLEMS

1. (Theory) Find the total number of additions/subtractions and multiplications/divisions required to evaluate:

(a) the matrix products $A(BC)$ and $(AB)C$ where A, B and C are, respectively, 3×2, 2×2 and 2×1;

[d] $\in$ means 'belonging to': see Appendix, p. 217.

(b) the product (performed in the usual way) of two $n \times n$ matrices. [When you find the answer to (b) marvel at the fact that the product can be evaluated in a non-standard way so that cn^{α} operations are required where c is some constant and α (which is known to be ≥ 2) can be assumed <2.5.]

2. Josephine used the matrix $\begin{bmatrix} 5 & 13 \\ 3 & 8 \end{bmatrix}$ to send to Napoleon a message which he receives as

$157,96|83,50|288,176|203,123|44,27|129,79|374,229|31,19|348,213|360,220|257,157$ $|136,83|300,184$. He replies using the matrix $\begin{bmatrix} 7 & 3 \\ 11 & 5 \end{bmatrix}$, sending the message $143,229$ $|200,320|147,235|84,134|116,188|115,185|148,234|136,216|105,169|62,100|197,315$ $|87,141|165,265|17,27|120,192|147,235|164,260|95,155|226,362$. By finding multiplicative inverses of the two given matrices (see Application (ii) above) decipher their messages. (And, having done so, explain the need for the 'X'!)

COMPUTER PACKAGE PROBLEMS

1. Families are classified as living in the country (C), the suburbs (S) and the town (T). If the transition matrix for each five-year period is given by

$$\begin{array}{c} \begin{matrix} \text{C} & \text{S} & \text{T} \end{matrix} \\ \begin{bmatrix} 0.70 & 0.5 & 0.6 \\ 0.15 & 0.45 & 0.3 \\ 0.15 & 0.05 & 0.1 \end{bmatrix} \begin{matrix} \text{C} \\ \text{S} \\ \text{T} \end{matrix} \end{array}$$

and if the present population is distributed amongst C, S and T in proportions 0.2, 0.3, 0.5, after how long (if at all) will there be no more than one-third of the population living in towns?

2. Investigate the powers of the 'nearly Markov' matrix

$$\begin{bmatrix} 0.70 & 0.5 & 0.6 \\ 0.14 & 0.44 & 0.3 \\ 0.15 & 0.05 & 0.09 \end{bmatrix}.$$

What do you notice here which does not happen in Problem 1?

3. Find the first power, if any, of the matrix

$$\begin{bmatrix} 0 & 0 & 0 & 1 & 1 & 0 \\ 1 & 0 & 0 & 1 & 0 & 0 \\ 0 & 0 & 0 & 0 & 1 & 0 \\ 0 & 1 & 1 & 0 & 0 & 0 \\ 0 & 0 & 1 & 0 & 1 & 1 \\ 0 & 0 & 1 & 1 & 1 & 0 \end{bmatrix}$$

which has no zero entries.

TUTORIAL PROBLEM 3.4

Can you find a 6×6 matrix with even fewer 1s (all other entries being 0) such that some power of it has no zero entries?

4. Let A be the matrix

$$\begin{bmatrix} \frac{55}{4} & 39 & \frac{353}{4} \\ \frac{13}{2} & -\frac{75}{4} & -44 \\ \frac{3}{4} & \frac{9}{4} & \frac{23}{4} \end{bmatrix}.$$

Investigate what happens to A^n as n 'tends to infinity', i.e. increases without bound? (We shall look at this again in Chapter 11.)

4 · Multiplicative Inverses of Matrices

In Chapter 3 we showed how matrix multiplication arises naturally both in succinctly describing linear systems and in constructing secret codes. Motivated by a desire to solve linear systems and to decode secret messages, we investigate the idea of (multiplicative) inverse of an $n \times n$ matrix (A, say). We shall show that inverses do not exist for every A but, if they do, how they can be found by using a succession of elementary row operations applied to the identity matrix I_n. We show that inverses, when they exist, are unique and, in Theorem 5, give four conditions each of which is equivalent to the existence of an inverse for A. Inverses (or their known non-existence) are perhaps more of theoretical than practical value here, but this changes somewhat in Chapter 11 which deals with so-called eigenvalue problems.

In Chapter 3 the linear system $A\mathbf{x} = \mathbf{b}$ suggested the 'solution' $\mathbf{x} = \mathbf{b}/A$. What are we to make of this? Let us look at the simple case where

$$A = \begin{bmatrix} a & b \\ c & d \end{bmatrix}, \quad \mathbf{x} = \begin{bmatrix} x \\ y \end{bmatrix} \quad \text{and} \quad \mathbf{b} = \begin{bmatrix} X \\ Y \end{bmatrix}$$

so that the linear system under consideration is

$$ax + by = X \tag{4.1}$$
$$cx + dy = Y. \tag{4.2}$$

(We use letters rather than numbers here so as more easily to keep track of what is going on.) The matrix $\mathbf{x}$ is then the solution of (4.1) and (4.2) giving x and y (if possible) in terms of X and Y. Applying Gaussian elimination to the system (4.1) and (4.2) we obtain, first,

$$cax + cby = cX \quad (c \times (4.1)) \tag{4.3}$$
$$acx + ady = aY \quad (a \times (4.2)) \tag{4.4}$$

and then, taking (4.3) – (4.4),

$$y = \frac{-c}{ad - bc} X + \frac{a}{ad - bc} Y. \tag{4.5}$$

(For the moment we ignore the possibility that $ad - bc$ might, in particular cases, be zero.) Then, by substituting for y in (4.3) or (4.4) we find that

$$x = \frac{d}{ad - bc} X - \frac{b}{ad - bc} Y. \tag{4.6}$$

The solution $\mathbf{x}$ is then expressible in matrix terms as $\mathbf{x} = B\mathbf{b}$ where B is the matrix

$$\begin{bmatrix} \frac{d}{ad-bc} & \frac{-b}{ad-bc} \\ \frac{-c}{ad-bc} & \frac{a}{ad-bc} \end{bmatrix}$$

which we may also write as

$$\frac{1}{ad-bc}\begin{bmatrix} d & -b \\ -c & a \end{bmatrix}.$$

Notice that the matrix product BA is (if we assume that $ad - bc \neq 0$) the identity matrix I_2 – which is not surprising since (see p. 37) BA is the matrix which, on substituting the expressions for X and Y in (4.1) and (4.2) into equations (4.5) and (4.6), expresses x and y in terms of . . . x and y(!) So, given $A\mathbf{x} = \mathbf{b}$, it appears that we should interpret $\mathbf{b}/A$ as $B\mathbf{b}$ where B is a matrix such that $BA = I_2$. [In brief: $A\mathbf{x} = \mathbf{b}$. Hence $BA\mathbf{x} = B\mathbf{b}$. Consequently $I_2\mathbf{x} = B\mathbf{b}$. That is $\mathbf{x} = B\mathbf{b}$.]

PROBLEM 4.1

For the above matrices work out the product AB. Are you surprised at the result? ●

Let us move up to the 3×3 case via a specific numerical example.

Example 1

The system of equations

$$\begin{aligned} u + 2v + 4w &= 0 \\ u + 3v + 5w &= 1 \\ 2u + 2v + 7w &= -1 \end{aligned}$$

may be written in the form $A\mathbf{x} = \mathbf{b}$ where

$$A = \begin{bmatrix} 1 & 2 & 4 \\ 1 & 3 & 5 \\ 2 & 2 & 7 \end{bmatrix}, \quad \mathbf{x} = \begin{bmatrix} u \\ v \\ w \end{bmatrix} \quad \text{and} \quad \mathbf{b} = \begin{bmatrix} 0 \\ 1 \\ -1 \end{bmatrix}.$$

According to the above we expect $\mathbf{x}$ to be equal to $B\mathbf{b}$ where B is a (the?) 3×3 matrix B such that $BA = I_3$. In fact, setting

$$B = \begin{bmatrix} 11 & -6 & -2 \\ 3 & -1 & -1 \\ -4 & 2 & 1 \end{bmatrix},$$

we find that $BA = I_3$ so that

$$\mathbf{x}(= BA\mathbf{x}) = \begin{bmatrix} 11 & -6 & -2 \\ 3 & -1 & -1 \\ -4 & 2 & 1 \end{bmatrix}\begin{bmatrix} 1 & 2 & 4 \\ 1 & 3 & 5 \\ 2 & 2 & 7 \end{bmatrix}\begin{bmatrix} u \\ v \\ w \end{bmatrix} = \begin{bmatrix} 11 & -6 & -2 \\ 3 & -1 & -1 \\ -4 & 2 & 1 \end{bmatrix}\begin{bmatrix} 0 \\ 1 \\ -1 \end{bmatrix}(= B\mathbf{b}), \quad \text{giving} \quad \mathbf{x} = \begin{bmatrix} -4 \\ 0 \\ 1 \end{bmatrix}.$$

The solution of the given system of equations is therefore, $u = -4$, $v = 0$ and $w = 1$.

Of course the big question is: given A can we find B (if it exists!) without (essentially) solving the system of equations first? (See Problem 4.4 below.)

PROBLEM 4.2

Work out the matrix AB. Are you surprised at the result? ●

[Now not all linear systems involve precisely the same number of equations as there are unknowns. In such cases the corresponding coefficient matrix A will be $m \times n$ where $m \neq n$. If there were a matrix B with $BA = I_n$, then B would be forced to be of shape $n \times m$ (why?). Admittedly the product AB would then exist – it would even be a (square) $m \times m$ matrix. However, we shall not consider this possibility in the main text; the exercises at the end of the chapter show the limitations of the idea.]

Finally let us look again at the coding application in Chapter 3. There the message which was encoded using the 2×2 matrix A can be decoded by finding a 2×2 matrix B such that $BA = I_2$.

The above example and problems give rise quite naturally to the following definition.

● *Definition I*

Let the $n \times n$ matrix A be given. If there exists a matrix B such that $AB = BA = I_n$ (so that B, too, is necessarily $n \times n$) we say that B is a **multiplicative inverse** for A and that A is **invertible** or, sometimes, **non-singular**. (Thus a square matrix is **singular** if and only if it has no multiplicative inverse.) ●

[Anyone worried by the insistence that both BA and AB must be equal to I_n should look for reassurance at Problems 4.1 and 4.2 above and, later, at Exercise 12 at the end of the chapter.]

Clearly the first question is: do all square matrices have multiplicative inverses? Equally clearly, the answer is 'no'! For example,

$$\begin{bmatrix} 0 & 0 \\ 0 & 0 \end{bmatrix}$$

has no such inverse. Of course, this is not too surprising since

$$\begin{bmatrix} 0 & 0 \\ 0 & 0 \end{bmatrix}$$

is the 2×2 matrix equivalent of the number 0. Indeed our opening remarks (and Problem 4.1) strongly suggest that

$$\begin{bmatrix} a & b \\ c & d \end{bmatrix}$$

will have an inverse iff $ad - bc \neq 0$.

PROBLEM 4.3

Find a multiplicative inverse B for the matrix

$$A = \begin{bmatrix} 4 & -19 \\ 7 & 11 \end{bmatrix}$$

by using the formula

$$B = \frac{1}{ad - bc} \begin{bmatrix} d & -b \\ -c & a \end{bmatrix}$$

obtained earlier. Check that AB and BA are both equal to I_3. Do you think that this B is the only one which will work? ●

PROBLEM 4.4

(a) Determine whether or not the matrix

$$A_1 = \begin{bmatrix} 2 & 3 & 5 \\ 7 & 11 & 13 \\ 17 & 19 & 23 \end{bmatrix}$$

has a multiplicative inverse by setting

$$B = \begin{bmatrix} a & b & c \\ d & e & f \\ g & h & j \end{bmatrix}$$

and then trying to solve $BA_1 = I_3$. (Don't forget to check that $A_1B = I_3$!)

(b) Do the same for the matrix

$$A_2 = \begin{bmatrix} 2 & 3 & 5 \\ 7 & 11 & 13 \\ 17 & 19 & 43 \end{bmatrix}.$$

●

Noting the amount of work involved in completing Problem 4.4 (and in dealing with even larger matrices), we raise one obvious question and reiterate another:

(i) Can we tell, in advance, which matrices have inverses – without attempting the (long?) task of trying to find them?

(ii) If we know a matrix has an inverse, can we find it easily – other than by brute force?

First, a nice result which (answering the question posed in Problem 4.3) tells us that once we have found one (multiplicative) inverse for a given matrix at least we need not look for any more!

• Theorem I

Let A, B_1 and B_2 be $n \times n$ matrices such that $AB_1 = B_1A = I_n$ and $AB_2 = B_2A = I_n$. Then $B_1 = B_2$. That is, if A has an inverse, then this inverse is unique.

[Here is a proof which clearly suffers from too much polishing! Proofs in mathematics are often like that. Furthermore, in presenting a proof in a logically ordered manner – in order to help the reader – any hint as to how the proof was originally conceived is often lost!]

PROOF[a]

$$B_1 = B_1 I_n \text{ (why?)} = B_1(AB_2) \text{ (why?)} = (B_1A)B_2 \text{ (why? etc)} = I_n B_2 = B_2.$$

This unique inverse, when it exists, we shall denote by A^{-1}. [One consequence of this theorem is that we may return to Definition 1 and replace the expression 'B is *a* multiplicative inverse for A' by 'B is *the* (unique) multiplicative inverse for A.']

We now answer the two questions raised above by showing you a method which appears to take much of the drudgery out of finding inverses. In order to illuminate the theory which follows, we show the way by means of a particular example.

Example 2

Find the inverse (if it exists!) of

$$A = \begin{bmatrix} 1 & 2 & 4 \\ 1 & 3 & 5 \\ 2 & 2 & 7 \end{bmatrix}.$$

(i) First write down

$$\begin{bmatrix} 1 & 2 & 4 & : & 1 & 0 & 0 \\ 1 & 3 & 5 & : & 0 & 1 & 0 \\ 2 & 2 & 7 & : & 0 & 0 & 1 \end{bmatrix},$$

which we write as $[A:I_3]$, the colons being just for convenience.

(ii) Now apply (elementary) row operations to this 3×6 matrix in order to change the left-hand 'half' to the matrix I_3 (if possible). We get

$$\begin{matrix} \rho_2 \to \rho_2 - \rho_1 \\ \rho_3 \to \rho_3 - 2\rho_1 \end{matrix} \quad \begin{bmatrix} 1 & 2 & 4 & : & 1 & 0 & 0 \\ 0 & 1 & 1 & : & -1 & 1 & 0 \\ 0 & -2 & -1 & : & -2 & 0 & 1 \end{bmatrix} \quad \text{(clearing the first column below the diagonal)}$$

$$\rho_3 \to \rho_3 + 2\rho_2 \quad \begin{bmatrix} 1 & 2 & 4 & : & 1 & 0 & 0 \\ 0 & 1 & 1 & : & -1 & 1 & 0 \\ 0 & 0 & 1 & : & -4 & 2 & 1 \end{bmatrix} \text{(I)} \quad \text{(clearing the second column below the diagonal)}$$

$$\rho_1 \to \rho_1 - 2\rho_2 \quad \begin{bmatrix} 1 & 0 & 2 & : & 3 & -2 & 0 \\ 0 & 1 & 1 & : & -1 & 1 & 0 \\ 0 & 0 & 1 & : & -4 & 2 & 1 \end{bmatrix} \quad \text{(clearing the second column above the diagonal)}$$

$$\begin{matrix} \rho_1 \to \rho_1 - 2\rho_3 \\ \rho_2 \to \rho_2 - \rho_3 \end{matrix} \quad \begin{bmatrix} 1 & 0 & 0 & : & 11 & -6 & -2 \\ 0 & 1 & 0 & : & 3 & -1 & -1 \\ 0 & 0 & 1 & : & -4 & 2 & 1 \end{bmatrix} \quad \text{(clearing the third column above the diagonal)}$$

and look what has happened in the right-hand half of this 3×6 matrix. We have the unique inverse of A. That is, A^{-1} exists and

[a] 'All theorems can be proved in one line . . . if the line is long enough!' (Anon.)

$$A^{-1} = \begin{bmatrix} 11 & -6 & -2 \\ 3 & -1 & -1 \\ -4 & 2 & 1 \end{bmatrix}.$$

[We have already checked that $A^{-1}A = I_3$ in Example 1. You should (unless you have completed Exercise 12) also check that $AA^{-1} = I_3$.]

Notice that, if we had started Example 2 with

$$A_1 = \begin{bmatrix} 1 & 2 & 4 \\ 1 & 3 & 5 \\ 2 & 2 & 6 \end{bmatrix}$$

line (I) above would have yielded

$$\begin{bmatrix} 1 & 2 & 4 & : & 1 & 0 & 0 \\ 0 & 1 & 1 & : & -1 & 1 & 0 \\ 0 & 0 & 0 & : & -4 & 2 & 1 \end{bmatrix}.$$

This would have told us that we cannot change the left-hand half of (I) to I_3 and, hence, that A_1 has no multiplicative inverse. (See Theorem 5(i) and (iv).)

PROBLEM 4.5

Repeat Problem 4.4 using the method of Example 2.

Of course the big question is: why does the algorithm in Example 2 work? To answer this it is perhaps tidiest (though certainly not necessary, as we shall see later) to introduce some special types of matrices.

• *Definition 2*

An **elementary** $n \times n$ **matrix** is one obtained from I_n by applying a single elementary row operation.

We have, for instance, the following examples.

Examples 3

$$\text{(i)} \begin{bmatrix} 1 & 0 & 0 \\ 0 & 1 & 0 \\ 0 & 0 & -\frac{3}{17} \end{bmatrix}, \quad \text{(ii)} \begin{bmatrix} 1 & 0 & 0 & 0 \\ 0 & 0 & 0 & 1 \\ 0 & 0 & 1 & 0 \\ 0 & 1 & 0 & 0 \end{bmatrix} \quad \text{and} \quad \text{(iii)} \begin{bmatrix} 1 & 0 & -\pi & 0 \\ 0 & 1 & 0 & 0 \\ 0 & 0 & 1 & 0 \\ 0 & 0 & 0 & 1 \end{bmatrix}$$

are all elementary matrices, the elementary row operations being (i) multiply row 3 by $-\frac{3}{17}$; (ii) interchange rows 2 and 4 and (iii) add $-\pi$ times row 3 to row 1. On the other hand, neither

$$\text{(iv)} \begin{bmatrix} 1 & 2 & 0 \\ 0 & 1 & 1 \\ 0 & 0 & 1 \end{bmatrix} \quad \text{nor} \quad \text{(v)} \begin{bmatrix} -1 & 0 & 0 \\ 0 & -1 & 0 \\ 0 & 0 & 1 \end{bmatrix}$$

is elementary as neither can be obtained from I_3 by use of a single elementary row operation.

NOTE 1 Each elementary matrix has a (multiplicative) inverse which is itself elementary. Indeed it is the elementary matrix which 'undoes' the original. (Cf. Exercise 16 in Chapter 2.)

Since examples are clearer to understand here than are words, we give the following.

Examples 4

The inverses of Examples 3(i), (ii) and (iii) are:

$$\text{(i)}\begin{bmatrix}1 & 0 & 0\\0 & 1 & 0\\0 & 0 & -\frac{17}{3}\end{bmatrix};\quad \text{(ii)}\begin{bmatrix}1 & 0 & 0 & 0\\0 & 0 & 0 & 1\\0 & 0 & 1 & 0\\0 & 1 & 0 & 0\end{bmatrix};\quad \text{(iii)}\begin{bmatrix}1 & 0 & \pi & 0\\0 & 1 & 0 & 0\\0 & 0 & 1 & 0\\0 & 0 & 0 & 1\end{bmatrix}.$$

(The proof of the first sentence of Note 1 is left to you. See Exercise 18.)

It is very useful to observe that elementary matrices can be used as a replacement for elementary row operations. For our present purposes we need only consider row operations on square matrices but, as we have previously discussed row operations on non-square matrices, we present the following theorem in the general case.

Theorem 2

Let A be an $m \times n$ matrix and E an elementary $m \times m$ matrix. If E is obtained from I_m by a particular elementary row operation, then EA is obtained from A by applying the same elementary row operation to A.

The truth of this theorem is fairly transparent once we have demonstrated what happens in one case by means of an example. We leave the full proof to Exercise 21.

Example 5

If

$$E = \begin{bmatrix}1 & \beta & 0\\0 & 1 & 0\\0 & 0 & 1\end{bmatrix} \quad \text{and} \quad A = \begin{bmatrix}a_{11} & a_{12} & a_{13} & a_{14}\\a_{21} & a_{22} & a_{23} & a_{24}\\a_{31} & a_{32} & a_{33} & a_{34}\end{bmatrix}$$

so that E is just I_3 changed by adding β times the second row to the first, then

$$EA = \begin{bmatrix}a_{11}+\beta a_{21} & a_{12}+\beta a_{22} & a_{13}+\beta a_{23} & a_{14}+\beta a_{24}\\a_{21} & a_{22} & a_{23} & a_{24}\\a_{31} & a_{32} & a_{33} & a_{34}\end{bmatrix}$$

which is just the matrix A changed by adding β times the second row to the first.

Theorem 2 is mainly of theoretical interest. It will be used for developing some results about matrices and linear systems. When actually using a computer it is better to perform the row operations directly.

We can now prove, fairly easily, the following nice theorem which answers the 'big' question raised above.

• Theorem 3

Suppose successive elementary row operations, performed on a given $n \times n$ matrix A, reduce it to the identity matrix I_n. Then exactly the same elementary row operations applied, in the same order, to I_n change I_n into A^{-1}. •

In the proof we shall find that we need a result the last bit of which seems somewhat curious if you have not seen its like before. (Cf. Exercise 24(a) of Chapter 3.)

• Theorem 4

Let A, B be invertible $n \times n$ matrices. Then AB is invertible. Furthermore $(AB)^{-1} = B^{-1}A^{-1}$.

PROOF

From Theorem 3 of Chapter 3 we can deduce that

$$(AB)(B^{-1}A^{-1}) = (A(BB^{-1}))A^{-1} = (AI_n)A^{-1} = AA^{-1} = I_n.$$

Similarly one can check that $(B^{-1}A^{-1})(AB) = I_n$. Hence $B^{-1}A^{-1}$ is an (and, hence, by Theorem 1 above, the unique) inverse for AB. Consequently we may write $B^{-1}A^{-1} = (AB)^{-1}$. •

PROBLEM 4.6

Let

$$A = \begin{bmatrix} 1 & 2 \\ 3 & 4 \end{bmatrix} \quad \text{and} \quad B = \begin{bmatrix} 5 & 6 \\ 7 & 8 \end{bmatrix}.$$

Confirm that A, B and AB all have multiplicative inverses and that $(AB)^{-1}$ is equal to $B^{-1}A^{-1}$ and not equal to $A^{-1}B^{-1}$. •

NOTE 2 Theorem 4 extends to products of more than two terms. That is: if $A, B, \ldots, K$ are invertible $n \times n$ matrices then $AB \ldots K$ is invertible and $(AB \ldots K)^{-1} = K^{-1} \ldots B^{-1}A^{-1}$.

PROOF OF THEOREM 3

Let $E_1, E_2, \ldots, E_t$ be the elementary matrices corresponding to the elementary row operations described in the theorem and let E denote $E_tE_{t-1} \ldots E_1$. Then $E_tE_{t-1} \ldots E_1A = EA = I_n$. Since each of $E_1, E_2, \ldots, E_t$ has an inverse so has E (by Note 2). But then we have

$$A = I_nA = (E^{-1}E)A = E^{-1}(EA) = E^{-1}I_n = E^{-1}$$

from which we deduce that $A^{-1} = E = EI_n = E_tE_{t-1} \ldots E_1I_n$.

You might compare the brevity (and the clarity) of this proof with that given just before the Applications, especially when written down for $n \times n$ matrices.

In fact our efforts above give us four statements each of which is equivalent to the invertibility of a given $n \times n$ matrix A.

• Theorem 5

Let A be an $n \times n$ matrix. Then, the following five statements are equivalent (that is, the truth of any one of them implies the truth of each – and hence every – other one):

(i) A is invertible;

(ii) For each $n \times 1$ (column) vector $\mathbf{b}$ the system of equations $A\mathbf{x} = \mathbf{b}$ has a unique solution for $\mathbf{x}$;

(iii) For the particular $n \times 1$ (column) vector $\mathbf{0}$ the system of equations $A\mathbf{x} = \mathbf{0}$ has the unique solution $\mathbf{x} = \mathbf{0}$;

(iv) A is row equivalent to I_n;

(v) A is a product of elementary $n \times n$ matrices.

NOTE ON THE PROOF As each statement above has to be proved on the assumption of each other, it would seem that we are about to embark on a sequence of 20 (is that right?) proofs of the form $X \Rightarrow Y$, $X \Rightarrow Z$, $Y \Rightarrow X$, etc. So it should please you that the task can be reduced to just five! Indeed we prove that (i) $\Rightarrow$ (ii), that (ii) $\Rightarrow$ (iii), that (iii) $\Rightarrow$ (iv), that (iv) $\Rightarrow$ (v) and, finally, that (v) $\Rightarrow$ (i). We may then deduce, for instance, that (ii) $\Rightarrow$ (i). For, assuming (ii), we can successively deduce (iii), then (iv), then (v) and finally (i).

PROOF

(i) $\Rightarrow$ (ii) Suppose A is invertible. Then, given $A\mathbf{x} = \mathbf{b}$, we obtain

$$\mathbf{x} = I_n\mathbf{x} = (A^{-1}A)\mathbf{x} = A^{-1}(A\mathbf{x}) = A^{-1}\mathbf{b}.$$

(ii) $\Rightarrow$ (iii) is immediate.

(iii) $\Rightarrow$ (iv) Consider the system of equations $A\mathbf{x} = \mathbf{0}$. In full, $A\mathbf{x} = \mathbf{0}$ is the system

$$\begin{array}{lllll} a_{11}x_1 + a_{12}x_2 + \ldots & & & + a_{1n}x_n & = 0 \\ a_{21}x_1 + a_{22}x_2 + \ldots & & & + a_{2n}x_n & = 0 \\ \vdots \quad\quad \vdots & \vdots & \vdots & \vdots & \\ a_{n1}x_1 + a_{n2}x_2 + \ldots & & & + a_{nn}x_n & = 0. \end{array}$$

Reducing these equations to echelon form we find, because (0, 0, . . ., 0) is their only solution, that the matrix to which A is reduced must have no rows full of zeros and, therefore, must be I_n. (See Exercise 13 of Chapter 2.) Thus A is indeed row equivalent to I_n.

(iv) $\Rightarrow$ (v) and (v) $\Rightarrow$ (i) both follow from Theorem 3. ●

We finish by giving an alternative presentation for the justification that the algorithm in Example 2 is valid. It may seem that this is a fairly convincing proof that that algorithm always works – indeed it is essentially the same idea dressed in slightly different clothing – but you must admit that it would be a bit messy to write it out for quite general $n \times n$ matrices.

If asked to solve the system of equations (given in matrix form):

$$\left[\begin{array}{ccc|c} 1 & 2 & 4 & 1 \\ 1 & 3 & 5 & 0 \\ 2 & 2 & 7 & 0 \end{array}\right],$$

we could apply elementary row operations to change the (sub)matrix

$$\begin{bmatrix}1&2&4\\1&3&5\\2&2&7\end{bmatrix} \text{ into } \begin{bmatrix}1&0&0\\0&1&0\\0&0&1\end{bmatrix} \text{ yielding the solution } \begin{bmatrix}a\\b\\c\end{bmatrix}=\begin{bmatrix}11\\3\\-4\end{bmatrix}.$$

Similarly, starting with

$$\left[\begin{array}{ccc|c}1&2&4&0\\1&3&5&1\\2&2&7&0\end{array}\right] \text{ and with } \left[\begin{array}{ccc|c}1&2&4&0\\1&3&5&0\\2&2&7&1\end{array}\right]$$

and applying the very same row operations, we again reduce

$$\begin{bmatrix}1&2&4\\1&3&5\\2&2&7\end{bmatrix} \text{ each time to } \begin{bmatrix}1&0&0\\0&1&0\\0&0&1\end{bmatrix},$$

obtaining solutions

$$\begin{bmatrix}\alpha\\\beta\\\gamma\end{bmatrix}=\begin{bmatrix}6\\-1\\2\end{bmatrix} \text{ and } \begin{bmatrix}A\\B\\C\end{bmatrix}=\begin{bmatrix}-2\\-1\\1\end{bmatrix}$$

respectively. But it is then immediate (see Exercise 9 of Chapter 3) that

$$\begin{bmatrix}1&2&4\\1&3&5\\2&2&7\end{bmatrix}\begin{bmatrix}11&-6&-2\\3&-1&-1\\-4&2&1\end{bmatrix}=\begin{bmatrix}1&0&0\\0&1&0\\0&0&1\end{bmatrix}.$$

The concept of matrix inverse first appears in a paper of Cayley's in 1855 in which he recasts a solution, given by Hermite, to a problem considered by Euler in the 18th century in the form of a product

$$\begin{bmatrix}a&h&g&\dots\\h&b&f&\dots\\g&f&c&\dots\\\dots&\dots&\dots&\dots\end{bmatrix}^{-1}\begin{bmatrix}a&h-\nu&g+\mu&\dots\\h+\nu&b&f-\lambda&\dots\\g-\mu&f+\lambda&c&\dots\\\dots&\dots&\dots&\dots\end{bmatrix}\begin{bmatrix}a&h+\nu&g-\mu&\dots\\h-\nu&b&f+\lambda&\dots\\g+\mu&f-\lambda&c&\dots\\\dots&\dots&\dots&\dots\end{bmatrix}^{-1}\begin{bmatrix}a&h&g&\dots\\h&b&f&\dots\\g&f&c&\dots\\\dots&\dots&\dots&\dots\end{bmatrix}.$$

Three years later, helped by the notion of addition, he can write this as $A^{-1}(A-S)(A+S)^{-1}A$. However, as with the definition of multiplication, the notion of inverse of a linear substitution had already been raised; Eisenstein had used the notation $1/S$ for a linear substitution S – provided the determinant associated with S (see Chapter 5) were non-zero.

Arthur Cayley (16 August 1821–26 January 1895) was born in England but, because of his father's business, spent his early years in Russia. In 1842 he was elected a fellow of Trinity College Cambridge, but left in 1846 rather than take the holy orders necessary for remaining there. He turned to the law, pursuing

mathematics and law simultaneously. Whilst at Lincoln's Inn he became friendly with Sylvester. These two wrote so many papers on the then popular topic of invariant theory that they were given the name 'invariant twins'. Cayley wrote 966 papers covering many areas of mathematics. In particular, in 1854, he introduced the idea (subsequently long neglected) of abstract group.

Applications

The idea of 'inverse' of a matrix arose, in Chapter 3, in connection with the solving of a system of equations $A\mathbf{x} = \mathbf{b}$ in the form $\mathbf{x} = A^{-1}\mathbf{b}$, Example 2 showing how this can be done. This method, however, is little used in real-life equation solving – at least one book on numerical linear algebra recommending that if the inverse of a matrix is not specifically needed it should not be evaluated since there are better methods of procedure. Nevertheless, inverses retain a theoretical usefulness: it is convenient to be able to express the solution to $A\mathbf{x} = \mathbf{b}$ in the form $\mathbf{x} = A^{-1}\mathbf{b}$ especially if one is more interested in the existence of a solution than (an approximation to) its actual value.

Later (Chapter 11) we shall be particularly interested in finding, for a given $n \times n$ matrix A, scalars λ and (non-zero) $n \times 1$ matrices (i.e. column vectors) $\mathbf{v}$ for which $A\mathbf{v} = \lambda\mathbf{v}$. Such a $\mathbf{v}$ will exist exactly when the matrix $A - \lambda I$ is singular. (A criterion for the (non-) invertibility of a square matrix, especially useful in this context, is given in Theorem 6 of Chapter 5.)

As a trifling, but amusing, example let

$$A = \begin{bmatrix} 3.3 & 4.8 & 10.2 \\ 0.8 & 2.1 & 3.1 \\ -1.0 & -1.8 & -3.2 \end{bmatrix}.$$

From computer evidence I might suspect that, as n increases indefinitely, the matrices A^n tend towards the 3×3 zero matrix. But computers can go only so far, oftentimes crediting as zero small numbers which are not. Can we check our suspicion conclusively? Yes! Simply write $A = M^{-1}DM$ where

$$M = \begin{bmatrix} 1 & 2 & 4 \\ 1 & 3 & 5 \\ 2 & 2 & 7 \end{bmatrix} \quad \text{and} \quad D = \begin{bmatrix} 0.9 & 0 & 0 \\ 0 & 0.7 & 0 \\ 0 & 0 & 0.6 \end{bmatrix}.$$

Exercise 7, below, shows you that, for all positive integers n, $A^n = M^{-1}D^nM$. But, clearly, as n increases, D^n (which is equal to

$$\begin{bmatrix} (0.9)^n & 0 & 0 \\ 0 & (0.7)^n & 0 \\ 0 & 0 & (0.6)^n \end{bmatrix}$$

– see Problem 3.4(iv) in Chapter 3) tends towards $0_{3\times3}$ and so A^n tends towards $M^{-1}0_{3\times3}M(= 0_{3\times3})$ as claimed. (Of course you are asking – are you not? – how I found the matrix M which so conveniently changed A into D. But that is a story for Chapter 11!)

Summary

The $n \times n$ matrix A has an **inverse** B if and only if $AB = BA = I_n$. B is then unique and is usually denoted by A^{-1}. [Non-square matrices do not have such inverses.] A^{-1} can be found (or shown not to exist!) by applying **elementary row operations** to the $n \times 2n$ matrix $[A{:}I_n]$ so that A is changed (if possible) into I_n. The matrix into which I_n is thereby changed is the required A^{-1}. This is most easily proved by using **elementary matrices**. These are matrices which are obtained from I_n by using just one elementary row operation. If E is an elementary ($n \times n$) matrix then EA is equal to the matrix obtained from A by applying to A the same row operation used to obtain E from I_n. One can prove that A is invertible if and only if A is a product of elementary matrices. [Other equivalent conditions are in Theorem 5.] In doing so we use the result that, if $A, B, \ldots, K$ are invertible $n \times n$ matrices, then so is $AB \ldots K$. Furthermore, then, $(AB \ldots K)^{-1} = K^{-1} \ldots B^{-1}A^{-1}$.

EXERCISES ON CHAPTER 4

1. Find the inverse (if it exists) for each of the following matrices:

$$\text{(i)}\begin{bmatrix} 1 & 0 \\ 1 & 1 \end{bmatrix}; \quad \text{(ii)}\begin{bmatrix} 7 & 2 \\ 1 & 6 \end{bmatrix}; \quad \text{(iii)}\begin{bmatrix} 12 & 28 \\ 27 & 63 \end{bmatrix}; \quad \text{(iv)}\begin{bmatrix} \cos\vartheta & -\sin\vartheta \\ \sin\vartheta & \cos\vartheta \end{bmatrix}.$$

Describe (iv) and its inverse geometrically. (See Exercise 15 of Chapter 3.)

2. Find $\begin{bmatrix} 1 & 0 \\ 1 & 1 \end{bmatrix}^{-3}$. That is, what is $\begin{bmatrix} 1 & 0 \\ 1 & 1 \end{bmatrix}^{-3}$ if

$$\begin{bmatrix} 1 & 0 \\ 1 & 1 \end{bmatrix}^3 \begin{bmatrix} 1 & 0 \\ 1 & 1 \end{bmatrix}^{-3} = \begin{bmatrix} 1 & 0 \\ 1 & 1 \end{bmatrix}^{-3} \begin{bmatrix} 1 & 0 \\ 1 & 1 \end{bmatrix}^3 = I_2 ?$$

3. Show that

$$\begin{bmatrix} a & b \\ -b & a \end{bmatrix} \left(\neq \begin{bmatrix} 0 & 0 \\ 0 & 0 \end{bmatrix} \right)$$

has a multiplicative inverse if a and b are real numbers but might not have a multiplicative inverse if a and b are allowed to be complex numbers.

4. (a) Find $A^{-1}B^{-1}AB$:

(i) if $A = \begin{bmatrix} 1 & \alpha \\ 0 & 1 \end{bmatrix}$ and $B = \begin{bmatrix} 1 & \beta \\ 0 & 1 \end{bmatrix}$;

(ii) if $A = \begin{bmatrix} 1 & a & c \\ 0 & 1 & b \\ 0 & 0 & 1 \end{bmatrix}$ and $B = \begin{bmatrix} 1 & d & f \\ 0 & 1 & e \\ 0 & 0 & 1 \end{bmatrix}$.

(b) Let C and D be (real) $n \times n$ matrices.

(i) If $CD = 0$ is DC necessarily also the zero matrix?

(ii) Can one have $CD \neq 0$ but $CD + DC = 0$?

5. (a) Under exactly what conditions is $(kA)^{-1} = k^{-1}A^{-1}$ (k being a scalar and A an $n \times n$ matrix)?

(b) Let A and B be invertible $n \times n$ matrices. Is $(A + B)^{-1} = A^{-1} + B^{-1}$. Proof or counterexample please.

(c) Show that, if A, B and $A + B$ are invertible then $A^{-1} + B^{-1}$ certainly is invertible.

(d) Let C and D bė $n \times n$ matrices. Show that if C or D is singular then so is CD [What is wrong with the argument: 'If CD is not singular then $(CD)^{-1}$ exists. But $(CD)^{-1} = D^{-1}C^{-1}$. Hence C and D are invertible.'?]

6. Find the inverses, if they exist, of:

(i) $\begin{bmatrix} 4 & 1 & 1 \\ 1 & -1 & 1 \\ 7 & 0 & 3 \end{bmatrix}$; (ii) $\begin{bmatrix} 1 & 3 & 5 \\ 0 & 7 & 9 \\ 0 & 0 & 11 \end{bmatrix}$; (iii) $\begin{bmatrix} 1 & 0 & 0 \\ 0 & -5 & 0 \\ 0 & 0 & 2 \end{bmatrix}$; (iv) $\begin{bmatrix} 0 & 0 & 1 \\ 0 & -5 & 0 \\ 2 & 0 & 0 \end{bmatrix}$;

(v) $\begin{bmatrix} a & 1 & 0 \\ 0 & a & 1 \\ 0 & 0 & a \end{bmatrix}$ $(a \neq 0)$; (vi) $\begin{bmatrix} 1 & 1 & 1 \\ 1 & 2 & 3 \\ 1 & 4 & 9 \end{bmatrix}$; (vii) $\begin{bmatrix} -1 & -1 & -1 \\ 1 & 2 & 3 \\ 1 & 4 & 9 \end{bmatrix}$;

(viii) $\begin{bmatrix} 1 & 1 & 1 \\ 1 & 2 & 4 \\ 1 & 3 & 9 \end{bmatrix}$; (ix) $\begin{bmatrix} 1 & 1 & 1 & 1 \\ 1 & 2 & 2 & 2 \\ 1 & 2 & 3 & 3 \\ 1 & 2 & 3 & 4 \end{bmatrix}$.

7. (a) Let A and P be $n \times n$ matrices, P being invertible. By writing the product out in full and then cancelling neighbouring P and P^{-1} show that $(P^{-1}AP)^3 = P^{-1}A^3P$. Prove more generally, by mathematical induction, that, for all positive integers n, $(P^{-1}AP)^n = P^{-1}A^nP$. Given that

$$\begin{bmatrix} -34 & -63 \\ 20 & 37 \end{bmatrix} = \begin{bmatrix} 4 & 7 \\ 5 & 9 \end{bmatrix}^{-1} \begin{bmatrix} 1 & 0 \\ 0 & 2 \end{bmatrix} \begin{bmatrix} 4 & 7 \\ 5 & 9 \end{bmatrix}$$

find (by hand, not by computer)

$$\begin{bmatrix} -34 & -63 \\ 20 & 37 \end{bmatrix}^{10}.$$

(b) The matrix B is said to be **similar** to the matrix A if there exists an invertible matrix P such that $P^{-1}AP = B$. Show that, if B is similar to A then A is similar to B. (We may then simply say that A and B are *similar.*)

8. Given that $\begin{bmatrix} a & b & c \\ d & e & f \\ g & h & j \end{bmatrix}^{-1} = \begin{bmatrix} A & D & G \\ B & E & H \\ C & F & J \end{bmatrix}$ write down $\begin{bmatrix} b & a & c \\ e & d & f \\ h & g & j \end{bmatrix}^{-1}$.

9. Let A, B and C be $n \times n$ matrices with A being invertible. Show that:
(i) A^{-1} is invertible and $(A^{-1})^{-1} = A$;
(ii) A^T is invertible and $(A^T)^{-1} = (A^{-1})^T$ (for the definition of A^T see Exercise 24(a) of Chapter 3);
(iii) if $AB = AC$ then $B = C$;
(iv) if A is symmetric then so is A^{-1}.

10. Solve the following systems of equations using matrix inverses. (The second can be done quickly once the first has been solved. Can you see why?)

$$\text{(i)}\ \begin{aligned} x+y+z&=1 \\ x+2y+3z&=2 \\ x+4y+9z&=3; \end{aligned} \qquad \text{(ii)}\ \begin{aligned} x+y+z&=1 \\ x+2y+4z&=2 \\ x+3y+9z&=3. \end{aligned}$$

[Hint: Exercise 9(ii) is helpful.]

11. (a) Let A, B and C be $n \times n$ matrices such that $AB = I_n = CA$. Prove that $B = C$.
(b) Let A, B and C be $n \times n$ matrices with $AB = CA$ and A invertible. Must $B = C$?

12. (A work saver! Did you, in Problem 4.5, check that $A^{-1}A = AA^{-1} = I_n$? If so, you worked too hard! – for the following is true. Try to prove it.)
Let A and B be $n \times n$ matrices such that $AB = I_n$. Then $BA = I_n$ (so that A is invertible with inverse B). [Hint: If $B\mathbf{x} = \mathbf{0}$ then $\mathbf{x} = I_n\mathbf{x} = AB\mathbf{x} = \mathbf{0}$. Now use Theorem 5.]

13. Let A be an $n \times n$ matrix such that: for all pairs of $n \times n$ matrices B and C we may deduce, from the equality $AB = AC$, that $B = C$. Prove that A must be invertible. [Hint: show that if $AB = 0_{n\times n}$ then $B = 0_{n\times n}$. Thus the only solution to $A\mathbf{x} = \mathbf{0}$ is $\mathbf{x} = \mathbf{0}$.]

14. Show that if

$$A = \begin{bmatrix} -1 & -1 & 1 \\ \frac{3}{2} & -\frac{1}{2} & 0 \end{bmatrix} \quad \text{and} \quad B = \begin{bmatrix} 1 & 2 \\ 3 & 4 \\ 5 & 6 \end{bmatrix}$$

then $AB = I_2$. Then A is a *left inverse* for B and B a *right inverse* for A. Find several more such inverses. Is A also a right inverse for B? That is, is $BA = I_3$?

15. (a) If $A_{m\times n}$ is not square show there cannot exist an $n \times m$ matrix C such that $AC = I_m$ and $CA = I_n$. [Hint: if $m < n$ choose $\mathbf{x} \neq \mathbf{0}$ such that $A\mathbf{x} = \mathbf{0}$. Then look at $CA\mathbf{x}$ $(= I_n\mathbf{x})$. If $n < m \ldots$]

(b) Show that if A, B are non-square matrices such that $AB = I_n$ then there must exist a matrix $C \neq B$ such that $AB = AC = I_n$.

16. Prove that a matrix with one row being a multiple of another cannot have an inverse.

17. Let A, B be $n \times n$ matrices. We know that $\mathrm{Tr}(AB) = \mathrm{Tr}(BA)$. (See Exercise 23 of Chapter 3.) Deduce that, if C is invertible then $\mathrm{Tr}(C^{-1}AC) = \mathrm{Tr}(A)$. [Hint: replace A by $C^{-1}A$ and B by C.]

18. Show that

$$\begin{bmatrix} 1 & 0 & 0 & 0 \\ 0 & 1 & 0 & k \\ 0 & 0 & 1 & 0 \\ 0 & 0 & 0 & 1 \end{bmatrix}^{-1} = \begin{bmatrix} 1 & 0 & 0 & 0 \\ 0 & 1 & 0 & -k \\ 0 & 0 & 1 & 0 \\ 0 & 0 & 0 & 1 \end{bmatrix}.$$

Now complete, for 4×4 matrices, the proofs of the statements in Note 1.

19. Let

$$A = \begin{bmatrix} 7 & -4 & -2 \\ 0 & -1 & 8 \\ 2 & 3 & 1 \end{bmatrix}.$$

Apply the following elementary row operations in succession. (i) Subtract 3 times row 3 from row 1; (ii) multiply row 3 by 4; (iii) interchange rows 2 and 1. Write down the new matrix B. Next write down the corresponding elementary matrices E_1, E_2, E_3. Finally evaluate $E_1E_2E_3A$. Do you get B? Explain!

20. $B = \begin{bmatrix} 5 & 1 & -2 \\ 0 & -1 & 1 \\ 6 & -9 & 11 \end{bmatrix}$ was obtained from the matrix A by first adding 5 times row 2 (of A) to row 3 and then dividing row 2 by 2. Write $B = E_2E_1A$ where E_1, E_2 are suitable elementary matrices. Find A (i) by using suitable elementary row operations; (ii) by finding E_1^{-1}, E_2^{-1} and then using $A = E_1^{-1}E_2^{-1}B$.

21. Prove Theorem 2 for each type of elementary row operation. [If you cannot do the $m \times n$ case, at least try the 4×3 case.]

22. Write the inverses of the matrices in Exercises 6(i) and (vi) as products of elementary matrices.

23. Show that A and B are row equivalent if and only if there exists an invertible matrix P, say, such that $A = PB$.

24. Let E be an elementary $n \times n$ matrix and A be an $n \times n$ matrix. Describe the relation between A and AE.

COMPUTING PROBLEM

1. Prove that if the invertible matrix A has rational number entries then so must A^{-1}. Why must you not necessarily expect a completely accurate answer from your computer when inverting such a 'rational' matrix?

COMPUTER PACKAGE PROBLEMS

1. Find the inverses, if they exist, of:

$$\text{(i)}\begin{bmatrix} 1 & 0 & 1 & -1 \\ 0 & 1 & -2 & 1 \\ 1 & -2 & 1 & 0 \\ -1 & 1 & 0 & 0 \end{bmatrix}; \quad \text{(ii)}\begin{bmatrix} 0 & 1 & 0 & 0 & 0 \\ 8 & 0 & 2 & 0 & 0 \\ 0 & 7 & 0 & 3 & 0 \\ 0 & 0 & 6 & 0 & 4 \\ 0 & 0 & 0 & 5 & 0 \end{bmatrix}; \quad \text{(iii)}\begin{bmatrix} -\frac{7}{16} & -\frac{1}{6} & \frac{19}{48} \\ \frac{1}{4} & \frac{1}{3} & -\frac{5}{12} \\ \frac{5}{16} & -\frac{1}{6} & \frac{7}{48} \end{bmatrix};$$

$$\text{(iv)}\begin{bmatrix} -3/5 & 0 & 0 & 0 & 2/5 \\ 0 & -7/13 & 0 & 6/13 & 0 \\ 0 & 0 & 1/9 & 0 & 0 \\ 0 & 8/13 & 0 & -5/13 & 0 \\ 4/5 & 0 & 0 & 0 & -1/5 \end{bmatrix}.$$

[If your package can work with fractions you should get interesting answers!]

2. Check to see if all the Hilbert matrices H_n ($n = 2, \ldots, 6$) of Computer Package Problem 1 of Chapter 2 have inverses. Do you think every H_n is invertible? Can you now explain the peculiar changes in solution obtained in Computer Package Problem 3 in Chapter 1?

3. Find the (formal) inverse of

$$\begin{bmatrix} 1 & 2 & 3 \\ 8 & x & 4 \\ 7 & 6 & 5 \end{bmatrix}, \quad \begin{bmatrix} 1 & 2 & x \\ 4 & x & 6 \\ x & 8 & 9 \end{bmatrix} \quad \text{and} \quad \begin{bmatrix} a & b & c \\ d & e & f \\ g & h & j \end{bmatrix}.$$

4. Let $A = [a_{ij}]$ be the $n \times n$ matrix defined by: $a_{ij} = 0$ if $i = j + 1$; $a_{ij} = 1$ otherwise. See if your package will construct this matrix from this formula. In any case find, for as many n as practical, A^{-1} (if it exists!). If computer evidence suggests a pattern, can you establish it theoretically for all n?

5 · Determinants

This chapter shows how, when solving systems of n equations in n unknowns, a number, the determinant of the coefficient matrix of the system, arises naturally. To evaluate a determinant directly from its definition can be quite messy so methods to help ease calculation are developed. Despite the (relative) ease with which the determinants of small matrices are evaluated by hand, determinants are, nowadays, probably more of theoretical than practical importance. In particular, Cramer's rule (Exercise 13(b)) is of little practical value. On the other hand, the criterion given in Theorem 6, for a matrix to be invertible, is useful in the study of eigenvalues in Chapter 11.

Readers who have previously studied determinants may well find this chapter rather appealing, because it merely confirms, for $n \times n$ matrices, what they probably know quite well already in the case of 2×2 and 3×3 matrices. However, whilst determinants may have played a more central role in readers' earlier studies, their role in algebra nowadays is mainly of a theoretical nature: many theorems can conveniently be stated using determinants – but as a computational tool their usefulness is somewhat limited.

Nevertheless, let us remind ourselves how determinants arise in the solving of systems of linear equations. As in Chapter 4 we start with the simplest case again working with letter (rather than actual number) coefficients so that we can better keep track of the various computations which arise.

Given the simultaneous equations

$$a_{11}x_1 + a_{12}x_2 = b_1 \tag{5.1}$$
$$a_{21}x_1 + a_{22}x_2 = b_2 \tag{5.2}$$

we may solve them for x_1 and x_2 (cf. equations $ax + by = X$ and $cx + dy = Y$ in Chapter 4) to obtain

$$x_1 = \frac{b_1a_{22} - b_2a_{12}}{a_{11}a_{22} - a_{21}a_{12}} \quad \text{and} \quad x_2 = \frac{a_{11}b_2 - a_{21}b_1}{a_{11}a_{22} - a_{21}a_{12}}.$$

(At this point we do not worry whether or not $a_{11}a_{22} - a_{21}a_{12}$ is zero.) Since $a_{11}a_{22} - a_{21}a_{12}$ is a number depending on the matrix

$$\begin{bmatrix} a_{11} & a_{12} \\ a_{21} & a_{22} \end{bmatrix}$$

we denote it, to remind us where it has come from, by

$$\begin{vmatrix} a_{11} & a_{12} \\ a_{21} & a_{22} \end{vmatrix}.$$

We call it the **determinant** of the (2×2) matrix

$$\begin{bmatrix} a_{11} & a_{12} \\ a_{21} & a_{22} \end{bmatrix}.$$

PROBLEM 5.1

Evaluate:

$$\text{(i)}\begin{vmatrix} 12 & 34 \\ 21 & -43 \end{vmatrix}; \qquad \text{(ii)}\begin{vmatrix} 187 & 209 \\ 221 & 247 \end{vmatrix}.$$

●

Note that introducing this notation permits us to write x_1 and x_2 above in the form

$$x_1 = \frac{\begin{vmatrix} b_1 & a_{12} \\ b_2 & a_{22} \end{vmatrix}}{\begin{vmatrix} a_{11} & a_{12} \\ a_{21} & a_{22} \end{vmatrix}}, \qquad x_2 = \frac{\begin{vmatrix} a_{11} & b_1 \\ a_{21} & b_2 \end{vmatrix}}{\begin{vmatrix} a_{11} & a_{12} \\ a_{21} & a_{22} \end{vmatrix}}.$$

Observe the position of $\begin{matrix} b_1 \\ b_2 \end{matrix}$, namely, in the first (respectively second) column of the numerator determinant in the case of x_1 (respectively x_2).

Moving to the case of three equations in three unknowns, that is,

$$\begin{aligned} a_{11}x_1 + a_{12}x_2 + a_{13}x_3 &= b_1 \\ a_{21}x_1 + a_{22}x_2 + a_{23}x_3 &= b_2 \\ a_{31}x_1 + a_{32}x_2 + a_{33}x_3 &= b_3 \end{aligned}$$

we find, likewise, that

$$x_1 = \frac{T_1}{D}, \qquad x_2 = \frac{T_2}{D}, \qquad x_3 = \frac{T_3}{D}$$

where

$$D = a_{11}a_{22}a_{33} - a_{11}a_{23}a_{32} - a_{12}a_{21}a_{33} + a_{12}a_{23}a_{31} + a_{13}a_{21}a_{32} - a_{13}a_{22}a_{31} \tag{5.3}$$

and

$$\begin{aligned} T_1 &= b_1a_{22}a_{33} - b_1a_{23}a_{32} - a_{12}b_2a_{33} + a_{12}a_{23}b_3 + a_{13}b_2a_{32} - a_{13}a_{22}b_3; \\ T_2 &= a_{11}b_2a_{33} - a_{11}a_{23}b_3 - b_1a_{21}a_{33} + b_1a_{23}a_{31} + a_{13}a_{21}b_3 - a_{13}b_2a_{31}; \\ T_3 &= a_{11}a_{22}b_3 - a_{11}b_2a_{32} - a_{12}a_{21}b_3 + a_{12}b_2a_{31} + b_1a_{21}a_{32} - b_1a_{22}a_{31}. \end{aligned}$$

The terms and minus signs in (5.3) are not as chaotically arranged as may, at first sight, appear. Indeed each term is of the form $a_{1x}a_{2y}a_{3z}$ where x, y, z are the numbers 1, 2, 3 in some order $\{$hence the six terms in $D\}$. This means that each term is a product involving exactly one element from each row and exactly one element from each column of the matrix

$$A=\begin{bmatrix} a_{11} & a_{12} & a_{1n} \\ a_{21} & a_{22} & a_{2n} \\ a_{31} & a_{32} & a_{3n} \end{bmatrix}.$$

Further, the sign preceding each product $a_{1x}a_{2y}a_{3z}$ is +1 or −1 according to whether x, y, z is what is called an even or odd permutation of 1, 2, 3. [For a discussion of permutations see Allenby, R. B. J. T. *Rings, Fields and Groups*.]

If we agree to denote D economically (but also to remind us of its origins) by

$$\begin{vmatrix} a_{11} & a_{12} & a_{13} \\ a_{21} & a_{22} & a_{23} \\ a_{31} & a_{32} & a_{33} \end{vmatrix} \tag{5.4}$$

we see that T_1, T_2 and T_3 may be written, similarly, as

$$T_1=\begin{vmatrix} b_1 & a_{12} & a_{13} \\ b_2 & a_{22} & a_{23} \\ b_3 & a_{32} & a_{33} \end{vmatrix}, \qquad T_2=\begin{vmatrix} a_{11} & b_1 & a_{13} \\ a_{21} & b_2 & a_{23} \\ a_{31} & b_3 & a_{33} \end{vmatrix}, \qquad T_3=\begin{vmatrix} a_{11} & a_{12} & b_1 \\ a_{21} & a_{22} & b_2 \\ a_{31} & a_{32} & b_3 \end{vmatrix}.$$

Now it is easy to check that D may be rewritten as

$$a_{11}(a_{22}a_{33}-a_{23}a_{32})-a_{12}(a_{21}a_{33}-a_{23}a_{31})+a_{13}(a_{21}a_{32}-a_{22}a_{31});$$

that is

$$D=a_{11}\begin{vmatrix} a_{22} & a_{23} \\ a_{32} & a_{33} \end{vmatrix}-a_{12}\begin{vmatrix} a_{21} & a_{23} \\ a_{31} & a_{33} \end{vmatrix}+a_{13}\begin{vmatrix} a_{21} & a_{22} \\ a_{31} & a_{32} \end{vmatrix} \tag{5.5}$$

Comparing this with (5.4) we see that the a_{11}, a_{12}, a_{13} are the elements of the first row of (5.4) whilst each such a_{1j} is multiplied by the determinant of the 2×2 (sub)matrix which is obtained from A by crossing out the elements of both row 1 and column j of A.

Example 1

Use (5.5) above to evaluate the following determinants:

$$\text{(i)}\begin{vmatrix} 1 & -2 & 3 \\ -4 & 5 & -6 \\ 7 & -8 & 9 \end{vmatrix}; \qquad \text{(ii)}\begin{vmatrix} 5 & 3 & 2 \\ 3 & 3 & 5 \\ 1 & 2 & 2 \end{vmatrix}; \qquad \text{(iii)}\begin{vmatrix} 10 & 6 & 4 \\ 3 & 3 & 5 \\ 1 & 2 & 2 \end{vmatrix}.$$

SOLUTIONS

$$\text{(i)}\begin{vmatrix} 1 & -2 & 3 \\ -4 & 5 & -6 \\ 7 & -8 & 9 \end{vmatrix}=+(1)\begin{vmatrix} 5 & -6 \\ -8 & 9 \end{vmatrix}-(-2)\begin{vmatrix} -4 & -6 \\ 7 & 9 \end{vmatrix}+(3)\begin{vmatrix} -4 & 5 \\ 7 & -8 \end{vmatrix}$$

$$=1.(45-48)+2(-36+42)+3(32-35)=0$$

$$\text{(ii)}\begin{vmatrix}5 & 3 & 2\\3 & 3 & 5\\1 & 2 & 2\end{vmatrix} = +(5)\begin{vmatrix}3 & 5\\2 & 2\end{vmatrix} - (3)\begin{vmatrix}3 & 5\\1 & 2\end{vmatrix} + (2)\begin{vmatrix}3 & 3\\1 & 2\end{vmatrix}$$

$$= 5(6-10) - 3(6-5) + 2(6-3) = -17$$

$$\text{(iii)}\begin{vmatrix}10 & 6 & 4\\3 & 3 & 5\\1 & 2 & 2\end{vmatrix} = +(10)\begin{vmatrix}3 & 5\\2 & 2\end{vmatrix} - (6)\begin{vmatrix}3 & 5\\1 & 2\end{vmatrix} + (4)\begin{vmatrix}3 & 3\\1 & 2\end{vmatrix} = -34$$

What relationship do you spot between (ii) and (iii)?

We call D, above, the **determinant** of the matrix A and the method of evaluating D, as in (5.5), **expansion of the determinant of A along its first row**. As there seems to be nothing particularly special about the first row of the matrix A this (and intellectual curiosity?) raises the obvious question as to whether or not a similar expansion along other rows – or even columns – of A will lead to the same result, namely D. If we denote by A_{ij} the 2×2 (sub)matrix obtained from A by deleting all the elements of row i and column j and its determinant by $|A_{ij}|$, we indeed find that

$$\left.\begin{aligned} D &= +a_{11}|A_{11}| - a_{12}|A_{12}| + a_{13}|A_{13}| \\ D &= -a_{21}|A_{21}| + a_{22}|A_{22}| - a_{23}|A_{23}| \\ D &= +a_{31}|A_{31}| - a_{32}|A_{32}| + a_{33}|A_{33}|. \end{aligned}\right\} \tag{5.6}$$

Note that the (plus/minus) signs attached to each a_{ij} can easily be remembered via the 'chess board' pattern

$$\begin{bmatrix} + & - & + \\ - & + & - \\ + & - & + \end{bmatrix},$$

the sign in place ij being $(-1)^{i+j}$.

Furthermore expansion along columns gives

$$\left.\begin{aligned} D &= +a_{11}|A_{11}| - a_{21}|A_{21}| + a_{31}|A_{31}| \\ D &= -a_{12}|A_{12}| + a_{22}|A_{22}| - a_{32}|A_{32}| \\ D &= +a_{13}|A_{13}| - a_{23}|A_{23}| + a_{33}|A_{33}|. \end{aligned}\right\} \tag{5.7}$$

PROBLEM 5.2

Confirm that the expression for D given in (5.3) can be written in each of the six forms in (5.6) and (5.7). ●

The above analysis shows two ways in which we might define the determinant of a general $n \times n$ matrix – namely by suitably generalising (5.3) or (5.5). We choose (5.5) because it relates more directly to the given matrix. The definition is made by 'induction', that is, we define the determinant of an $n \times n$ matrix in terms of $(n-1) \times (n-1)$ matrices. The 'bottom line', so to speak, then involves 2×2 matrices which we evaluate directly.

• Definition 1

Let

$$A=\begin{bmatrix} a_{11} & a_{12} & \cdots & a_{1n} \\ a_{21} & a_{22} & \cdots & a_{2n} \\ \vdots & \vdots & \vdots & \vdots \\ a_{n1} & a_{n2} & \cdots & a_{nn} \end{bmatrix}.$$

The **determinant of A**, written Det(A) or $|A|$, is the number

$$a_{11}|A_{11}|-a_{12}|A_{12}|+a_{13}|A_{13}|-\ldots+(-1)^{1+n}a_{1n}|A_{1n}|$$

where each A_{1j} is the $(n-1)\times(n-1)$ (sub)matrix of A obtained by deleting all the elements of the 1st row and the jth column of A. ●

Example 2

$$\begin{vmatrix} 2 & 6 & -4 & -5 \\ 1 & -3 & 5 & 7 \\ 0 & 1 & 8 & 4 \\ 9 & 5 & -7 & -8 \end{vmatrix} = +2\begin{vmatrix} -3 & 5 & 7 \\ 1 & 8 & 4 \\ 5 & -7 & -8 \end{vmatrix} - 6\begin{vmatrix} 1 & 5 & 7 \\ 0 & 8 & 4 \\ 9 & -7 & -8 \end{vmatrix} + (-4)\begin{vmatrix} 1 & -3 & 7 \\ 0 & 1 & 4 \\ 9 & 5 & -8 \end{vmatrix} - (-5)\begin{vmatrix} 1 & -3 & 5 \\ 0 & 1 & 8 \\ 9 & 5 & -7 \end{vmatrix}$$

PROBLEM 5.3

Using Definition 1 evaluate

$$\begin{vmatrix} 3 & 7 & 0 & -6 \\ 2 & -4 & 0 & 8 \\ 1 & 2 & 0 & 5 \\ 4 & 6 & 0 & -9 \end{vmatrix}.$$

When you have completed Problem 5.3 you might wonder if you could have obtained the answer by using, instead, a column expansion down column 3, like those given above in (5.7) for D. The good news is that you *may*. The bad news is that you may *not* – at least not until we have proved that such a calculation is legally valid! First, in order to save ourselves from being overwhelmed by minus signs, we introduce a convenient idea and notation.

• Definition 2

Let A be an $n\times n$ matrix and (for each i and j) let A_{ij} be the $(n-1)\times(n-1)$ (sub)matrix obtained from A by deleting all the elements of row i and all the elements of column j of A. The number $(-1)^{i+j}|A_{ij}|$ is called the **cofactor** of a_{ij} in A. We denote it by C_{ij}. ●

Example 3

(i) The cofactor of a_{23} $(=-4)$ in

$$\begin{bmatrix} 6 & -7 & -3 \\ -1 & 4 & -4 \\ 1 & 2 & 8 \end{bmatrix} \text{ is } (-1)^{2+3}\begin{vmatrix} 6 & -7 \\ 1 & 2 \end{vmatrix} = -19.$$

(ii) Definition 1 may be rewritten $\text{Det}(A) = a_{11}C_{11} + a_{12}C_{12} + \ldots + a_{1n}C_{1n}$.
Our 'big' theorem is then as follows.

• Theorem 1

Let A be as in Definition 1. For each row i $(1 \le i \le n)$

$$\text{Det}(A) = a_{i1}C_{i1} + a_{i2}C_{i2} + \ldots + a_{in}C_{in}$$

and for each column j $(1 \le j \le n)$

$$\text{Det}(A) = a_{1j}C_{1j} + a_{2j}C_{2j} + \ldots + a_{nj}C_{nj}.$$

These expressions are called, respectively, the **expansion of the determinant of A along row i** and **down column j**. •

We have seen how such expansions might be useful when evaluating the determinant of a matrix with many zeros together in one row or column. Here is another consequence.

• Corollary 1

Let A be an **upper triangular** $n \times n$ matrix (that is, $a_{ij} = 0$ whenever $i > j$). Then $\text{Det}\, A = a_{11}a_{22} \ldots a_{nn}$, is the product of the diagonal elements of A. •

We leave a proper proof to you. To show you what is involved consider the following example.

Example 4

Expanding

$$\begin{vmatrix} 2 & 6 & -4 & -5 \\ 0 & -3 & 5 & 7 \\ 0 & 0 & 2 & 4 \\ 0 & 0 & 0 & -8 \end{vmatrix}$$

down its first column yields

$$\begin{vmatrix} 2 & 6 & -4 & -5 \\ 0 & -3 & 5 & 7 \\ 0 & 0 & 2 & 4 \\ 0 & 0 & 0 & -8 \end{vmatrix} = +2\begin{vmatrix} -3 & 5 & 7 \\ 0 & 2 & 4 \\ 0 & 0 & -8 \end{vmatrix} - 0\begin{vmatrix} 6 & -4 & -5 \\ 0 & 2 & 4 \\ 0 & 0 & -8 \end{vmatrix} + 0\begin{vmatrix} 6 & -4 & -5 \\ -3 & 5 & 7 \\ 0 & 0 & -8 \end{vmatrix} - 0\begin{vmatrix} 6 & -4 & -5 \\ -3 & 5 & 7 \\ 0 & 2 & 4 \end{vmatrix}$$

(now expanding down the first column of the only non - zero summand)

$$= 2\begin{vmatrix} -3 & 5 & 7 \\ 0 & 2 & 4 \\ 0 & 0 & -8 \end{vmatrix}$$

$$= 2\left\{+(-3)\begin{vmatrix} 2 & 4 \\ 0 & -8 \end{vmatrix} - 0\begin{vmatrix} 5 & 7 \\ 0 & -8 \end{vmatrix} + 0\begin{vmatrix} 5 & 7 \\ 2 & 4 \end{vmatrix}\right\}$$

$$= 2.(-3).2.(-8) = 96.$$

But this is only a start! The usefulness of Theorem 1 goes much further in often enabling us to 'simplify' a determinant even before we use Definition 1. Because a proof of Theorem 1 is a bit messy, we ignore our implied stricture that we should prove it immediately, deducing from it, instead, some consequences which are useful in effecting the quick evaluation of certain determinants. We begin with the following theorem.

• Theorem 2

Let A be as in Definition 1 and let c be any scalar (i.e. number).

(i) Let B be the matrix obtained from A by multiplying each of the elements of the ith row by c. Then Det $B = c$. Det A. In particular, a matrix with a row full of 0s has determinant 0;

(ii) $\text{Det}(cA) = c^n$. Det A;

(iii) Let C be the $n \times n$ matrix obtained from A by changing (just) the rth row (from $a_{r1}, a_{r2}, \ldots, a_{rn}$ to $c_{r1}, c_{r2}, \ldots, c_{rn}$). Then $\text{Det}(A + C) = \text{Det } A + \text{Det } C$.

PROOF

(i) The ith row expansion of Det(A) is $a_{i1}C_{i1} + a_{i2}C_{i2} + \ldots + a_{in}C_{in}$. Likewise the ith row expansion of Det(B) is $(ca_{i1})C_{i1} + (ca_{i2})C_{i2} + \ldots + (ca_{in})C_{in}$ which is clearly equal to c times Det(A).

(ii) This follows from (i) on noting that cA is just A with every one of the n rows of A multiplied by c.

(iii) We leave this to you after observing that the rth rows of A, C and $A + C$ comprise the elements $a_{r1}, a_{r2}, \ldots, a_{rn}$; $c_{r1}, c_{r2}, \ldots, c_{rn}$ and $a_{r1} + c_{r1}, a_{r2} + c_{r2}, \ldots, a_{rn} + c_{rn}$, respectively. ●

Example 5

(i) $\begin{vmatrix} 2 & 7 & 1 \\ 8 & 2 & 8 \\ 4 & 5 & 9 \end{vmatrix} = -\frac{1}{3}\begin{vmatrix} -6 & -21 & -3 \\ 8 & 2 & 8 \\ 4 & 5 & 9 \end{vmatrix}$; (ii) $\begin{vmatrix} 4 & -6 & 8 & -2 \\ -10 & 14 & 2 & 0 \\ 8 & 2 & 4 & -4 \\ 0 & 12 & 12 & 2 \end{vmatrix} = 2^4\begin{vmatrix} 2 & -3 & 4 & -1 \\ -5 & 7 & 1 & 0 \\ 4 & 1 & 2 & -2 \\ 0 & 6 & 6 & 1 \end{vmatrix}$;

(iii) $\begin{vmatrix} 1 & -6 & 5 \\ 2 & 1 & -4 \\ 1 & -1 & 9 \end{vmatrix} + \begin{vmatrix} 1 & -6 & 5 \\ 6 & -4 & 4 \\ 1 & -1 & 9 \end{vmatrix} = \begin{vmatrix} 1 & -6 & 5 \\ 8 & -3 & 0 \\ 1 & -1 & 9 \end{vmatrix}$.

Many questions of a similar nature now raise themselves quite naturally. For example: what (if any) is the relationship between the determinants of two matrices A and B which differ only in that two of their rows are interchanged?

The answer is given in the following theorem

• Theorem 3

Let the $n \times n$ matrix B be obtained from A, above, by interchanging any two rows. Then Det $B = -$Det A.

PROOF

We proceed using mathematical induction beginning with the case of 2×2 matrices. Direct calculation easily shows that the result claimed is true in this case. So we may suppose A and B to be of size $n \times n$ where $n \geq 3$ and that the desired result is true for all $(n-1) \times (n-1)$ matrices.

Let us suppose that B is obtained from A by interchanging rows r and s (where $r \neq s$). We expand Det(A) and Det(B) along (any) row t where $t \neq r$ and $t \neq s$. (How do I know such a row can be chosen?) Thus

$$\text{Det}(A) = a_{t1}C_{t1} + a_{t2}C_{t2} + \ldots + a_{tn}C_{tn} \quad \text{and} \quad \text{Det}(B) = a_{t1}D_{t1} + a_{t2}D_{t2} + \ldots + a_{tn}D_{tn}$$

where the rows of the matrix B_{tk} corresponding to D_{tk} are identical to those of the matrix A_{tk} corresponding to C_{tk} except that two of the rows (those which are part of rows r and s in A) are interchanged. But each A_{tk} and B_{tk} is obtained from an $(n-1) \times (n-1)$ matrix and so, using the inductive hypothesis, we may assume that $D_{tk} = -C_{tk}$. That Det(A) = $-$Det(B) then follows immediately. ●

Example 6

(i) $\begin{vmatrix} 3 & 1 & 4 \\ 1 & 5 & 9 \\ 2 & 6 & 5 \end{vmatrix} = -\begin{vmatrix} 3 & 1 & 4 \\ 2 & 6 & 5 \\ 1 & 5 & 9 \end{vmatrix} = \begin{vmatrix} 1 & 5 & 9 \\ 2 & 6 & 5 \\ 3 & 1 & 4 \end{vmatrix}$.

(ii) (Cf Example 2 above)

$$\begin{vmatrix} 2 & 6 & -4 & -5 \\ 9 & 5 & -7 & -8 \\ 0 & 1 & 8 & 4 \\ 1 & -3 & 5 & 7 \end{vmatrix} = +2\begin{vmatrix} 5 & -7 & 8 \\ 1 & 8 & 4 \\ -3 & 5 & 7 \end{vmatrix} - 6\begin{vmatrix} 9 & -7 & -8 \\ 0 & 8 & 4 \\ 1 & 5 & 7 \end{vmatrix} + (-4)\begin{vmatrix} 9 & 5 & -8 \\ 0 & 1 & 4 \\ 1 & -3 & 7 \end{vmatrix} - (-5)\begin{vmatrix} 9 & 5 & -7 \\ 0 & 1 & 8 \\ 1 & -3 & 5 \end{vmatrix}$$

$$= -2\begin{vmatrix} -3 & 5 & 7 \\ 1 & 8 & 4 \\ 5 & -7 & 8 \end{vmatrix} + 6\begin{vmatrix} 1 & 5 & 7 \\ 0 & 8 & 4 \\ 9 & -7 & -8 \end{vmatrix} - (-4)\begin{vmatrix} 1 & -3 & 7 \\ 0 & 1 & 4 \\ 9 & 5 & -8 \end{vmatrix} + (-5)\begin{vmatrix} 1 & -3 & 5 \\ 0 & 1 & 8 \\ 9 & 5 & -7 \end{vmatrix}$$

$$= -\begin{vmatrix} 2 & 6 & -4 & -5 \\ 1 & -3 & 5 & 7 \\ 0 & 1 & 8 & 4 \\ 9 & 5 & -7 & -8 \end{vmatrix}.$$

Two immediate consequences of Theorem 3, the second being the key in most calculations, are given in the next theorem.

• Theorem 4

Let A be as in Definition 1 and let c be a scalar.

(i) If A has two identical rows then Det $A = 0$.
(ii) Fix on distinct rows r and s of A. If B is obtained from A by replacing each a_{rj} by $a_{rj} + ca_{sj}$ $(1 \leq j \leq n)$ then Det B = Det A.

PROOF

(i) On switching the identical rows, Theorem 3 shows that Det A = –Det A.
(ii) By Theorem 2(iii) Det B = Det A + Det C where C is identical to A except that its rth row is $ca_{s1}, ca_{s2}, \ldots, ca_{sn}$. By Theorem 2(i) we see that Det $C = c$ Det D where D is a matrix with rth and sth rows both being $a_{s1}, a_{s2}, \ldots, a_{sn}$. Hence, by part (i) Det $D = 0$ and so Det B = Det A (+ 0), as claimed. ●

Example 7

(i) $\begin{vmatrix} 2 & 7 & 1 \\ 8 & 2 & 8 \\ 2 & 7 & 1 \end{vmatrix} = 0;$

(ii) $\begin{vmatrix} 9 & 3 & 7 \\ 5 & 8 & 8 \\ 1 & 5 & 3 \end{vmatrix} = \begin{vmatrix} 9 & 3 & 7 \\ 79 & 378 & 230 \\ 1 & 5 & 3 \end{vmatrix}$ (on adding ... what to what?)

NOTE I

(i) Theorems 3, 2(i) and 4(ii) show the change in the value of the determinant of an $n \times n$ matrix A after applying an elementary row operation to A. Since elementary row operations can reduce a matrix to one whose determinant is more easily worked out, these particular results are of especial importance.
(ii) Theorems 3, 2(i) and 4(ii) apply, in particular, to I_n and tell us, at a glance, the determinant of any elementary matrix. In particular, each elementary matrix has non-zero determinant. In fact, we can deduce something even more useful: let E be any $n \times n$ elementary matrix. Then Det(EA) = Det E Det A. (See Exercise 23(b).) This fact is central in proving Theorem 7.

Before we can evaluate determinants more efficiently we need another theorem which follows from the (alleged) equality of the expansions of Det(A) along the first row and the first column of A.

• Theorem 5

Let A be as in Definition 1 and let A^T be its transpose. Then Det(A^T) = Det A.

PROOF

Expanding Det(A) along its first column gives

$$\text{Det}(A) = a_{11}C_{11} + a_{21}C_{21} + \ldots + a_{n1}C_{n1}$$

where each C_{j1} comes from the $(n-1) \times (n-1)$ matrix A_{j1}. On the other hand, expanding A^T along its first row shows that

$$\text{Det}(A^T) = a_{11}D_{11} + a_{21}D_{12} + \ldots + a_{n1}D_{1n}$$

where each D_{1j} arises from A_{j1}^T. Since, for each j, the obvious induction hypothesis implies that we have $D_{1j} = C_{j1}$. we see that $\text{Det}(A) = \text{Det}(A^T)$, as claimed. ●

Example 8

$$\begin{vmatrix} 3 & 7 & 1 \\ 4 & -6 & 5 \\ 1 & 2 & -3 \end{vmatrix} = \begin{vmatrix} 3 & 4 & 1 \\ 7 & -6 & 2 \\ 1 & 5 & -3 \end{vmatrix}.$$

Theorem 5 together with Theorems 3, 2(i) and 4(ii) immediately produces the following corollary.

● Corollary 2

Let A be as in Definition 1 and let c be a scalar.

(i) If B is obtained from A by interchanging two columns then $\text{Det } B = -\text{Det } A$.

(ii) If B is obtained from A by multiplying one column by c then $\text{Det } B = c \text{ Det } A$.

(iii) If B is obtained from A by adding c times column s to column r then $\text{Det } B = \text{Det } A$. ●

Example 9

(i) $$\begin{vmatrix} 3 & 1 & 4 \\ 1 & 5 & 9 \\ 2 & 6 & 5 \end{vmatrix} = -\begin{vmatrix} 3 & 4 & 1 \\ 1 & 9 & 5 \\ 2 & 5 & 6 \end{vmatrix} = \begin{vmatrix} 4 & 3 & 1 \\ 9 & 1 & 5 \\ 5 & 2 & 6 \end{vmatrix};$$

(ii) $$\begin{vmatrix} 2 & 7 & 1 \\ 8 & 2 & 8 \\ 4 & 5 & 9 \end{vmatrix} = -\frac{1}{3}\begin{vmatrix} 2 & 7 & -3 \\ 8 & 2 & -24 \\ 4 & 5 & -27 \end{vmatrix};$$ (iii) $$\begin{vmatrix} 9 & 3 & 7 \\ 5 & 8 & 8 \\ 1 & 5 & 3 \end{vmatrix} = \begin{vmatrix} 485 & 3 & 7 \\ 549 & 8 & 8 \\ 205 & 5 & 3 \end{vmatrix}.$$

Note 1(ii) can be used to obtain yet another criterion for an $n \times n$ matrix to have a multiplicative inverse as follows. (Cf. Theorem 5 of Chapter 4.)

● Theorem 6

The $n \times n$ matrix A is invertible if and only if $\text{Det } A \neq 0$.

PROOF

Since A can be changed to reduced (row) echelon form by a succession of elementary row operations, there exist elementary matrices $E_1, E_2, \ldots, E_t$ and a reduced (row) echelon matrix R, say, such that $R = E_t E_{t-1} \ldots E_1 A$. It follows, from Note 1(ii), that

$$\begin{aligned} \text{Det R} &= \text{Det}\left(E_t\left\{E_{t-1}\ldots E_1 A\right\}\right) = \text{Det } E_t \text{ Det}\left\{E_{t-1}\ldots E_1 A\right\} \\ &= \ldots = \text{Det } E_t \text{ Det } E_{t-1}\ldots \text{ Det } E_1 \text{ Det } A. \end{aligned} \tag{5.8}$$

Since A is $n \times n$, either R must have at least one row full of zeros – in which $\text{Det } R = 0$ and R has no inverse – or R must be the identity matrix I_n, in which case $\text{Det } R \neq 0$ and R has an inverse!!

Since, for each E_i, (i) $\text{Det } E_i \neq 0$ (see Note 1(ii)) and (ii) E_i is invertible, we see from

(5.8) that (i) Det $A = 0$ if and only if Det $R = 0$ and (ii) (by Note 2 of Chapter 4) A is invertible if and only if R is. This does it!

Example 10

$$\begin{bmatrix} 1 & 1 & 1 & 1 \\ 1 & 1 & 1 & 2 \\ 1 & 1 & 2 & 3 \\ 1 & 2 & 3 & 4 \end{bmatrix} \text{ is invertible since } \begin{vmatrix} 1 & 1 & 1 & 1 \\ 1 & 1 & 1 & 2 \\ 1 & 1 & 2 & 3 \\ 1 & 2 & 3 & 4 \end{vmatrix} = \begin{vmatrix} 1 & 1 & 1 & 1 \\ 0 & 0 & 0 & 1 \\ 0 & 0 & 1 & 1 \\ 0 & 1 & 1 & 1 \end{vmatrix} = -\begin{vmatrix} 1 & 1 & 1 & 1 \\ 0 & 1 & 1 & 1 \\ 0 & 0 & 1 & 1 \\ 0 & 0 & 0 & 1 \end{vmatrix}$$

on using Theorem 4(ii) three times to get the second determinant from the first and Theorem 3 once to get the third determinant from the second. The last determinant is not 0 by Corollary 1 and so

$$\begin{bmatrix} 1 & 1 & 1 & 1 \\ 1 & 1 & 1 & 2 \\ 1 & 1 & 2 & 3 \\ 1 & 2 & 3 & 4 \end{bmatrix}$$

is invertible by Theorem 6.

We have certainly not run out of questions which natural inquisitiveness would have us ask about matrices. To name just a couple: do we have (for appropriately shaped square matrices) (i) $\text{Det}(A + B) = \text{Det } A + \text{Det } B$; (ii) $\text{Det}(AB) = \text{Det } A \text{ Det } B$? Exercise 4 below deals with (i); for (ii) we have the following theorem.

Theorem 7

Let A and B be $n \times n$ matrices. Then $\text{Det}(AB) = \text{Det } A \text{ Det } B$.

PROOF

If A is invertible then, by Theorem 5(v) of Chapter 4, (or Theorem 6 above) A is expressible as a product $F_1F_2 \dots F_t$, say, of elementary matrices. Then, as in Theorem 6, we have

$$\text{Det}(AB) = \text{Det}(F_1\{F_2 \dots F_tB\}) = \text{Det } F_1 \text{ Det}\{F_2 \dots F_tB\} = \text{Det } F_1 \text{ Det } F_2 \dots \text{ Det } F_t \text{ Det } B.$$

But, likewise,

$$\text{Det } F_1 \text{ Det } F_2 \dots \text{ Det } F_t = \text{Det}(F_1F_2 \dots F_t) = \text{Det } A.$$

If A is not invertible then neither is AB (Exercise 5(d) at the end of Chapter 4). But then by Theorem 6 we know that Det A and Det(AB) are both zero. Hence Det A Det B = Det(AB) (since both are 0).

(Exercise 3 below shows that, geometrically, Theorem 7 is not so great a surprise.)

PROBLEM 5.4

(a) Confirm Theorem 7 for the matrices

$$A=\begin{bmatrix}1 & 2\\3 & 4\end{bmatrix} \quad \text{and} \quad B=\begin{bmatrix}5 & 6\\7 & 8\end{bmatrix}.$$

(b) Deduce that for all pairs of $n \times n$ matrices A and B we have $\mathrm{Det}(AB) = \mathrm{Det}(BA)$ – even though $AB \neq BA$ in general. ●

Finally we offer a sketch of part of the following proof.

PROOF OF THEOREM 1

Once again we take A_{ij} and C_{ij} as in Definition 2. We prove that, for each i, where $1 < i \leq n$:

(i) $a_{11}C_{11} + a_{12}C_{12} + \ldots a_{1n}C_{1n} = a_{i1}C_{i1} + a_{i2}C_{i2} + \ldots + a_{in}C_{in}$;

(ii) $a_{11}C_{11} + a_{12}C_{12} + \ldots a_{1n}C_{1n} = a_{11}C_{11} + a_{21}C_{21} + \ldots + a_{n1}C_{n1}$. (Why does this suffice?)

We prove (i) by showing that, for each j (where $1 \leq j \leq n$) and for each k (where $1 \leq k \leq n$, $k \neq j$) each side of (i) is a sum of identical terms of the form $a_{1j}a_{ik}d$. First suppose that $j < k$. Then a_{ik} lies in the $(i-1)$st row and the $(k-1)$st column of A_{1j}. Thus in the first expression in (i) we find that $a_{1j}a_{ik}$ is multiplied by $(-1)^{1+j}$ (coming from the factor C_{1j}) and by $(-1)^{(i-1)+(k-1)}M$ where M is the determinant obtained from A_{1j} by deleting its $(i-1)$st row and its $(k-1)$st column [these being parts of row i and column j, respectively, of A itself]. In a similar manner, we observe, looking at the second expression in (i), that a_{1j} lies in row 1 and column j (since $j < k$) of A_{ik}. Hence, in $a_{i1}C_{i1} + a_{i2}C_{i2} + \ldots + a_{in}C_{in}$, the term $a_{ik}a_{1j}$ is multiplied by $(-1)^{i+k}$ {coming from the factor C_{ik}} and by $(-1)^{1+j}N$ where N is the determinant obtained from A_{ik} by deleting its first row and jth column. But, deleting the $(i-1)$st row and the $(k-1)$st column of A_{1j} and deleting the 1st row and jth column of A_{ik}, we obtain the same $(n-2) \times (n-2)$ submatrix of A. This means that $M = N$ which completes the proof of the case $j < k$ since $(-1)^{1+j}.(-1)^{(i-1)+(k-1)} = (-1)^{i+k}.(-1)^{1+j}$. We leave the reader to cover the case $j > k$ and find a full proof of Theorem 1 in Fraleigh, B. and Beauregard, R. A. *Linear Algebra*. ●

The founder of the theory of determinants is usually regarded as being G. W. Leibniz. Of course there were the usual precursors; Cardano (1545) gave a rule for solving pairs of equations in two unknowns and the Japanese Takakazu Seki (1642–1708) used determinant-like quantities in 1683. And, naturally, Leibniz did not have all the information given in this chapter, nor did he prove the results he did have. Nevertheless, Leibniz did know about the 24 terms (and their signs) in the 4×4 array (see Exercise 9) – calling their (signed) sum the 'resultant' and, by 1684, he knew Cramer's rule (see Exercise 13(b)) for solving linear systems. Cramer's rule (1750) derived from Cramer's attempt to determine the equation of a curve $P(x, y) = 0$ of the nth degree passing through $[n(n + 3)]/2$ given points. Laplace, in 1772, rejected Cramer's method as impractical and made improvements of his own. (Theorem 1 is due to him.) Determinants occur naturally in Gauss's theory of forms. This led Cauchy (1812) to his systematic development of the general theory in which he introduced the name 'determinant' to mean what it does today. This work went unnoticed for 30 years and it was Jacobi (1841) who brought the theory to the mathematicians' attention. The vertical parallel lines at each side of a determinant

were introduced by Cayley also in 1841; the double suffix notation had been employed by Leibniz and Vandermonde much earlier.

Gottfried Wilhelm Leibniz was born in Leipzig on 1 July 1646. He is best known as the co-inventor, along with Newton, of the calculus. A precocious child from an academic background, he did not start out as a mathematician. Indeed, aged 15, he entered the University of Leipzig to study law. In 1666 he obtained a doctorate for a work concerning a universal symbolic method of reasoning by which he hoped all disputes could be settled by logic rather than by fighting! In 1671 he invented a calculating machine superior to Pascal's. Instead of accepting a professorship (in law) offered by the University of Altdorf he became an ambassador (Bell, a writer not averse to a little embellishment of the facts, prefers the term 'glorifed commercial traveller') working for the Elector of Mainz. This brought him into contact with many influential people and prominent scientists, including Huygens, whose researches awakened his interest in mathematics. Kline lists his fields of expertise as logic, mechanics, optics, mathematics, hydrostatics, pneumatics, nautical science and calculating machines! But this 'universal genius' spent much of his life trying to sort out the family tree of the Duke of Brunswick. The man who, at one stage, was offered the librarianship of the Vatican died, apparently neglected and forgotten, in Hanover, on 14 November 1716.

We now give a selection of examples showing the evaluation of some determinants.

Examples 11

(i) $\begin{vmatrix} 2 & 3 \\ 5 & 7 \end{vmatrix} = 2.7 - 5.3 = -1;$

(ii) $$\begin{vmatrix} 2 & 3 & 7 \\ 1 & 6 & 5 \\ 8 & 9 & 2 \end{vmatrix} = 2\begin{vmatrix} 6 & 5 \\ 9 & 2 \end{vmatrix} - 3\begin{vmatrix} 1 & 5 \\ 8 & 2 \end{vmatrix} + 7\begin{vmatrix} 1 & 6 \\ 8 & 9 \end{vmatrix}$$

$$= 2.(12-45) - 3.(2-40) + 7.(9-48) = -225,$$

by expanding along the first row. This can be done a bit more elegantly by noting

$$\begin{vmatrix} 2 & 3 & 7 \\ 1 & 6 & 5 \\ 8 & 9 & 2 \end{vmatrix} = -\begin{vmatrix} 1 & 6 & 5 \\ 2 & 3 & 7 \\ 8 & 9 & 2 \end{vmatrix} = -\begin{vmatrix} 1 & 6 & 5 \\ 0 & -9 & -3 \\ 0 & -39 & -38 \end{vmatrix} = -3\begin{vmatrix} 1 & 6 & 5 \\ 0 & 3 & 1 \\ 0 & 39 & 38 \end{vmatrix}$$

$$= -3\begin{vmatrix} 1 & 6 & 5 \\ 0 & 3 & 1 \\ 0 & 0 & 25 \end{vmatrix} = (-3).1.3.25 = -225.$$

What steps have I used here? I leave it up to you to find out since they are fairly transparent. In more involved cases it is often helpful to indicate to the reader – and to remind yourself! – which elementary row (or column) operations you have applied by making use of the ρ (or κ for column?) notation we employed earlier.

Why did I take those particular steps? The whole point of Theorems 2(i), 4(ii) (and 3) as far as calculating determinants is concerned, is that they enable determinants to be evaluated more easily, perhaps by taking out large factors or by introducing many zero entries. [However, for 'small' matrices it is perhaps not worth spending too long on trying to find crafty labour-saving row operations. See Exercise 16(c).]

Slightly different kinds of examples are included in the following.

Examples 12

(i) Evaluate

$$\begin{vmatrix} x & y & z \\ y & z & x \\ z & x & y \end{vmatrix}.$$

(Here x, y, z are just 'unknowns' so the result will be a polynomial expression in x, y and z.) Starting by adding rows 2 and 3 to row 1 we get

$$\begin{vmatrix} x & y & z \\ y & z & x \\ z & x & y \end{vmatrix} = \begin{vmatrix} x+y+z & x+y+z & x+y+z \\ y & z & x \\ z & x & y \end{vmatrix} = (x+y+z)\begin{vmatrix} 1 & 1 & 1 \\ y & z & x \\ z & x & y \end{vmatrix}$$

$$= (x+y+z)\left\{\begin{vmatrix} z & x \\ x & y \end{vmatrix} - \begin{vmatrix} y & x \\ z & y \end{vmatrix} + \begin{vmatrix} y & z \\ z & x \end{vmatrix}\right\}$$

$$= (x+y+z)\left\{zy - x^2 - y^2 + xz + yx - z^2\right\}.$$

Alternatively we may write

$$\begin{vmatrix} 1 & 1 & 1 \\ y & z & x \\ z & x & y \end{vmatrix} \quad \text{as} \quad \begin{vmatrix} 1 & 0 & 0 \\ y & z-y & x-y \\ z & x-z & y-z \end{vmatrix} = \{(z-y)(y-z)-(x-y)(x-z)\}$$

by expanding along the first row.

(ii) Evaluate

$$\begin{vmatrix} 1 & x & y & z \\ 1 & x^2 & y^2 & z^2 \\ 1 & x^3 & y^3 & z^3 \\ 1 & x^4 & y^4 & z^4 \end{vmatrix}$$

where, again, x, y, z are unknowns. Here one could proceed by a long-winded expansion but we can exploit some of the general results obtained above in a rather nice way as follows. By expanding down the first column we see (cf. the expression (5.3) given earlier for D) that the value of this determinant will be a sum of terms of the form $x^r y^s z^t$ where $r + s + t \leq 9$. Let us call this expression $f(x, y, z)$.

If x takes the value 1 we have a determinant with two identical columns. The determinant would therefore be 0. But this implies (cf. the factor theorem in the Appendix), that $(x - 1)$ is a factor of $f(x, y, z)$. Likewise, so are $(y - 1)$ and $(z - 1)$, and $(x - y)$, $(y - z)$ and $(z - x)$ – since if x and y or if x and z or if y and z are equal the determinant has two equal columns. Thus

$$f(x,y,z) = xyz(x-1)(y-1)(z-1)(x-y)(y-z)(z-x)g(x,y,z)$$

for some suitable $g(x, y, z)$. [The x, y and z factors are obtained immediately from columns 2, 3 and 4.] Since $f(x, y, z)$ is a polynomial of degree 9, we see that $g(x, y, z)$ must have degree 0, i.e. $g(x, y, z)$ must be a constant, c, say. To find c, note that the coefficient of $x^2y^3z^4$ in $f(x, y, z)$ is precisely c. However, the only way of getting the product $x^2y^3z^4$ in the determinant expansion is from the main diagonal – hence it appears +1 times. Consequently,

$$c = 1 \quad \text{and} \quad f(x,y,z) = xyz(x-1)(y-1)(z-1)(x-y)(y-z)(z-x).$$

To end with, some amusing problems for discussion.

TUTORIAL PROBLEMS 5.1

1. Let A be an $n \times n$ matrix whose entries are 0s and 1s. What is the maximum possible value of Det(A)? (Of course lots of variants on this theme will occur to you! See, for example, the computer package problems.)
2. Define the concept of $\infty \times \infty$ determinant. (Such things first arose in solving infinite linear systems in infinitely many unknowns which themselves arose in the study of series solutions of differential equations.) What problems do you envisage? Which of the above theorems will also be true for infinite determinants? For example, what about Theorems 2, 3, and 4?

Finally, a lovely problem due to Gilbert Strang (p. 178 of *Linear Algebra and its Applications*, 2nd edition, Academic Press, 1980) reproduced with his permission.

3. Let A be an $n \times n$ matrix with positive determinant. Show that the elements of A can be changed continuously so that A changes into I_n *without becoming singular on the way*. Note that as t changes continuously from 0 to 1, $A(t) = A + t(I_n - A)$ changes continuously from A to I_n – but Det $A(t)$ might be 0 for some value of t.

Applications

Like matrix inverses, determinants are rather more of a theoretical than a practical tool. Indeed whilst it is nice to be able to give an exact formula for the solution of a system of equations by using Cramer's rule (see Exercise 13(b) below), the rule itself is useless in practice, taking about n times as long as Gaussian elimination. (Actually Cramer's rule does indicate why, if the determinant of the coefficient matrix of a system of equations is close to zero and the coefficients themselves are somewhat unreliable, the solution of the system may be extremely unreliable.) Nevertheless, a practical use of determinants does arise if, as in the Applications in Chapter 4, we seek λ, $\mathbf{v}$, for which $A\mathbf{v} = \lambda\mathbf{v}$. For, by Theorem 6, $A - \lambda I$ is invertible if and only if $\text{Det}(A - \lambda I) \neq 0$. This leads to a polynomial equation seemingly giving all possible such λ. (Unfortunately these λ may also be computationally unreliable, as pointed out in Ciarlet, P. G. *Introduction to Numerical Linear Algebra and Optimization* and Hager, W. W. *Applied Numerical Linear Algebra*. On the other hand, if A is $n \times n$, it does tell us that there can be no more than n such λ, a fact which is trickier to establish by non-determinantal means (see Theorem 1 of Chapter 11).

Determinants are useful in changing variables in multiple integrals and as a testing device (via the Wronskian) for the linear independence of functions (see Exercise 23 of Chapter 8).

As a mildly amusing 'theoretical' application consider the following problem. (If you believe what follows you'll believe anything!)

The other day I wanted to find a prime number value for x so that the matrix

$$A = \begin{bmatrix} 3 & 5 & 7 \\ 11 & 13 & 17 \\ 19 & 23 & x \end{bmatrix}$$

had an inverse with integer entries. After some huffing and puffing I saw why it could not have one – in one line! (Can you see why not?)

Summary

With each square matrix

$$A = \begin{bmatrix} a_{11} & a_{12} & \cdots & a_{1n} \\ a_{21} & a_{22} & \cdots & a_{2n} \\ \vdots & \vdots & \vdots & \vdots \\ a_{n1} & a_{n2} & \cdots & a_{nn} \end{bmatrix}$$

we can associate a number,

$$\text{Det}(A) = a_{11}|A_{11}| - a_{12}|A_{12}| + a_{13}|A_{13}| - \ldots + (-1)^{1+n} a_{1n}|A_{1n}|$$

called the **determinant of** A. In this form we say that the determinant of A is found by **expansion along the first row of** A. Det A may also be evaluated for each i, $1 \le i \le n$ (by **expansion along the** i**th row**) as $\sum_{j=1}^{n} a_{ij}C_{ij}$, where C_{ij} is the ij**th cofactor** of A and, by similar formulae, down each column. Elementary operations, applied to the rows of a matrix, may change the value of its determinant, but they do not change whether or not the determinant is zero. (Similar remarks apply to the columns of A since Det A = Det(A^T).) This is useful since A is invertible if and only if Det $A \ne 0$, a fact proved by showing that the determinant of each elementary matrix is non-zero.

For all $n \times n$ matrices A, B we have Det(AB) = Det A Det B, a proof of which uses the fact that Det $A \ne 0$ if and only if A is invertible.

EXERCISES ON CHAPTER 5

1. Evaluate $\begin{vmatrix} 1 & 2 \\ 3 & 4 \end{vmatrix}$ and $\begin{vmatrix} 3 & 4 \\ 1 & 2 \end{vmatrix}$.

2. (a) Find x such that (i) $\begin{vmatrix} 1 & 2 \\ 3 & x \end{vmatrix} = 0$; (ii) $\begin{vmatrix} 1-x & -2 \\ 3 & 4-x \end{vmatrix} = 0$.

   (b) Find all x such that $\begin{vmatrix} 1-x & 1 \\ a & 3-x \end{vmatrix} = 0$ if a equals (i) 0; (ii) -1; (iii) -2.

3. (a) Let A be the matrix

$$\begin{bmatrix} a & b \\ c & d \end{bmatrix}.$$

   Assuming, for simplicity, that each of a, b, c and d are positive real numbers, determine the points $A\mathbf{x}$ where

$$\mathbf{x} = \begin{bmatrix} 0 \\ 0 \end{bmatrix}, \begin{bmatrix} 1 \\ 0 \end{bmatrix}, \begin{bmatrix} 1 \\ 1 \end{bmatrix} \text{ and } \begin{bmatrix} 0 \\ 1 \end{bmatrix}.$$

   Find the ratio of the area of the square whose vertices are the four given points to the area of the shape whose vertices are the four values of $A\mathbf{x}$. Relate this ratio to Det(A).

   (b) Let B be the matrix

$$\begin{bmatrix} e & f \\ g & h \end{bmatrix}$$

where e, f, g and h are positive. Find the areas of the shapes determined by (i) the points $AB\mathbf{x}$; (ii) the points $BA\mathbf{x}$, as $\mathbf{x}$ ranges over the points

$$\begin{bmatrix}0\\0\end{bmatrix}, \quad \begin{bmatrix}1\\0\end{bmatrix}, \quad \begin{bmatrix}1\\1\end{bmatrix} \quad \text{and} \quad \begin{bmatrix}0\\1\end{bmatrix}.$$

What does this tell you about Det(AB) and Det(BA)?

4. Find two $n \times n$ matrices A and B – for whatever n (≥ 1) you wish – such that Det($A + B$) $\neq$ Det A + Det B. (Does this not contradict Theorem 2(iii)?)

5. Evaluate, using (5.3),

$$\text{(i)} \begin{vmatrix}1 & 2 & 3\\4 & 5 & 6\\7 & 8 & 9\end{vmatrix}; \qquad \text{(ii)} \begin{vmatrix}1 & 2 & 3\\1 & 4 & 9\\1 & 8 & 27\end{vmatrix}.$$

Check your results by easier means when you can.

6. Using Theorem 1, evaluate $\begin{vmatrix}17 & -19 & 31\\0 & 0 & 1\\62 & 47 & -15\end{vmatrix}$ along two different rows.

7. For which x is $\begin{vmatrix}7 & x & -1\\2 & 6 & 4\\4 & -7 & 5\end{vmatrix} = 0$?

8. Exercise 13 of Chapter 3 defined the vector product $\mathbf{a}\wedge\mathbf{b}$ of the vectors $\mathbf{a} = (a_1, a_2, a_3)$ and $\mathbf{b} = (b_1, b_2, b_3)$. If we define $\mathbf{i} = (1, 0, 0)$, $\mathbf{j} = (0, 1, 0)$ and $\mathbf{k} = (0, 0, 1)$ – so that each (a, b, c) in $\mathbb{R}^3$ can be written as $a\mathbf{i} + b\mathbf{j} + c\mathbf{k}$ – show that $\mathbf{a}\wedge\mathbf{b}$ may be represented formally by

$$\begin{vmatrix}\mathbf{i} & \mathbf{j} & \mathbf{k}\\a_1 & a_2 & a_3\\b_1 & b_2 & c_3\end{vmatrix}.$$

Now find a vector perpendicular to the (plane containing) the two vectors (1, 2, 3) and (4, 5, 6).

9. (i) Take $n = 4$ in Definition 1. If the determinant is expanded fully (as in equation (5.3) for the 3×3 case) as a sum of products of the form $a_{1\alpha}a_{2\beta}a_{3\gamma}a_{4\delta}$, how many terms are there in the expression for Det A? Can you identify three of these terms whose associated signs (I) are positive; (II) are negative? How many terms are there in the full expansion of the determinant of a 5×5 matrix?

(ii) One way of remembering the full expansion (5.3) of a 3×3 determinant is to draw the picture

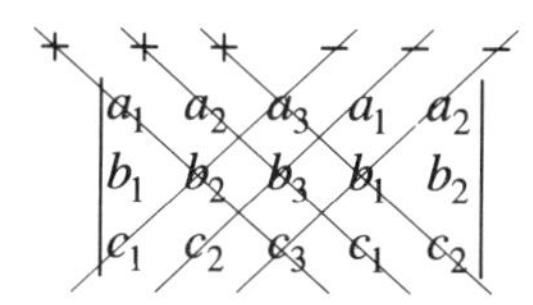

Explain why this method works here but cannot possibly work for the $n \times n$ case if $n \geq 4$.

10. Given that $\begin{vmatrix} 1-x & 2 & 3 \\ 4 & 5-x & 6 \\ 7 & 8 & 9-x \end{vmatrix} = 0,$ find x.

11. (i) Arrange the numbers, 1, 1, 1, 2, 2, 3, 4, 5, 6 into a 3×3 matrix whose determinant is 0.

(ii) Arrange seven 1s and two 0s to form a 3×3 matrix with non-zero determinant.

12. (a) For the matrix

$$A = \begin{bmatrix} -3 & 1 & 5 \\ 4 & -2 & -2 \\ -5 & 1 & 10 \end{bmatrix}$$

evaluate $a_{13}C_{13} + a_{23}C_{23} + a_{33}C_{33}$ and $a_{13}C_{11} + a_{23}C_{21} + a_{33}C_{31}$.

(b) For the general $n \times n$ matrix show that, for $i \neq k$,

$$\sum_{j=1}^{n} a_{ij}C_{kj} = a_{i1}C_{k1} + a_{i2}C_{k2} + \ldots + a_{in}C_{kn}$$

(a so-called *expansion by wrong cofactors*) has value 0.

13. (a) Let A be as in Definition 1 and let C be the matrix

$$\begin{bmatrix} C_{11} & C_{21} & \cdots & C_{n1} \\ C_{12} & C_{22} & \cdots & C_{n2} \\ \vdots & \vdots & \vdots & \vdots \\ C_{1n} & C_{2n} & \cdots & C_{nn} \end{bmatrix},$$

where the Cij are as in Definition 2. Use Theorem 1 to deduce that $AC = |A|I_n$. Deduce further that, if $\text{Det}(A) \neq 0$ then

$$A^{-1} = \frac{1}{|A|}\begin{bmatrix} C_{11} & C_{21} & \cdots & C_{n1} \\ C_{12} & C_{22} & \cdots & C_{n2} \\ \vdots & \vdots & \cdots & \vdots \\ C_{1n} & C_{2n} & \cdots & C_{nn} \end{bmatrix} \quad (\text{if } n \geq 2).$$

(b) Use part (a) to show that if

$$A\begin{bmatrix} x_1 \\ \vdots \\ x_n \end{bmatrix} = \begin{bmatrix} b_1 \\ \vdots \\ b_n \end{bmatrix}$$

where $A_{n \times n}$ is invertible, then, for each i, $x_i = \frac{1}{|A|}\{b_1C_{1i} + b_2C_{2i} + \ldots + b_nC_{ni}\}$.

[This is **Cramer's rule**. In brief each $x_i = \text{Det } A(i)/\text{Det } A$ where $A(i)$ is obtained from A by replacing column i by

$$\begin{bmatrix} b_1 \\ \vdots \\ b_n \end{bmatrix}.$$

For practical computation the rule is almost useless because of the 'cost' of evaluating so many determinants. It can, however be convenient theoretically (as in part (b)) since it gives a specific formula for each x_i.]

(c) Suppose A has integer coefficients and that $\text{Det}(A) = \pm 1$. Prove that $A\mathbf{x} = \mathbf{b}$ has a unique solution in which all the x_i are integers.

14. (i) Solve, for x, the equation $\begin{vmatrix} 1 & 1 & x \\ 1 & x & 1 \\ x & 1 & 1 \end{vmatrix} = 0.$ [Hint: add the rows.]

(ii) Solve for x (if you can!) $\begin{vmatrix} 1 & 2 & x \\ 3 & x & 4 \\ x & 5 & 6 \end{vmatrix} = 0.$

(Thus relying on easily solving a cubic may indicate too much optimism!)

15. Evaluate

$$\text{(i)} \begin{vmatrix} 4 & 3 & 2 & 1 \\ 3 & 3 & 2 & 1 \\ 2 & 2 & 2 & 1 \\ 1 & 1 & 1 & 1 \end{vmatrix}; \quad \text{(ii)} \begin{vmatrix} a+1 & b & c & d \\ a & b+1 & c & d \\ a & b & c+1 & d \\ a & b & c & d+1 \end{vmatrix}.$$

16. Find

$$\text{(a)} \begin{vmatrix} 0 & 0 & 0 & 0 & 4 \\ 0 & -2 & 0 & 0 & 0 \\ 0 & 0 & 0 & 1 & 0 \\ -5 & 0 & 0 & 0 & 0 \\ 0 & 0 & -3 & 0 & 0 \end{vmatrix}; \quad \text{(b)} \begin{vmatrix} 1 & 11 & 31 & 0 & 0 \\ 17 & 18 & -7 & -14 & 16 \\ 0 & 2 & 0 & 0 & 0 \\ 0 & 1 & 19 & 0 & 0 \\ 15 & 11 & -14 & 28 & 15 \end{vmatrix}; \quad \text{(c)} \begin{vmatrix} 17 & 4 & 12 \\ -11 & 6 & 4 \\ 7 & 9 & -5 \end{vmatrix}.$$

17. (a) Let U_n denote the $n \times n$ matrix with 0s down the main diagonal and 1s in every other position. Evaluate $\text{Det}(U_n)$ for as many n as you can.

(b) Evaluate the $n \times n$ determinant

$$\begin{vmatrix} 2 & -1 & 0 & \dots & 0 & 0 & 0 \\ -1 & 2 & -1 & \dots & 0 & 0 & 0 \\ \vdots & \vdots & \vdots & & \vdots & \vdots & \vdots \\ 0 & 0 & 0 & \dots & -1 & 2 & -1 \\ 0 & 0 & 0 & \dots & 0 & -1 & 2 \end{vmatrix}$$

for $n = 3$, 4 and 5. Can you find (and prove it correct by mathematical induction) a formula which gives the value of the determinant for each value of n?

18. Think of the 4×4 matrix

$$Z = \begin{bmatrix} A & B \\ C & D \end{bmatrix}$$

as being 'partitioned' into 2×2 submatrices A, B, C and D. Is Det Z = Det A Det D – Det C Det B?) (Proof or counterexample please!)

19. (a) Prove that the determinant of an elementary matrix is non-zero.
(b) Prove from first principles, for each $n \times n$ elementary matrix E and each $n \times n$ matrix A, that Det(EA) = Det E Det A.

20. Let A be a **skew symmetric** $n \times n$ matrix (i.e. for all i, j, $a_{ji} = -a_{ij}$). Show that Det $A = (-1)^n$ Det A. What may you deduce if n is odd?

21. (a) The invertible $n \times n$ matrix A has integer coefficients and is such that $A^{10} = A^{13}$. Show that A^{-1} has integer entries.
(b) Given that the matrix

$$\Pi = \begin{bmatrix} 3 & 1 & 4 & 1 \\ 5 & 9 & 2 & 6 \\ 5 & 3 & 5 & 8 \\ 9 & 7 & 9 & 3 \end{bmatrix}$$

is invertible, how can I tell, immediately, that not all the entries in Π^{-1} are integers?

22. Let A, B be $n \times n$ matrices with A invertible. Prove that (i) $\text{Det}(A^{-1}) = (\text{Det } A)^{-1}$ – and explain why, here the -1 is not the same as the -1 (!?); (ii) $\text{Det}(A^{-1}BA) = \text{Det } B$; (iii) $\text{Det}(A^{-1}BA - \lambda I_n) = \text{Det}(B - \lambda I_n)$, λ being any scalar.

23. Are there 2×2 matrices A, B such that $A^{-1}B^{-1}AB = \begin{bmatrix} 7 & 5 \\ 2 & 1 \end{bmatrix}$?

24. The matrix

$$\begin{bmatrix} C_{11} & C_{21} & \cdots & C_{n1} \\ C_{12} & C_{22} & \cdots & C_{n2} \\ \vdots & \vdots & \cdots & \vdots \\ C_{1n} & C_{2n} & \cdots & C_{nn} \end{bmatrix}$$

in Exercise 13 is called the **adjoint** of A, Adj(A). Find Det(Adj(A)). Show that A is singular if and only if Det(Adj(A)) = 0 and that, if A is invertible, then $(\text{Adj } A)^{-1} = \text{Adj}(A^{-1}) = A/|A|$.

25. What is wrong with the following proof that ALL determinants are zero?
'Proof' Let $a_1, a_2, \ldots, a_n$ and $b_1, b_2, \ldots, b_n$ be the first two rows of the $n \times n$ matrix A. Applying, in succession, the row operations $\rho_1 \to \rho_1 - \rho_2$ (= ρ_1', say) and $\rho_2 \to \rho_2 - \rho_1$ (= ρ_2', say) we see (very easily) that $\rho_2' = -\rho_1'$. This does it.

26. Write the so-called **Vandermonde**[a] **determinant**

$$\begin{vmatrix} 1 & 1 & 1 \\ x & y & z \\ x^2 & y^2 & z^2 \end{vmatrix}$$

[a]Alexandre-Théophile Vandermonde, 28 February 1735–1 January 1796.

as a product of linear factors (i.e. terms of degree 1).

Questions 27–30 show some uses of determinants in geometry.

27. Show that the equation of the straight line drawn through the points (x_1, y_1), (x_2, y_2) in the x–y plane is given by

$$\begin{vmatrix} 1 & 1 & 1 \\ x & x_1 & x_2 \\ y & y_1 & y_2 \end{vmatrix}.$$

28. Draw a triangle with all its vertices in the first quadrant of the x–y plane. From each vertex drop a perpendicular onto the x-axis. Use this picture to show that the area of this triangle is given by the modulus of

$$\frac{1}{2}\begin{vmatrix} 1 & 1 & 1 \\ x_1 & x_2 & x_3 \\ y_1 & y_2 & y_3 \end{vmatrix}.$$

If the triangle is 'lowered' partially into the fourth quadrant, what determinantal fact shows that its area remains unchanged?

29. In a similar way one may show that the equation of the circle through the points (x_1, y_1), (x_2, y_2), (x_3, y_3) is given by

$$\begin{vmatrix} x^2+y^2 & x & y & 1 \\ x_1^2+y_1^2 & x_1 & y_1 & 1 \\ x_2^2+y_2^2 & x_2 & y_2 & 1 \\ x_3^2+y_3^2 & x_3 & y_3 & 1 \end{vmatrix}.$$

Use this to find the equation of the circle through the points (1, 2), (3, 4) and (5, 7).

30. The volume of the parallelepiped (a figure like a sheared rectangular box but with parallelograms rather than rectangles for sides) defined by the vectors (a_{11}, a_{12}, a_{13}), (a_{21}, a_{22}, a_{23}) and (a_{31}, a_{32}, a_{33}) is given by the positive value of $\pm$Det A where

$$A = \begin{vmatrix} a_{11} & a_{12} & a_{13} \\ a_{21} & a_{22} & a_{23} \\ a_{31} & a_{32} & a_{33} \end{vmatrix}.$$

Find the volume of the parallelepiped three of whose sides join the point (3, 1, –2) to each of (3, –1, 4), (–1, 5, –9), (2, –6, 5). Can you determine the volume of the tetrahedron defined by these vertices?

HAND CALCULATOR PROBLEM

1. Find

$$\begin{vmatrix} 5.74 & 6.33 & 7.92 \\ 3.44 & 2.47 & 1.50 \\ 6.39 & 8.26 & 11.13 \end{vmatrix}.$$

If you read your calculator display upside down you will probably realise immediately if you have got the correct result!

COMPUTER PACKAGE PROBLEMS

1. Find the values of the determinants of several of the Hilbert matrices H_n as defined in Computer Package Problem 1 of Chapter 2. Do you see any pattern to these answers?

2. For which of x are the following determinants zero? It might be nice to guess before you find out!

$$\text{(a)} \begin{vmatrix} x & 2 & 3 & x \\ 12 & 13 & 14 & 5 \\ 11 & 16 & 15 & 6 \\ x & 9 & 8 & x \end{vmatrix}; \quad \text{(b)} \begin{vmatrix} 1 & 2 & 3 & 4 \\ 12 & x & x & 5 \\ 11 & x & x & 6 \\ 10 & 9 & 8 & 7 \end{vmatrix}; \quad \text{(c)} \begin{vmatrix} 1 & 2 & 3 & 4 & 5 \\ 16 & 17 & 18 & 19 & 6 \\ 15 & 24 & x & 20 & 7 \\ 14 & 23 & 22 & 21 & 8 \\ 13 & 12 & 11 & 10 & 9 \end{vmatrix}.$$

3. Is either

$$\text{(a)} \begin{vmatrix} 1 & 11 \\ 111 & 1111 \end{vmatrix} \quad \text{or} \quad \text{(b)} \begin{vmatrix} 1 & 11 & 111 \\ 1111 & 11111 & 111111 \\ 1111111 & 11111111 & 111111111 \end{vmatrix}$$

equal to 0?

Are you prepared to go to the 'next' two cases? The next twenty? All cases?

4. Find the value of

$$\begin{vmatrix} 7 & 1 & 3 \\ 2 & 8 & 5 \\ 4 & 6 & 9 \end{vmatrix}.$$

Can the first nine integers be rearranged so that a 3×3 matrix with larger determinant will result? (Actually I have no idea! I have not checked it!)

5. Evaluate

$$\begin{vmatrix} \sqrt{1} & \sqrt{2} & \sqrt{3} \\ \sqrt{4} & \sqrt{5} & \sqrt{6} \\ \sqrt{7} & \sqrt{8} & \sqrt{9} \end{vmatrix} \quad \text{and} \quad \begin{vmatrix} \sqrt{1} & \sqrt{2} & \sqrt{3} & \sqrt{4} \\ \sqrt{5} & \sqrt{6} & \sqrt{7} & \sqrt{8} \\ \sqrt{9} & \sqrt{10} & \sqrt{11} & \sqrt{12} \\ \sqrt{13} & \sqrt{14} & \sqrt{15} & \sqrt{16} \end{vmatrix}$$

etc. in decimal form. How many places of decimals will your package work to?

6 · Real Vector Spaces

In Chapter 3 we showed how to introduce addition and (scalar) multiplication into sets of matrices and we noted some of the rules that these operations satisfy. In this chapter we describe many other sets which can be treated similarly and which obey the same rules. To organise this information into a coherent whole we introduce the notion of (abstract) vector space. By adding and multiplying 'elements' (rather than specific things such as matrices, functions, etc.), we obtain an improved (purer?) clarity of view as well as saving ourselves the work of repeating, for each particular example, identical calculations to those already carried out in essentially similar situations. (This idea of abstracting the important – and discarding the peripheral – will be new to most readers but operating at this somewhat higher intellectual level will prove useful – and exhilarating if you work at it!)

Often, in mathematics, things which appear different on the surface have an underlying sameness. This sameness is frequently of a structural kind. We illustrate what we mean by considering a number of different-looking examples.

Example 1

By direct substitution we find that (2, −1, 0, −1) [i.e. $w = 2$, $x = -1$, $y = 0$, $z = -1$] and (2, 0, −1, 0) are both solutions of the homogeneous linear system

$$\begin{aligned} w + 3x + 2y - z &= 0 \\ 2w + x + 4y + 3z &= 0 \\ w + x + 2y + z &= 0 \end{aligned} \tag{6.1}$$

Furthermore, (4, −1, −1, −1), obtained by adding together, component by component, the two solutions above, is also a solution of (6.1) (since

$$\begin{aligned} 4 + 3(-1) + 2(-1) - (-1) &= 2(4) + (-1) + 4(-1) + 3(-1) \\ &= 4 + (-1) + 2(-1) + (-1) = 0). \end{aligned}$$

In fact the sum of any two solutions of (6.1) (where 'sum' means adding together the two w values, the two x values, etc.) is again a solution of (6.1). We also note that multiplying each component w, x, y and z of a solution by the same constant again produces a solution of (6.1). For example, (6, −3, 0, −3) {= 3(2, −1, 0, −1)} and $(2, -\frac{1}{2}, -\frac{1}{2}, -\frac{1}{2})$ $\{= \frac{1}{2}(4, -1, -1, -1)\}$ are also solutions of (6.1) as you may readily check.

It is clear that similar remarks apply to any homogeneous system of linear equations.

PROBLEM 6.1

Given that $\mathbf{u} = (8, 13, 0, 6)$ and $\mathbf{v} = (2, 4, -2, 6)$ are solutions of the system

$$4x-2y-3z-t=0$$
$$5x-4y+3z+2t=0$$

show that (6, 9, 2, 0) is also a solution. Find a solution with each of x, y, z and $t > 100$.

Almost identical remarks apply to differential equations.

Example 2

Consider the differential equation

$$\frac{d^2y}{dx^2}+4\frac{dy}{dx}+3y=0. \tag{6.2}$$

It is not difficult to verify (check it!) that both e^{-3x} and e^{-x} satisfy (6.2) and, consequently, that for any constants A and B, the functions $A\,e^{-3x}$ and $B\,e^{-x}$ and their sum $A\,e^{-3x} + B\,e^{-x}$ satisfy (6.2). Indeed $A\,e^{-3x} + B\,e^{-x}$ is called the *general solution* of the (*homogeneous* – because of the 0 on the right-hand side) equation of (6.2).

Let us introduce some convenient terminology. Suppose we are given any set S on which it is possible to define a *sum* $s_1 + s_2$ of any two elements s_1 and s_2, and a *scalar product* λs of any element s of S by any (real) number λ. We say that S is **closed under addition** if $s_1 + s_2$ is always in S and we say that S is **closed under multiplication by scalars** if λs is always a member of S.

For example, if E denotes the set of all solutions in Example 1 and if D denotes the set of all solutions in Example 2, we see that both E and D are closed under addition and scalar multiplication. We employ this terminology repeatedly below.

Here are two somewhat more frivolous (but fascinating!) examples.

Definition 1

An **arithmetic progression** (AP) is a sequence of numbers in which each term after the first differs from its predecessor by the same constant.

Example 3

$$2, 5, 8, 11, 14, 17, 20, \ldots \tag{6.3}$$

and

$$19, 15, 11, 7, 3, -1, -5, \ldots \tag{6.4}$$

are APs whereas 1, 3, 6, 10, 15, 21, 28, . . . is not.

PROBLEM 6.2

Write down any two APs of your choice, then add corresponding terms. [Doing this with (6.3) and (6.4) gives

$$21, 20, 19, 18, 17, 16, 15, \ldots \tag{6.5}$$

another AP!] We naturally call (6.5) the ‘sum’ of (6.3) and (6.4). [Is the sum of your chosen APs also an AP? Try a few more. Can you explain why things work out as they do?] As you will have guessed, we can also multiply APs by constants. For example, from (6.4) we obtain the sequence –57, –45, –33, –21, –9, 3, 15, . . ., which is yet another AP. Can you explain why it is?

Example 4

You may be familiar with the idea of magic squares. A **magic square** is just a square array of numbers in which the numbers in each row, each column and both diagnonals add up to the same total (the 'magic constant' of the given magic square).

We shall restrict ourselves here to 3×3 magic squares:

$$\begin{array}{ccc} 4 & 3 & 8 \\ 9 & 5 & 1 \\ 2 & 7 & 6 \end{array} \quad \text{is one example;} \quad \begin{array}{ccc} 0 & 0 & 0 \\ 0 & 0 & 0 \\ 0 & 0 & 0 \end{array}$$

is a slightly less interesting(?) one. Multiplying each entry in a magic square by the same constant again gives us a magic square, as does adding corresponding entries in any two magic squares. (Example:

$$\begin{array}{ccc} 2 & -5 & 0 \\ -3 & -1 & 1 \\ -2 & 3 & -4 \end{array} \text{`+'} \begin{array}{ccc} 4 & 3 & 8 \\ 9 & 5 & 1 \\ 2 & 7 & 6 \end{array} = \begin{array}{ccc} 6 & -2 & 8 \\ 6 & 4 & 2 \\ 0 & 10 & 2 \end{array}$$

is magic. Question: how is the magic constant of the sum related to those of its summands?)

We therefore see that the set A, say, of all APs and the set M, say, of all magic squares share with Examples 1 and 2 the properties of being closed under the (quite naturally defined) actions of addition and scalar multiplication.

And now for some more examples of sounder mathematical pedigree.

Example 5

In the set (which we denote by $M_{m,n}(\mathbb{R})$) of all $m \times n$ matrices with real entries, we have already defined (in Chapter 3) the concept of sum and scalar multiple of matrices, the result being, in all cases, yet another element of $M_{m,n}(\mathbb{R})$. Thus $M_{m,n}(\mathbb{R})$ is closed under addition and scalar multiplication.

Example 6

Consider the set of all n-tuples $(r_1, r_2, \ldots, r_n)$ of real numbers. We may denote this set by $\mathbb{R}^n$ since it may be identified as ordinary n-dimensional space by associating with each point in that space its n-tuple of coordinates. (See p. 41.) Once again we define addition componentwise:

$$(r_1,\ldots,r_n)+(s_1,\ldots,s_n)=(r_1+s_1,\ldots,r_n+s_n)$$

and multiplication by a constant c by $c(r_1, \ldots, r_n) = (cr_1, \ldots, cr_n)$. Clearly $\mathbb{R}^n$ is closed under addition and multiplication as just defined.

Example 7

Let P be the set of all triples (x, y, z) in $\mathbb{R}^3$ for which $2x - 3y + 11z = 0$. [Geometrically this is a plane in 3-dimensional space which passes through the origin of coordinates. Algebraically it is the full set of solutions of a homogeneous linear system comprising a single equation.]

With addition and scalar multiplication defined as in Examples 1 and 6, we see that sums and scalar multiples of elements in P are again in P.

Example 8

Let $\mathbb{R}[x]$ denote the set of all polynomials in the indeterminate x and with real coefficients. If we add two members of $\mathbb{R}[x]$ or multiply a member of $\mathbb{R}[x]$ by any real number the result is again a member of $\mathbb{R}[x]$. [Note that the degree of a sum of two polynomials cannot be greater than the degrees of each of the polynomials and may be less. For example, the sum of $3x^3 - \pi x - 7$ and $-3x^3 + 7.1x^2 - \frac{1}{3}x + 8.5$ is $7.1x^2 - (\pi + \frac{1}{3})x + 1.5$ which has a degree less than the degrees (namely 3) of each of the summands. Thus $\mathbb{R}[x]$ is closed under the (naturally) defined sum and (scalar) multiple of polynomials. [Of course the product of two polynomials is again a polynomial – but we choose to ignore that fact here!]

Example 9

In 2-dimensional (or in 3-dimensional) space we may consider the set G, say, of all geometrical vectors ('arrows'). The reader will (no doubt many times) have added two such vectors together and multiplied such vectors by scalars (i.e. numbers) each time producing another vector. (Notice that, as with polynomials above, we are happy here, when working in three dimensions, to 'forget' about the possibility of combining two vectors by the vector product method.) Thus G is closed under vector addition and scalar multiplication of vectors.

Example 10

Let $\mathfrak{F}$ denote the set of all functions from the real numbers $\mathbb{R}$ to $\mathbb{R}$ and let $\mathfrak{D}$ denote the subset of all those functions from $\mathbb{R}$ to $\mathbb{R}$ which are differentiable. For functions f and g in either of these sets we may define their sum $f \oplus g$ and for each real scalar α we may define the scalar multiple αf of f by α. (See the Appendix.) Note that, if $f, g \in \mathfrak{F}$ then so do $f \oplus g$ and αf whilst if $f, g \in \mathfrak{D}$ then so do $f \oplus g$ and αf (the latter since sums and multiples of differentiable functions are differentiable).

Example 11

Let $\mathbb{C}$ denote the set of all complex numbers. If we add any two complex numbers or multiply any complex number by a real(!) number the result is again a complex number. (Note – cf. Examples 8 and 9 – that we choose to ignore(!) the fact that the product of two elements in $\mathbb{C}$ is still in $\mathbb{C}$.]

From the above examples we see that instances of entities being combined by a kind of 'addition' and modified by a kind of 'constant multiplication' (the result always being an entity of the same type) are not uncommon. Now whenever pure mathematicians find a plethora of examples all of apparently similar nature, if not actually of the same outward appearance, they like to abstract their common properties and then work with these properties rather than with each individual example separately.

This abstracting has several advantages. One is that, working with general 'elements' rather than with solutions of linear systems or with magic squares etc., we can develop a single theory which is applicable to all specific instances at once. Another is that, by developing the common infrastructure which underlies the surface 'gloss' we expose the real reason why these examples behave as they do.

Let us now list the more obvious properties of addition and (scalar) multiplication which the above examples have in common. These properties will be very reminiscent of many which we listed for matrices. This is scarcely surprising since sets of matrices form important instances of our theory!

• *Axioms V.S.*[a]

Let V be any of the sets in Examples 1 to 11. Let **a, b**, and $\mathbf{c}$ be typical elements of V and let λ and μ be any real scalars. Let $\mathbf{a} \oplus \mathbf{b}$ denote the appropriate 'sum' of **a** and **b** and let $\lambda \circ \mathbf{a}$ denote the 'scalar multiple' of **a** by λ. Then V satisfies the following rules (also called **axioms**). Concerning addition:

Axiom A(i) $\mathbf{a} \oplus \mathbf{b} \in V$ [the **sum** of **a** and **b** is in V]

A(ii) $(\mathbf{a} \oplus \mathbf{b}) \oplus \mathbf{c} = \mathbf{a} \oplus (\mathbf{b} \oplus \mathbf{c})$ [the **associative law** holds for $\oplus$]

A(iii) There exists in V an element, which we denote by $\mathbf{z}$, which is such that $\mathbf{a} \oplus \mathbf{z} = \mathbf{z} \oplus \mathbf{a} = \mathbf{a}$ [V contains a '**zero type**' element]

A(iv) Corresponding to each $\mathbf{a} \in V$ there exists an element $\bar{\mathbf{a}}$ in V such that $\mathbf{a} \oplus \bar{\mathbf{a}} = \bar{\mathbf{a}} \oplus \mathbf{a} = \mathbf{z}$ [the same $\mathbf{z}$ as in A(iii)]. [Each **a** in V has an **additive inverse** or **negative**.]

A(v) $\mathbf{a} \oplus \mathbf{b} = \mathbf{b} \oplus \mathbf{a}$ [the **commutative law** holds for $\oplus$]

And concerning scalar multiplication:

M(i) $\lambda \circ \mathbf{a} \in V$ [**scalar multiples** are in V]

M(ii) $\lambda \circ (\mathbf{a} \oplus \mathbf{b}) = \lambda \circ \mathbf{a} \oplus \lambda \circ \mathbf{b}$ [the **distributive law** holds for $\circ$ over $\oplus$]

M(iii) $(\lambda + \mu) \circ \mathbf{a} = \lambda \circ \mathbf{a} \oplus \mu \circ \mathbf{a}$ [the **distributive law** holds for + over $\circ$]

M(iv) $\lambda \circ (\mu \circ \mathbf{a}) = (\lambda\mu) \circ \mathbf{a}$ [the **associative law** holds for $\circ$]

M(v) $1 \circ \mathbf{a} = \mathbf{a}$ [the scalar 1 acts as expected with respect to $\circ$] •

We have claimed that Example 7 satisfies these axioms. Let us check some of them. To help with this we shall suppose that $\mathbf{a} = (a_1, a_2, a_3)$ and $\mathbf{b} = (b_1, b_2, b_3)$ are any two elements of P. We therefore know that $2a_1 - 3a_2 + 11a_3 = 0$ and that $2b_1 - 3b_2 + 11b_3 = 0$. Now, by definition of $\oplus$, we have $\mathbf{a} \oplus \mathbf{b} = (a_1 + b_1, a_2 + b_2, a_3 + b_3)$. Furthermore

$$\begin{aligned} 2(a_1+b_1)-3(a_2+b_2)+11(a_3+b_3) &= (2a_1-3a_2+11a_3)+(2b_1-3b_2+11b_3) \\ &= 0+0=0. \end{aligned}$$

This shows that Axiom A(i) holds for P.

Axiom A(iii) holds since, first, (0, 0, 0) clearly belongs to P and, second,

[a]V.S., vector space. See Definition 2.

$$(a_1,a_2,a_3)\oplus(0,0,0)=(0,0,0)\oplus(a_1,a_2,a_3)=(a_1,a_2,a_3).$$

Thus, here, (0, 0, 0) plays the role of the **z** element.

Axiom A(iv) holds since, if we set $\bar{\mathbf{a}}=(-a_1,-a_2,-a_3)$, we clearly have

$$\mathbf{a}\oplus\bar{\mathbf{a}}=\bar{\mathbf{a}}\oplus\mathbf{a}=(0,0,0)=\mathbf{z}.$$

PROBLEM 6.3

Show that, if $\mathbf{a} \in P$, then $\bar{\mathbf{a}} \in P$. ●

Finally we check Axiom M(i), leaving the rest to you. So, let $\mathbf{a}(\in P)$ be as above and let λ be any real number. Now $\lambda \circ \boldsymbol{a} = (\lambda a_1, \lambda a_2, \lambda a_3)$. Thus we look at $2(\lambda a_1) - 3(\lambda a_2) + 11(\lambda a_3)$. But this is equal to $\lambda(2a_1 - 3a_2 + 11a_3) = \lambda 0 = 0$. Thus $\lambda \circ \boldsymbol{a} \in P$, that is, Axiom M(i) holds.

PROBLEM 6.4

Check the remaining six axioms hold for Example 7. (Methods introduced in Chapter 7 will reduce this task by about 80%!) ●

For further insight let us consider just one axiom, Axiom A(iii), for each of Examples 1–11. It is not too difficult to see that the role of the element **z** is played, in Example 1, by the solution (0, 0, 0, 0), that is $w = x = y = z = 0$. In Example 2 it is played by the **zero function** Z [defined by $Z(x) = 0$ for each real number x]. The zero type elements in Examples 3 and 4 are the AP 0, 0, 0, 0, . . . and the magic square

$$\begin{matrix} 0 & 0 & 0 \\ 0 & 0 & 0 \\ 0 & 0 & 0 \end{matrix}$$

respectively. The remaining zero type elements are: in Example 5, the $m \times n$ zero matrix (see Theorem 1(iii) of Chapter 3); in Example 6, the n-tuple (0, 0, . . ., 0); in Example 7 the solution (0, 0, 0), as we have just seen; in Example 8 the zero polynomial ($0 + 0x + 0x^2 + \ldots$) usually denoted, simply, by 0, in which all coefficients are zero; in Example 9, the zero vector **0**; in Example 10, the zero function (for $\mathfrak{F}$ and for $\mathfrak{D}$); in Example 11, the zero complex number.

TUTORIAL PROBLEM 6.1

> Do you agree that, for polynomials $p(x)$ and $q(x)$ in $\mathbb{R}[x]$, we have deg(ree)$\{p(x)q(x)\} = \deg\{p(x)\} + \deg\{q(x)\}$? If so, what degree should one ascribe to the zero polynomial? (Surely deg{0} = deg{any constant polynomial} = 0? Well, . . .)

[For those who have seen a little group theory, A(i), . . ., A(v) merely state that, under $\oplus$, V is an abelian group.

Examples 1–11 are far from being the only examples which satisfy all the rules A(i)–A(v) and M(i)–M(v): and if you study mathematical physics or mathematical analysis to any depth you will come across even more important ones! Instead of calling all of them 'examples which satisfy rules A(i) to M(v)'(!) we say, in deference to the basic Example 9, that each is an example of a *vector space* which we define as follows.

• Definition 2

Let V be any non-empty set in which it is possible to define (for each pair of elements) an 'addition' which we denote by $\oplus$ and (for each element) a 'multiplication', which we denote by $\circ$, by a real scalar so that all the axioms in the list V.S. hold. Then V is called a **real vector space**, or **vector space over** $\mathbb{R}$, the elements of V being called (in deference to the basic Example 9), **vectors**. •

[Pedants (I believe I am one!) may wish to observe that, technically, it is the triple $\langle V, \oplus, \circ\rangle$ which is the vector space. V itself is only a set!]

Anyway, Examples 1–11 supply specific examples of real vector spaces.

On occasions it is very convenient to allow the scalars in Definition 2 to be complex numbers. The definition of **complex vector space** (or **vector space over** $\mathbb{C}$) is then identical to that given above – once the two occurrences of the word 'real' in Definition 2 have been changed to 'complex'. Indeed $\mathbb{R}$ can be replaced by any *field*, that is, any system of numbers in which addition, subtraction, multiplication and division (except by zero) are always possible within the system. (In particular the set $\mathbb{Q}$ of **rational numbers** forms a field and may replace $\mathbb{R}$ in Definition 1.)

We shall sometimes denote the set of scalars by $\mathbb{F}$ (for 'field') if we do not want to be specific as to which set of scalars we are using.

Given the number of different specific instances of the concept, later called vector space, which had occurred in the mathematical literature, the formal (abstract) definition of vector space, as given above, took a long time to arrive. For example, Euler's observation, that each of the solutions of a (homogeneous) differential equation could be expressed as (linear) combinations of only some (finite number) of them meant that the set of solutions satisfied the vector space axioms. At a more primitive level it had long been known that forces and velocities could be represented using directed line segments and that the 'sum' of two such line segments should be defined by the parallelogram law. In these examples it is not sensible (or, at best, not obvious how) to multiply two of the objects concerned together and expect the result to be another one. [In the first the product of two solutions to Euler's equation is rarely another.] In some spaces, however, the vectors themselves can be multiplied to produce new vectors. That this is true for the 3-dimensional physical/geometrical vectors emerged from the work of Hamilton and Grassmann (though the latter's *Ausdehnungslehre* in 1844 was so impenetrable that it – and its 1862 successor – were largely ignored). Hamilton's quaternions, resulting from his attempt to push to three dimensions a successful representation of motion in the plane using complex numbers (which, eliminating the problematic symbol $\sqrt{-1}$, he had equated with number pairs and hence points in the plane) involve 'numbers' $a + bi + cj + dk$ where a, b, c, d are real numbers and $i^2 = j^2 = k^2 = -1$ are quantities such that $ij = -ji = k$, $jk = -kj = i$ and $ki = -ik = j$. The i, j, k part was extracted by Maxwell and Gibbs to form the familiar physical/geometrical vectors we use today. In this context the term 'vector' was coined by Hamilton.

About the same time the identification of points in the plane and in space with their coordinate pairs and triples led Cayley to the obvious suggestion that sets of n-tuples of real numbers could be regarded as lying in 'n-dimensional space'. (Some mathematicians were unable to make this psychological jump because of the impossibility of geometrical interpretation.) These ideas developed in several different directions which we have no space to report on here.

Later, in number theory and Galois theory, examples occurred where, although a product of 'vectors' was readily available, it was ignored! Still the need for a formal abstract definition of vector space just did not seem worth while. Nevertheless, Peano, working with differential equations, and always keen to rid mathematics of the intuitive element, had thought it prudent to give a formal definition – but this was still a system of n-tuples in which the axes of reference chosen were the analogues of the usual x–y–z ones. In any case this work also went unnoticed for some time. Around this time other problems in differential equations led to interest in infinte linear systems (as they had earlier interested Fourier when developing his theory of heat flow in solid bodies) and hence infinite dimensional vector spaces. Perhaps the subtleties of such systems, together with the processes of abstraction which were in vogue at the turn of the century, helped persuade mathematicians that the time was ripe for a formal definition. Such was given, in a work on the theory of relativity, by Weyl in 1918. However, the 1924 edition of the very influential algebra book of Maxime Bocher does not contain a formal definition of vector space.

Herman Gunther Grassmann was born in Stettin (now Szczecin) on 15 April 1809, the third of 12 children. Following quite closely in the path of his father, Grassmann first studied theology for six semesters at the University of Berlin. Returning to Stettin in 1830 he undertook a study of mathematics and physics, eventually qualifying to enter the teaching profession. As well as mathematics he taught Latin and German and even wrote schoolbooks on these topics. In 1832 he began to work on a new geometric calculus out of which came the *Ausdehnungslehre*. Grassmann tried to popularise the *Ausdehnungslehre* by describing numerous applications, but to little avail. Greatly disappointed he turned to linguistics where he did important work. The *Ausdehnungslehre* eventually gained some belated recognition with Peano's 1888 attempted elaboration and Gibbs's preference for it over the quaternions of Hamilton. Grassmann died, in Stettin, on 26 September 1877.

At this point it might be quite instructive to give some examples which look as if they might be vector spaces – but are not! First, one which 'fails' at the outset.

Example 12

It is easy to check that (–1, 1, 0, 6) and (3, –2, 1, 3) are solutions of the system

$$\begin{aligned} w+3x+2y-z&=-4\\ 2w+x+4y+3z&=17\\ w+x+2y+z&=6 \end{aligned} \qquad (6.6)$$

However, their sum is not a solution (since, for example, (2, –1, 1, 9) does not satisfy even the first equation). Further, no multiples $(-c, c, 0, 6c)$, $(3d, -2d, d, 3d)$ (except for $c=d=1$) are solutions of (6.6). Thus axioms A(i) and M(i) both fail.

PROBLEM 6.5

Is there any solution of the form $c(-1, 1, 0, 6) + d(3, -2, 1, 3)$ where neither c nor d is 0?

Next, an example which fails at a deeper level.

Example 13

Let $\mathbb{R}^2$ denote, as usual, the set $\{(x, y){:}x, y \in \mathbb{R}\}$. Define addition, as usual (i.e. componentwise), but define 'scalar multiplication', $\circ$, by: for all $\lambda \in \mathbb{R}$ and $(x, y) \in \mathbb{R}^2$, $\lambda \circ (x, y) = (\lambda y, \lambda x)$.

Now it is not too difficult to see that all the axioms A(i)–A(v) are certain to hold for this example (for each such axiom involves only the addition symbol and here the definition of addition is the usual one). If anything is to stop our example from being a vector space it will be that one of the axioms M(i)–M(v) will fail. In fact, if we examine them each in turn, we find that M(iv) will, in general (meaning 'usually', though, for a mathematician, just once is enough!) fail. We show this most convincingly by exhibiting a . . . what? . . . Answer: a counterexample. Indeed, from our definition of $\circ$, we find: (Please check I am right – do not just assume I am!)

$$1\circ\{1\circ(1,0)\} \overset{\text{(why?)}}{=} 1\circ(0,1) \overset{\text{(why?)}}{=} (1,0) \quad \text{whereas} \quad \{1.1\}\circ(1,0) \overset{\text{(why?)}}{=} 1\circ(1,0) \overset{\text{(why?)}}{=} (0,1).$$

Thus axiom M(iv) fails to hold and our example is barred from joining the vector space club.

We now demonstrate the efficacy, described above, of 'going abstract' by making a few deductions (based on rules of logic alone rather than on hearsay, intuition or guesswork!) from the Axioms V.S. Because each of Examples 1–11 satisfies all 10 of the axioms, each will automatically satisfy every logical deduction we make from the axioms. Thus, even with our restricted list of examples of vector spaces, we get 12 theorems for the price of one!

For each of Examples 1–11 we easily identified an element which satisfies the requirements imposed on it by Axiom A(iii). You are probably quite certain that in none of Examples 1–11 does a second such zero element exist. You may even be fairly sure that, in no vector space yet studied, nor in any still to be invented or discovered, can two 'zero type' elements satisfying Axiom A(iii) sit side by side. But are you *absolutely certain*?

Rather than prove the uniqueness of the **z** element separately for each of the examples above (and then again each time a 'new' vector space of interest arises) we show how we can deduce **z**'s uniqueness from the Axioms V.S. Whilst doing this we also take the opportunity to prove three other little consequences of the axioms; first, because we shall need them in Chapter 7 and, second, because we wish to show you something of the spirit in which the pure mathematician frequently operates. (These 'little consequences' might strike you as 'obvious' – but on what grounds? There is no shortage in mathematics (or in real life for that matter!) of 'obviously true' statements which, on closer examination, turn out to be false! Hence the pure mathematician's caution.)

We prove the following theorem.

• *Theorem I*

The following statements are true in every vector space:

(A) The **z** element defined in Axiom A(iii) is unique.
(B) If $\mathbf{a} \oplus \mathbf{x} = \mathbf{a} \oplus \mathbf{y}$ then $\mathbf{x} = \mathbf{y}$ in V.
(C) For all $\lambda \in \mathbb{R}, \lambda \circ \mathbf{z} = \mathbf{z}$; for all $\mathbf{a} \in V$, $0 \circ \mathbf{a} = \mathbf{z}$.
(D) For all $\mathbf{a} \in V, (-1) \circ \mathbf{a} = \bar{\mathbf{a}}$.

[Advice to the reader: To begin with just check, slowly but carefully, that each of the assertions made below is logically acceptable to you. If not, find a colleague or teacher with whom to argue it out. Also whilst you are reading, try to get a feeling for the style of presention and the method of attack. But do not despair that you would 'never have thought of doing it like that'. After all, one of the reasons for your reading this book is to learn to master things like proofs and their construction.]

PROOF

(A) Suppose, then, that there are two elements, $\mathbf{z}_1$ and $\mathbf{z}_2$, say, in V which satisfy Axiom A(iii). Then, because $\mathbf{z}_1$ satisfies A(iii), we have, on substituting $\mathbf{z}_2$ for **a** in A(iii), $\mathbf{z}_1 \oplus \mathbf{z}_2 = \mathbf{z}_2$. But, since $\mathbf{z}_2$ also satisfies A(iii), we see that, on substituting $\mathbf{z}_1$ for **a** in A(iii), $\mathbf{z}_1 \oplus \mathbf{z}_2 = \mathbf{z}_1$. Thus $\mathbf{z}_1 = \mathbf{z}_2$ (since each is equal to $\mathbf{z}_1 \oplus \mathbf{z}_2$).

(B) Since we are given that $\mathbf{a} \oplus \mathbf{x} = \mathbf{a} \oplus \mathbf{y}$ we may deduce (from which axiom?) that $\bar{\mathbf{a}} \oplus (\mathbf{a} \oplus \mathbf{x}) = \bar{\mathbf{a}} \oplus (\mathbf{a} \oplus \mathbf{y})$. It then follows (which axiom?) that $(\bar{\mathbf{a}} \oplus \mathbf{a}) \oplus \mathbf{x} = (\bar{\mathbf{a}} \oplus \mathbf{a}) \oplus \mathbf{y}$, that is, $\mathbf{z} \oplus \mathbf{x} = \mathbf{z} \oplus \mathbf{y}$ (why?). But, then, we may infer (which axiom?) that $\mathbf{x} = \mathbf{y}$, as claimed.

(C) [Somewhat more quickly.] Directly from A(iii) we deduce that $\mathbf{z} \oplus \mathbf{z} = \mathbf{z}$(!) Multiplying through by λ gives

$$\lambda \circ (\mathbf{z} \oplus \mathbf{z}) = \lambda \circ \mathbf{z} \tag{6.7}$$

But the left-hand side is $\lambda \circ (\mathbf{z} \oplus \mathbf{z}) = \lambda \circ \mathbf{z} \oplus \lambda \circ \mathbf{z}$ (by M(ii)) whilst the right-hand side may be rewritten $\lambda \circ \mathbf{z} = \lambda \circ \mathbf{z} \oplus \mathbf{z}$ (by A(iii)). Substituting these in (6.7) gives $\lambda \circ \mathbf{z} \oplus \lambda \circ \mathbf{z} = \lambda \circ \mathbf{z} \oplus \mathbf{z}$. Then, using (B) on this last equality, we may cancel out the first $\lambda \circ \mathbf{z}$ term on each side to deduce that $\lambda \circ \mathbf{z} = \mathbf{z}$, as claimed. {I hope you enjoyed that! I did!]

We leave the other part of (C) to the exercises.

(D) From $1 + (-1) = 0$ in $\mathbb{R}$, we deduced that $(1 + (-1)) \circ \mathbf{a} = 0 \circ \mathbf{a}$. As in (C) we find, using M(iii) on the left-hand side and the second part of (C) on the right-hand side, that $1 \circ \mathbf{a} \oplus (-1) \circ \mathbf{a} = \mathbf{z}$. Using M(v) to write $1 \circ \mathbf{a} = \mathbf{a}$, we get $\mathbf{a} \oplus (-1) \circ \mathbf{a} = \mathbf{z}$. But, according to A(iv) – and Exercise 24 – $\bar{\mathbf{a}}$ is the unique element of V satisfying the equality $\mathbf{a} \oplus \bar{\mathbf{a}} = \mathbf{z}$. Consequently $(-1) \circ \mathbf{a}$ must be the unique element $\bar{\mathbf{a}}$, as required. ●

COMMENT When we chose our list of (desirable?) Axioms V.S., we were aided by a large slice of hindsight (experience, if you like). We did not, for instance, include, as an axiom, the requirement $\lambda \circ \mathbf{z} = \mathbf{z}$, since we know, from experience, that it can be deduced from those we have chosen. The reader may justifiably ask if we might reduce our axiom list still further so that, if nothing else, we shall have less axioms to check if someone asks us if what (s)he has to hand is a vector space or not. In fact some of the listed axioms are redundant: they can be deduced from the rest! I leave you to ferret in the exercises below to find some of these – and to prove, in particular, that axiom M(v) cannot be so eliminated: it is not a logical consequence of the remaining axioms. (See Exercise 22.)

A NOTATIONAL POINT In the above the symbols $\mathbf{z}$ and $\bar{\mathbf{a}}$ were used to represent elements of V with properties akin to (i) the number 0 and (ii) the additive inverse of the number a. The symbols 0 and $-a$ were eschewed in order to help the reader avoid the trap of attributing to $\mathbf{z}$ and $\bar{\mathbf{a}}$, uncritically, special properties of the numbers 0 and $-a$. However, to maintain the use of $\mathbf{z}$ and, in particular, $\bar{\mathbf{a}}$, is somewhat overdoing things (especially as we are also using bold type) and so, unless it is crucial not to do so, we shall replace $\mathbf{z}$ by $\mathbf{0}$, $\bar{\mathbf{a}}$ by $-\mathbf{a}$ and, indeed, $\mathbf{a} \oplus \mathbf{b}$ by $\mathbf{a} + \mathbf{b}$ and $\mathbf{a} \oplus \bar{\mathbf{b}}$ by $\mathbf{a} - \mathbf{b}$ on the understanding that the reader will not attribute to these, nor any other symbols used, any intuitive meaning or property which has not been formally established or listed.

As already implied, it is quite often necessary, in practice, to study vector spaces in which the real scalars are replaced by complex (or even rational) numbers. For example, the set of real number $\{a + b\sqrt{2} : a, b \in \mathbb{Q}\}$ can be shown to satisfy Axioms V.S. if we take, as scalars, the set $\mathbb{Q}$ of rational numbers. Likewise the set $\mathbb{C}$ of complex numbers can be regarded as a vector space over $\mathbb{C}$ with $\mathbb{C}$ playing the roles of vectors and scalars simultaneously! Similarly, the set $\mathbb{R}$ can be regarded as a vector space over the set of scalars $\mathbb{Q}$. Examples similar to the first of these are made frequent use of in the theory of numbers.

Applications

As Examples 1–11 help to show, there are many different instances of sets of elements which satisfy all the Axioms V.S. Consequently, a major usefulness of the vector space concept is that it expresses, succinctly, the qualities of many common mathematical systems. In brief, the language and notation of vector spaces is the right language in which to express and investigate many and diverse real-life problems. In particular, the vector space setting is the most appropriate setting for discussing the theory of differential equations. Furthermore, the act of abstraction often suggests ideas which might otherwise have remained dormant, hidden by a mass of superfluous and irrelevant detail.

Summary

Several diverse examples were given of sets whose elements could (sensibly) be added in pairs and multiplied by 'scalars' (usually from $\mathbb{R}$) in such a way that many of the algebraic properties possessed by $m \times n$ matrices were seen to hold. Isolating these properties led us to the (abstract) concept of **vector space**. The advantages of making this abstraction include (i) the illumination of the common underlying algebraic structure of those examples and (ii) the necessity of deriving logical consequences of the underlying rules only once rather than for every 'new' vector space.

EXERCISES ON CHAPTER 6

1. For the linear system of Example 1 show that, for each pair of real numbers s and t, both $(2s, -s, 0, -s)$ and $(2t, 0, -t, 0)$ and their sum, namely $(2s + 2t, -s, -t, -s)$, are solutions.

2. Are the following sets of integers closed under (ordinary) addition?
(i) The set of all powers of 2;
(ii) The set of all integers of the form $5r + 7s$ (where r and s range over all the integers.)

3. (a) Find a 3×3 magic square with magic constant $\frac{1}{2}$;
(b) Show that the sum of any two 3×3 magic squares is again magic;
(c) Is there a magic square with distinct integer coefficients and with (i) magic constant 21; (ii) magic constant 22? [Hint for (c)(ii): by adding entries along various rows/columns, etc. deduce that the middle entry in a magic square is one-third of the magic constant.]

4. Are there real numbers λ, μ and ν such that $\lambda(1, 2, 3) + \mu(4, 5, 6) + \nu(7, 8, 9)$ is (i) equal to $(0, 0, 0)$; (ii) equal to $(0, 0, 1)$?

5. For each of the following write down its additive inverse (see Axioms A(iii) and A(iv)):
(i) $(8, -9, 5, -9)$ in Example 1;
(ii) $2\,e^{-3x} - 5\,e^{-x}$ in Example 2;
(iii) $\begin{bmatrix} 3 & -3 & -5 \\ -1 & 2 & -2 \end{bmatrix}$ in $M_{2,3}$ (see Example 5).

In Exercises 6–16 determine whether or not the given sets are real vector spaces (where addition and multiplication by scalars will be the usual ones unless otherwise stated). For those which are not vector spaces give the first of the 10 axioms in the list V.S. which is not satisfied. (Later [Theorem 1 of Chapter 7] we shall show – by use of more abstraction – how the somewhat tedious checking of all 10 axioms – in those cases which are vector spaces – can frequently be avoided.)

6. The set of all solutions of the system in Example 1. [Hint: let (w_1, x_1, y_1, z_1) and (w_2, x_2, y_2, z_2) be any two solutions. Show that their sum and each scalar multiple of (w_1, x_1, y_1, z_1) are again solutions etc.]

7. The set of all solutions of the equation in Example 2.

8. (a) The set of all $n \times n$ upper triangular matrices. [See Corollary 1 of Chapter 5.];
(b) the set of all skew symmetric matrices in $M_2(\mathbb{R})$. [See Exercise 20 of Chapter 5.]

9. The set $C(A)$ defined as follows. Let A be a fixed $n \times n$ matrix in $M_n(\mathbb{R})$. Then $C(A)$ denotes the subset of all matrices B in $M_n(\mathbb{R})$ for which $AB = BA$.

10. (i) The subset $\{(x, y, z)\colon 3x - 4y + z = 0\}$ of $\mathbb{R}^3$.
(ii) The subset $\{(x, y, z)\colon 3x - 4y + z = 1\}$ of $\mathbb{R}^3$.
(iii) The subset $\{(x, y, z)\colon 3x - 4y + z \geq 0\}$ of $\mathbb{R}^3$.
Describe the sets (i), (ii) and (iii) geometrically.

11. The subset $\{(u, v, 4v - 3u)\colon u, v \in \mathbb{R}\}$ of $\mathbb{R}^3$.

12. The subset of $\mathbb{R}^3$ comprising the single element (0, 0, 0).

13. The set of all polynomials $a + bx + cx^2 + dx^3 + ex^4 + fx^5$ in x with real coefficients which have degree exactly 5 (together with the zero polynomial). [Hint: is the sum of every pair of polynomials of degree 5 also of degree 5?]

14. The set of all polynomials in x with real coefficients which have degree at most 5 (together with the zero polynomial).

15. $\mathcal{S}$ is the set of all infinite sequences $(x_1, x_2, \ldots)$ of real numbers; the sum of two such sequences is defined componentwise, that is,

$$(a_1, a_2, \ldots) + (b_1, b_2, \ldots) = (a_1 + b_1, a_2 + b_2, \ldots)$$

and scalar multiplication is defined by $\lambda(a_1, a_2, \ldots) = (\lambda a_1, \lambda a_2, \ldots)$.

16. The subset of $\mathbb{R}^3$ comprising all elements of $\mathbb{R}^3$ of the form $a(1, 0, 1) + b(5, 1, 2)$ where $a, b \in \mathbb{R}$. Does (7, 7, 7) belong to this set?

17. Let V and W be real vector spaces and let S denote the set $\{(\mathbf{v}, \mathbf{w})\colon \mathbf{v} \in V, \mathbf{w} \in W\}$ of all pairs. For $(\mathbf{v}_1, \mathbf{w}_1)$ and $(\mathbf{v}_2, \mathbf{w}_2)$ in S and for $\alpha \in \mathbb{R}$ define $(\mathbf{v}_1, \mathbf{w}_1) \boxplus (\mathbf{v}_2, \mathbf{w}_2) = (\mathbf{v}_1 + \mathbf{v}_2, \mathbf{w}_1 + \mathbf{w}_2)$ and $\alpha \boxdot (\mathbf{v}_1, \mathbf{w}_1) = (\alpha\mathbf{v}_1, \alpha\mathbf{w}_1)$. Show that, with respect to $\boxplus$ and $\boxdot$, S forms a (real) vector space called the *direct sum* of V and W and denoted by $S = V \oplus W$.

18. Show that the difference $\mathbf{d} = (w_2 - w_1, x_2 - x_1, y_2 - y_1, z_2 - z_1)$ of the two solutions $\mathbf{s}_2 = (w_2, x_2, y_2, z_2)$ and $\mathbf{s}_1 = (w_1, x_1, y_1, z_1)$ of the linear system

$$\begin{aligned} w + 3x + 2y - z &= -4 \\ 2w + x + 4y + 3z &= 17 \\ w + x + 2y + z &= 6 \end{aligned}$$

is a solution of the system

$$\begin{aligned} w+3x+2y-z&=0\\ 2w+x+4y+3z&=0\\ w+x+2y+z&=0 \end{aligned}.$$

[Having done this note that from the equality $\mathbf{d}=\mathbf{s}_2-\mathbf{s}_1$ we get $\mathbf{s}_2=\mathbf{d}+\mathbf{s}_1$. This shows that once we know one solution ($\mathbf{s}_1$) of the given system of equations and a full set of solutions of the corresponding homogeneous system, then we shall know the full set of solutions of the given (non-homogeneous) linear system. Similar remarks apply to all linear systems. Since each homogeneous system has either one or infinitely many solutions we can deduce that each (general) linear system has either (i) 0; (ii) 1 or (iii) infinitely many solutions. This answers the question in Chapter 1.]

Because the following exercises directly concern Axioms V.S. we return, for the moment, to our use of the symbols $\oplus$ and $\circ$.

19. In $\mathbb{R}^2$ define a 'new' addition, namely, $(x_1, y_1) \oplus (x_2, y_2) = (x_1 + y_2, y_1 + x_2)$. Does $\mathbb{R}^2$ form a vector space with respect to this new addition and the usual (scalar) multiplication?

20. Show that one-half of each of Axioms A(iii) and A(iv) follows from the other half together with A(v).

21. Expand $\{1 + 1\} \circ (\mathbf{u} + \mathbf{v})$ using (i) M(ii) then M(iii); (ii) M(iii) then M(ii). Then use Theorem 2(B) and its left-handed equivalent to deduce that A(v) is a consequence of the other nine axioms. [Warning: do not use A(v) in this deduction!]

22. On $\mathbb{R}^2$ define $\oplus$ as usual but $\lambda \circ (x, y)$ to be $(0, 0)$ for every $\lambda \in \mathbb{R}$. Show that, with this definition of addition and scalar multiplication, $\mathbb{R}^2$ satisfies all the Axioms V.S. except M(v). [So we would have to include this silly example in our list of vector spaces if it were not for Axiom M(v).]

23. On $\mathbb{R}^2$ define $\oplus$ as usual but $\lambda \circ (x, y)$ to be (x, y) for every $\lambda \in \mathbb{R}$. Show that, with this definition of addition and scalar multiplication, $\mathbb{R}^2$ satisfies all the Axioms V.S. except M(iii).

24. Prove, directly from Axioms V.S., that for each $\mathbf{a}$ in the vector space V, the element $\bar{\mathbf{a}}$ is unique. [Hint: if $\mathbf{a}^0$ also satisfies A(iv) then $\mathbf{a}^0 \oplus (\mathbf{a} \oplus \bar{\mathbf{a}}) = (\mathbf{a}^0 \oplus \mathbf{a}) \oplus \bar{\mathbf{a}}$.]

25. Prove, directly from Axioms V.S., that if λ and $\mathbf{v}$ are such that $\lambda \circ \mathbf{v} = \mathbf{z}$, then either $\lambda = 0$ or $\mathbf{v} = \mathbf{z}$ (or both!) [Hint: if $\lambda \neq 0$ then $\mathbf{v} = 1 \circ \mathbf{v} = (\lambda^{-1}\lambda) \circ \mathbf{v} = \lambda^{-1} \circ (\lambda \circ \mathbf{v}) = \lambda^{-1} \circ \mathbf{z} = \mathbf{z}$. (Why?)]

26. Prove, directly from Axioms V.S., that for each λ and $\mathbf{v}$ we have $\ominus(\lambda \circ \mathbf{v}) = (-\lambda) \circ \mathbf{v} = \lambda \circ (\ominus \mathbf{v})$. [Hint: cf. Theorem 2(D).]

27. Prove that, for all $\mathbf{v} \in V$, $0 \circ \mathbf{v} = \mathbf{z}$.

28. Let V be a vector space over $\mathbb{C}$. Show the V may be regarded as a vector space over $\mathbb{R}$.

7 • Subspaces and Linear Combinations

In any mathematical system M those systems of similar type which are contained in M (subsystems of M) usually play an important role. Here we look at the concept of subspace of a vector space. Many examples are given; each is easily checked using the simple criteria of Theorem 1. Examples of subspaces may be given by specifying a spanning (or generating) set of vectors. Of particular importance are those subspaces of $\mathbb{R}^n$ and $\mathbb{R}^m$ spanned by the rows and columns of a given $m \times n$ matrix A, say, as well as the set of all $n \times 1$ matrices x satisfying the equation $Ax = 0$. The subspace concept has many applications – one particuarly nice one is concerned with the real-life problem of finding the 'best solution' to an inconsistent system of linear equations.

Consider, once again, Examples 6 and 7 of Chapter 6. Each was seen to be a vector space over $\mathbb{R}$. Furthermore, if we take $n = 3$ in Example 6, we see that the set P of triples of Example 7 forms a subset of the set, $\mathbb{R}^3$, of Example 6. (Geometrically we identified P as a plane passing through the origin in $\mathbb{R}^3$.) Likewise, the solution set E of Example 1 is a vector space which sits naturally inside the vector space $\mathbb{R}^4$. (Geometrically it could be described as a plane – passing through the origin – in 4-dimensional space!) Yet again, the set of all real 2×2 matrices with trace zero (see p. 53) is easily checked to be a real vector space which is a subset of the vector space $M_2(\mathbb{R})$.

This situation, of one vector space lying inside another (it occurs frequently but – for the moment – we spare you more instances) leads us to introduce the following general definition which covers all these examples.

• *Definition 1*

Let V be a vector space (over the set of scalars[a] $\mathbb{F}$) and let S be a non-empty subset of V. Then S is called a **subspace of V** if and only if S is itself a vector space *under the same definition of addition and scalar multiplication as in* V. (In particular we use the same scalars $\mathbb{F}$ for S as for V.) •

We leave comment on the italics to Exercise 9(a) and continue with more examples, below. But, first, we offer a theorem which will save a great deal of labour when looking at these examples, so the small amount of effort given to proving the theorem now will be rewarded many times over.

It seems that, to decide whether or not the (non-empty) subset S of the vector space V is a vector (sub)space, we must check all 10 vector space axioms. Fortunately this is not so, for we have the following theorem.

[a]The reader may continue to let $\mathbb{F}$ stand for $\mathbb{R}$ (or $\mathbb{C}$ or $\mathbb{Q}$ or any other field of numbers).

• Theorem 1

Let S be a non-empty subset of the vector space V. Then S is a subspace of V if and only if the following conditions (I) and (II) both hold: (I) If $\mathbf{s}, \mathbf{t} \in S$ then[b] $\mathbf{s} \oplus \mathbf{t} \in S$; (II) If $\mathbf{s} \in S$ and $\alpha \in \mathbb{F}$ then $\alpha \circ \mathbf{s} \in S$.

PROOF

(only if) Let S be a subspace of V. Then (I) and (II) necessarily hold since S is given to be a vector space in its own right with respect to the operations $\oplus$ and $\circ$. Thus, in particular, $\mathbf{s} \oplus \mathbf{t}$ and $\alpha \circ \mathbf{s}$ must belong to S.

(if) Here we must prove that, if S satisfies conditions (I) and (II), then S is a vector space with respect to $\oplus$ and $\circ$. So, on this one occasion, we must indeed check that all 10 vector space axioms V.S. of Chapter 6 hold for S.

Now A(i) and M(i) are given – they are the conditions (I) and (II). What about A(ii)? Well, since V is a vector space with respect to $\oplus$ and $\circ$, A(ii) holds for all $\mathbf{a}, \mathbf{b}, \mathbf{c}$ in V and hence, certainly, for all $\mathbf{a}, \mathbf{b}, \mathbf{c} \in S$. In fact axioms A(v), M(ii), M(iii), M(iv) and M(v) all hold in S for similar reasons. Thus we only have to check that axioms A(iii) and A(iv) hold for S.

To prove A(iii) holds, let $\mathbf{s}$ be any element of S. (How do we know any such $\mathbf{s}$ exists?) Then, by (II), $(-1) \circ \mathbf{s}$ (that is $\ominus\mathbf{s}$) belongs to S. We then use (I) to deduce that $\mathbf{s} \oplus (\ominus\mathbf{s})$ (that is, $\mathbf{z}_V$, the zero element of V) belongs to S. And then, of course, for all $\mathbf{s} \in S$, we have $\mathbf{s} \oplus \mathbf{z}_V = \mathbf{z}_V \oplus \mathbf{s} = \mathbf{s}$ (because these equalities hold in V itself). We have therefore shown that: S has an element (namely $\mathbf{z}_V$) satisfying axiom A(iii) of Axioms V.S. and, to each $\mathbf{s} \in S$, an element (namely $\ominus\mathbf{s}$) such that $\mathbf{s} \oplus (\ominus\mathbf{s}) = (\ominus\mathbf{s}) \oplus \mathbf{s} = \mathbf{z}_V$. Consequently, all the 10 vector space axioms hold for S (with respect to $\oplus$ and $\circ$) thus showing that S is a vector space in its own right and, hence, a subspace of V. ●

If (I) and (II) hold we say, as in Chapter 6, that 'S is **closed** with respect to (i) addition and (ii) multiplication by scalars'. Thus we may restate Theorem 1 as:

'A non-empty subset S of a vector space V is a subspace of V if and only if S is closed under addition and under multiplication by scalars.'

Actually Theorem 1 is often useful for checking whether or not a given set S is a vector space – provided it happens to be contained as a subset of another *known* vector space V, say (and involves the same set of scalars). We then only have to check that S is (or is not) a subspace of V.

We now show Theorem 1 in action.

• Examples 1

(i) The subset $S = \{(x, y, z): 2x + 4y - 3z = 0\}$ is a subspace of $\mathbb{R}^3$. We prove that conditions (I) and (II) hold, as follows. (Cf. the check we carried out on Example 7 of Chapter 6 – after listing Axioms V.S.)

For (I): Suppose that $\mathbf{u} = (u_1, u_2, u_3)$ and $\mathbf{v} = (v_1, v_2, v_3) \in S$. Then, necessarily, $2u_1 + 4u_2 - 3u_3 = 0$ and $2v_1 + 4v_2 - 3v_3 = 0$. Adding, we obtain

$$2(u_1 + v_1) + 4(u_2 + v_2) - 3(u_3 + v_3) = 0.$$

[b]We retain the $\oplus$, $\circ$ and $\mathbf{z}$ notation in preference to $+$, . and $\mathbf{0}$ wherever it seems helpful.

But this implies that $\mathbf{u} \oplus \mathbf{v}$ $\{= (u_1 + v_1, u_2 + v_2, u_3 + v_3)\} \in S$.

For (II): Suppose that $\mathbf{u} = (u_1, u_2, u_3) \in S$ and that $\alpha \in \mathbb{R}$. Again $2u_1 + 4u_2 - 3u_3 = 0$ and then, trivially $\alpha(2u_1 + 4u_2 - 3u_3) = 0$ so that $2(\alpha u_1) + 4(\alpha u_2) - 3(\alpha u_3) = 0$. But this means that $(\alpha u_1, \alpha u_2, \alpha u_3)$ $\{= \alpha \circ \mathbf{u}\} \in S$. Thus (provided S is non-empty – is it?) S has been shown to be a subspace of $\mathbb{R}^3$ – and, consequently, a vector space in its own right, over $\mathbb{R}$. (Geometrically S is a plane passing through the origin in $\mathbb{R}^3$.)

(ii) The subset $S = \{(x, y, z): 2x + 4y - 3z = 1\}$ is not a subspace of $\mathbb{R}^3$.

Recall that, when wishing to assert that some (mathematical) statement is false, one needs only give a single instance of failure. In this case, taking $\mathbf{u} = (0, 1, 1)$ and $\mathbf{v} = (1, 2, 3)$, say, we see that $\mathbf{u}$ and $\mathbf{v}$ do belong to S (since

$$2.0 + 4.1 - 3.1 = 2.1 + 4.2 - 3.3 = 1)$$

and yet $\mathbf{u} \oplus \mathbf{v}$ $(= (1, 3, 4))$ does not (since $2.1 + 4.3 - 3.4 \neq 1$). Consequently, condition (I) in Theorem 1 does not hold. Hence S cannot be a subspace of $\mathbb{R}^3$. (Note that it is now irrelevant as to whether or not condition (II) holds for S. [As it happens, it does not!])

QUESTION Geometrically this S also represents a plane in $\mathbb{R}^3$. Thus not all planes in $\mathbb{R}^3$ correspond to subspaces of $\mathbb{R}^3$ – indeed only those through the origin do. Nevertheless S is parallel to a (unique) plane which corresponds to a subspace of $\mathbb{R}^3$. Which?

(iii) Is $S = \{(x, y, z): xy + yz + zx = 0\}$ a subspace of $\mathbb{R}^3$?

[It is perhaps a little early for you to have an intuitive feeling about this, but the presence of the nonlinear terms xy, yz and zx might make you suspect that S will not be a subspace. In any case we still have to confirm our suspicions, so let us try for a counterexample.] Choosing $\mathbf{u}$, $\mathbf{v}$ almost at random – but with an eye on simple arithmetic – we note that $\mathbf{u} = (1, 0, 0)$ and $\mathbf{v} = (0, 1, 0)$ lie in S (since $1.0 + 0.0 + 0.1 = 0$ and $0.1 + 1.0 + 0.0 = 0$). However, $\mathbf{u} \oplus \mathbf{v} = (1, 1, 0) \notin S$. (Am I right?) Thus S fails to satisfy condition (I) of Theorem 1 and so cannot be a subspace of $\mathbb{R}^3$.

(iv) Let $S = \{(x, y, z, t): y \in \mathbb{Z}\}$ [where $\mathbb{Z}$ denotes the set $\{\ldots, -2, -1, 0, 1, 2, 3, \ldots\}$ of whole numbers or **integers**]. Then S is not a subspace of R^4. Here condition (I) is satisfied (essentially because the sum of two integers is again an integer). However condition (II) fails. Surely you can offer an instance of this failure, i.e. a counterexample? (Exercise 7(a).)

(v) The set M_0 of all 3×3 magic squares with centre component zero is a vector space. For, clearly: (a) M_0 is a non-empty subset of the set (vector space) M of Example 4 of Chapter 6

[example ?: $\begin{matrix} 0 & 0 & 0 \\ 0 & 0 & 0 \\ 0 & 0 & 0 \end{matrix}$ (!)]

and (b) conditions (I) and (II) of Theorem 1 hold. [As an example of (I) holding, but not, of course, a full proof that it holds universally, we offer

$$\begin{matrix} 4 & -2 & -2 & -5 & 2 & 3 & -1 & 0 & 1 \\ -6 & 0 & 6 \oplus 8 & 0 & -8 = 2 & 0 & -2.] \\ 2 & 2 & -4 & -3 & -2 & 5 & -1 & 0 & 1 \end{matrix}$$

Since M_0 is a subspace of a known vector space it is a vector space in its own right.

(vi) Let

$$A = \begin{bmatrix} 1 & 3 & 1 \\ 2 & 1 & 1 \end{bmatrix}, \quad \mathbf{b} = \begin{bmatrix} -2 \\ 3 \end{bmatrix}, \quad \mathbf{c} = \begin{bmatrix} 0 \\ 0 \end{bmatrix}.$$

Then the full set of solutions of $A\mathbf{x} = \mathbf{b}$ {respectively of $A\mathbf{x} = \mathbf{c}$} is not {respectively is} a subspace of $\mathbb{R}^3$. Much more generally, if A is any $m \times n$ matrix, the set S, say, of all solutions $\mathbf{x}$ (in $\mathbb{R}^n$) of the linear system $A\mathbf{x} = \mathbf{0}$ is a vector space. For, if $\mathbf{x}$ and $\mathbf{y} \in S$ then $A\mathbf{x} = A\mathbf{y} = \mathbf{0}$. Consequently $A(\mathbf{x} \oplus \mathbf{y}) = \mathbf{0} \oplus \mathbf{0} = \mathbf{0}$ and, for all real numbers α, $A(\alpha\mathbf{x}) = \alpha A(\mathbf{x}) = \alpha\mathbf{0} = \mathbf{0}$. Hence $\mathbf{x} \oplus \mathbf{y} \in S$ and $\alpha\mathbf{x} \in S$, showing that S (which is clearly non-empty) is a subspace of $\mathbb{R}^n$ and, hence, a vector space in its own right. S is called the **null space** or **kernel** of A, denoted by ker(A). As a particular example, suppose

$$A = \begin{bmatrix} 1 & 2 & -1 & 4 \\ 3 & 5 & 2 & 8 \\ 2 & 5 & -7 & 12 \end{bmatrix}.$$

Then

$$\ker A = \left\{ c\begin{bmatrix} -9 \\ 5 \\ 1 \\ 0 \end{bmatrix} + d\begin{bmatrix} 4 \\ -4 \\ 0 \\ 1 \end{bmatrix} : c, d \in \mathbb{R} \right\}$$

since the general solution of the system of equations

$$\begin{bmatrix} 1 & 2 & -1 & 4 \\ 3 & 5 & 2 & 8 \\ 2 & 5 & -7 & 12 \end{bmatrix}\begin{bmatrix} x \\ y \\ z \\ t \end{bmatrix} = \begin{bmatrix} 0 \\ 0 \\ 0 \\ 0 \end{bmatrix} \quad \text{is} \quad \left\{ \begin{bmatrix} -9c + 4d \\ 5c - 4d \\ c \\ d \end{bmatrix} : c, d \in \mathbb{R} \right\}.$$

PROBLEM 7.1

Find the null space of the matrix

$$\begin{bmatrix} 1 & 3 & 1 \\ 2 & 1 & 1 \end{bmatrix}.$$

Example 1 (cont.)

(vii) For each positive integer d let $\mathbb{R}_d[x]$ denote the set of polynomials of degree at most d in the indeterminate x and with real coefficients, together with the zero polynomial. Then, for each d, $\mathbb{R}_d[x]$ is a vector space and for each pair of positive integers m and n with $m \leq n$, $\mathbb{R}_m[x]$ is a subspace of $\mathbb{R}_n[x]$.

(viii) The set $\mathfrak{F}$ in Example 10 of Chapter 6 is a vector space. Furthermore $\mathfrak{D}$ is a subspace of $\mathfrak{F}$. (Why is it?)

(ix) Two (somewhat degenerate) examples of subspaces to be found in every vector space V are: (a) the whole set V itself; (b) the subset $\{\mathbf{0}_v\}$ comprising the zero vector of V alone. [This latter is called the **zero** or **trivial subspace** of V. Indeed *every* subspace S of a vector space V must contain the vector $\mathbf{0}_v$ of V – for if $\mathbf{s} \in S$ then $0 \circ \mathbf{s} (= \mathbf{0}_v) \in S$.]

PROBLEM 7.2

(i) What geometrical shape corresponds to the subset $S = \{(x, y, z): z = x^2 + y^2\}$? Is S a subspace of $\mathbb{R}^3$?

(ii) Show that each straight line through the origin in $\mathbb{R}^3$ corresponds to a subspace of $\mathbb{R}^3$. What about straight lines which do not pass through the origin? Apart from the infinitely many straight lines and planes through the origin $\mathbb{R}^3$ has just two other subspaces. What are they? ●

PROBLEM 7.3

(i) Let M_t denote the set of magic squares with centre component equal to the real number t. Given that M_t is a subspace of M (the vector space of all real 3×3 magic squares) is it necessarily the case that $t = 0$?

(ii) Let $M_{\times 3}$ denote the subset of M in which the centre component is some multiple of 3. Is $M_{\times 3}$ a vector space? (i.e. is it a subspace of M?) ●

It is easy to see geometrically that, in $\mathbb{R}^3$, any two distinct non-parallel planes which pass through the origin meet in a straight line which again passes through the origin (and hence is a subspace of $\mathbb{R}^3$). In higher dimensions the above geometric conclusion is not so clear: if we define a *flat* in $\mathbb{R}^n$ to be a subspace of $\mathbb{R}^n$ (necessarily passing through the origin) of $\mathbb{R}^n$ is it true that the set-theoretic intersection of two flats F_1 and F_2 (that is, the set $F_1 \cap F_2$ of all points common to F_1 and F_2) is again a flat? The (positive) answer is a consequence of the following very general result.

• *Theorem 2*

Let S_1 and S_2 be any two subspaces of the vector space V. Then their set-theoretic intersection, $S_1 \cap S_2$ (= S, say), is also a subspace of V.

PROOF

First note that S is not the empty set. (For, by Example 1(ix), the vector . . .? belongs to both S_1 and S_2.) We now check that conditions (I) and (II) of Theorem 1 hold. To that end, let $\mathbf{s}_1, \mathbf{s}_2 \in S$ and let $\alpha \in \mathbb{R}$. Then $\mathbf{s}_1, \mathbf{s}_2 \in S_1$. Hence $\mathbf{s}_1 \oplus \mathbf{s}_2 \in S_1$ and $\alpha \circ \mathbf{s}_1 \in S_1$ (since S_1 is a subspace of V.) Likewise $\mathbf{s}_1 \oplus \mathbf{s}_2 \in S_2$ and $\alpha \circ \mathbf{s}_1 \in S_2$ (since S_2 is a . . .?) Consequently $\mathbf{s}_1 \oplus \mathbf{s}_2 \in S_1 \cap S_2$ and $\alpha \circ \mathbf{s}_1 \in S_1 \cap S_2$, showing that S is indeed a subspace of V. ●

(Incidentally, there is nothing special here about there being two intersecting subspaces. The same result holds for any number of given subspaces – see Exercise 33(a).)

Having just seen how to produce a 'new' subspace from the set-theoretical intersection of given ones it is quite natural to enquire if the same is true for unions. In general it is not! Let us consider an example. (For effect we do not choose the simplest example possible!)

Example 2

Let S_1 and S_2 denote the subsets $\{(\lambda, 2\lambda, -3\lambda): \lambda \in \mathbb{R}\}$, $\{(4\mu, -2\mu, 5\mu): \mu \in \mathbb{R}\}$ of $\mathbb{R}^3$. Thus S_1 and S_2 correspond to lines l_1 and l_2 through the origin passing, respectively, through the points $(1, 2, -3)$ and $(4, -2, 5)$. Let S denote the set-theoretic union $S_1 \cup S_2$. Now it is easy to check that any scalar multiple of any triple in S is again in S. However, the sum of two elements in S is not, in general, in S. For example, $(1, 2, -3) + (4, -2, 5)$ $\{= (5, 0, 2)\}$ is neither in S_1 nor S_2. (Geometrically the point $(5, 0, 2)$ lies in neither the line l_1 nor l_2.) In fact, if T is any subspace of $\mathbb{R}^3$ containing all the elements of S_1 and of S_2 then, by Theorem 1, T must also contain all elements of the form $\lambda(1, 2, -3) + \mu(4, -2, 5)$ where $\lambda, \mu \in \mathbb{R}$. But this set of triples corresponds precisely to the set of points lying on the (unique) plane P, say, determined by the lines l_1 and l_2, and so is a subspace of $\mathbb{R}^3$.

This example suggests a way of generating subspaces in any vector space. First we need the following definition.

Definition 2

Let $\mathbf{v}_1, \mathbf{v}_2, \ldots, \mathbf{v}_r$ be elements of a vector space V. Each element $\mathbf{v}$ of the form[c] $\mathbf{v} = \alpha_1\mathbf{v}_1 + \alpha_2\mathbf{v}_2 + \ldots + \alpha_r\mathbf{v}_r$ (the α_i being scalars) is called a **linear combination** of the $\mathbf{v}_1, \mathbf{v}_2, \ldots, \mathbf{v}_r$. The set of all such linear combinations is called the **span** of $\mathbf{v}_1, \mathbf{v}_2, \ldots, \mathbf{v}_r$, written $\text{sp}\{\mathbf{v}_1, \mathbf{v}_2, \ldots, \mathbf{v}_r\}$.

(Below we shall confirm the suggestion made above, namely that $\text{sp}\{\mathbf{v}_1, \mathbf{v}_2, \ldots, \mathbf{v}_r\}$ is a subspace of V.)

First some examples.

Example 3

(i) $\mathbf{v} = (-\frac{4}{3}, 0, \frac{47}{24}, 1)$ is a linear combination of $\mathbf{v}_1 = (2, 1, \frac{1}{2}, 0)$, $\mathbf{v}_2 = (\frac{1}{4}, \frac{1}{2}, \frac{2}{3}, 1)$ and $\mathbf{v}_3 = (\frac{1}{3}, \frac{1}{4}, -1, 1)$. In fact $\mathbf{v} = -\frac{3}{4}\mathbf{v}_1 + 2\mathbf{v}_2 - 1\mathbf{v}_3$. On the other hand, $(\frac{7}{3}, \frac{5}{2}, \frac{31}{6}, 1)$ is not such a linear combination. How can I prove these assertions? Well,

$$\begin{aligned}\text{sp}\{\mathbf{v}_1, \mathbf{v}_2, \mathbf{v}_3\} &= \{\alpha_1\mathbf{v}_1 + \alpha_2\mathbf{v}_2 + \alpha_3\mathbf{v}_3 : \alpha_1, \alpha_2, \alpha_3 \in \mathbb{R}\}\\ &= \{\alpha_1(2,1,\tfrac{1}{2},0) + \alpha_2(\tfrac{1}{4},\tfrac{1}{2},\tfrac{2}{3},1) + \alpha_3(\tfrac{1}{3},\tfrac{1}{4},-1,1) : \alpha_1, \alpha_2, \alpha_3 \in \mathbb{R}\}\\ &= \{(2\alpha_1 + \tfrac{1}{4}\alpha_2 + \tfrac{1}{3}\alpha_3, \alpha_1 + \tfrac{1}{2}\alpha_2 + \tfrac{1}{4}\alpha_3, \tfrac{1}{2}\alpha_1 + \tfrac{2}{3}\alpha_2 - \alpha_3, \alpha_2 + \alpha_3):\\ &\qquad \alpha_1, \alpha_2, \alpha_3 \in \mathbb{R}\}.\end{aligned}$$

So, all I need to do is to check that the systems of equations

$$\begin{aligned} 2\alpha_1 + \tfrac{1}{4}\alpha_2 + \tfrac{1}{3}\alpha_3 &= -\tfrac{4}{3} \\ \alpha_1 + \tfrac{1}{2}\alpha_2 + \tfrac{1}{4}\alpha_3 &= 0 \\ \tfrac{1}{2}\alpha_1 + \tfrac{2}{3}\alpha_2 - \alpha_3 &= \tfrac{47}{24} \\ \alpha_2 + \alpha_3 &= 1 \end{aligned} \qquad \begin{aligned} 2\alpha_1 + \tfrac{1}{4}\alpha_2 + \tfrac{1}{3}\alpha_3 &= \tfrac{7}{3} \\ \alpha_1 + \tfrac{1}{2}\alpha_2 + \tfrac{1}{4}\alpha_3 &= \tfrac{5}{2} \\ \tfrac{1}{2}\alpha_1 + \tfrac{2}{3}\alpha_2 - \alpha_3 &= \tfrac{31}{6} \\ \alpha_2 + \alpha_3 &= 1 \end{aligned}$$

[c]We resort, once again, to writing $\mathbf{u} + \mathbf{v}$ and $\alpha\mathbf{u}$ instead of $\mathbf{u} \oplus \mathbf{v}$ and $\alpha \circ \mathbf{u}$.

are, respectively, consistent (with solution $(\alpha_1, \alpha_2, \alpha_3) = (-\frac{3}{4}, 2, -1)$) and inconsistent. (In particular the subspace $\mathrm{sp}\{\mathbf{v}_1, \mathbf{v}_2, \mathbf{v}_3\}$ of $\mathbb{R}^4$ is not the whole of $\mathbb{R}^4$.)

(ii) Let A be an $m \times n$ matrix. If we regard the rows, $\mathbf{r}_1, \mathbf{r}_2, \ldots, \mathbf{r}_m$ (respectively, columns $\mathbf{c}_1, \mathbf{c}_2, \ldots, \mathbf{c}_n$) of A as vectors in $\mathbb{R}^n$ (respectively $\mathbb{R}^m$) then $\mathrm{sp}\{\mathbf{r}_1, \mathbf{r}_2, \ldots, \mathbf{r}_m\}$ (respectively $\mathrm{sp}\{\mathbf{c}_1, \mathbf{c}_2, \ldots, \mathbf{c}_n\}$), is a subspace of $\mathbb{R}^n$ (respectively $\mathbb{R}^m$) called the **row space** (respectively **column space**) of A.

PROBLEM 7.4

By using linear systems, as in Example 3(i), you may readily check that (8, 21, 7, 9) does belong whereas (3, 2, 0, 2) does not belong to the row space of

$$A = \begin{bmatrix} 1 & 1 & -1 & 2 \\ 2 & 1 & 1 & -1 \\ -1 & 2 & 1 & 1 \end{bmatrix}.$$

[A somewhat quicker way to check these statements will follow from Theorem 2 of Chapter 8.]

A slightly more interesting point is that (using the same matrix A) the system of equations $A\mathbf{x} = \mathbf{b}$ (where $\mathbf{x} = [x_1\, x_2\, x_3\, x_4]^T$ and $\mathbf{b} = [b_1\, b_2\, b_3]^T$) has a solution $\mathbf{x}$ if and only if

$$\begin{aligned} 1x_1 + 1x_2 - 1x_3 + 2x_4 &= b_1 \\ 2x_1 + 1x_2 + 1x_3 - 1x_4 &= b_2, \\ -1x_1 + 2x_2 + 1x_3 + 1x_4 &= b_3 \end{aligned}$$

that is, if and only if $\mathbf{b}$ belongs to the column space of A. (For $\mathbf{x}$ is such a solution if and only if

$$x_1\begin{bmatrix} 1 \\ 2 \\ -1 \end{bmatrix} + x_2\begin{bmatrix} 1 \\ 1 \\ 2 \end{bmatrix} + x_3\begin{bmatrix} -1 \\ 1 \\ 1 \end{bmatrix} + x_4\begin{bmatrix} 2 \\ -1 \\ 1 \end{bmatrix} = \mathbf{b}.$$

See Exercise 9(c) of Chapter 3.) In particular, the column space of A can be identified as the set $\{A\mathbf{x}: \mathbf{x} \in \mathbb{R}^4\}$. Consequently, regarding A as a function from $\mathbb{R}^4$ to $\mathbb{R}^3$, the column space is seen to coincide with the **image** or **range** of A (see the Appendix).

PROBLEM 7.5

(a) Show that the magic square

5	1	3
1	3	5
3	5	1

is a linear combination of

$$\begin{matrix} 1 & 1 & 1 \\ 1 & 1 & 1 \\ 1 & 1 & 1 \end{matrix} \quad \text{and} \quad \begin{matrix} -1 & 1 & 0 \\ 1 & 0 & -1 \\ 0 & -1 & 1 \end{matrix} \quad \text{whilst} \quad \begin{matrix} 4 & 3 & 8 \\ 9 & 5 & 1 \\ 2 & 7 & 6 \end{matrix} \quad \text{is not.}$$

(b) Show that the AP (–4, 5, 14, 23, . . .) is a linear combination of the APs (0, 1, 2, 3, . . .) and (1, 1, 1, . . .). Indeed . . . (see Exercise 14).

Let us now return to the claim we made following Definition 2.

• Theorem 3

Let $\mathbf{v}_1, \mathbf{v}_2, \ldots, \mathbf{v}_r$ be vectors of the vector space V over the scalars $\mathbb{F}$. Then $\text{sp}\{\mathbf{v}_1, \mathbf{v}_2, \ldots, \mathbf{v}_r\}$ (= S, say) is a subspace of V.

PROOF

We must check that conditions (I) and (II) of Theorem 1 hold. So let $\mathbf{u} = \alpha_1\mathbf{v}_1 + \alpha_2\mathbf{v}_2 + \ldots + \alpha_r\mathbf{v}_r$ and $\mathbf{w} = \beta_1\mathbf{v}_1 + \beta_2\mathbf{v}_2 + \ldots + \beta_r\mathbf{v}_r$ be any two elements of S. Then

$$\mathbf{u} + \mathbf{w} = (\alpha_1 + \beta_1)\mathbf{v}_1 + (\alpha_2 + \beta_2)\mathbf{v}_2 + \ldots + (\alpha_r + \beta_r)\mathbf{v}_r$$

whilst, for each $\gamma \in \mathbb{F}$, we have

$$\gamma\mathbf{u} = (\gamma\alpha_1)\mathbf{v}_1 + (\gamma\alpha_2)\mathbf{v}_2 + \ldots + (\gamma\alpha_r)\mathbf{v}_r.$$

Clearly $\mathbf{u} + \mathbf{w}$ and $\gamma\mathbf{u}$ belong to S and so S (being non-empty – why?) is a subspace of V. ●

Following Theorem 3 it is quite convenient to introduce the next definition.

• Definition 3

If V is a vector space and $\mathbf{v}_1, \mathbf{v}_2, \ldots, \mathbf{v}_r$ are vectors in the subspace S of V such that $\text{sp}\{\mathbf{v}_1, \mathbf{v}_2, \ldots, \mathbf{v}_r\} = S$ we say that $\{\mathbf{v}_1, \mathbf{v}_2, \ldots, \mathbf{v}_r\}$ **spans** (or **generates**) S or that S is **spanned** (or **generated**) **by** $\{\mathbf{v}_1, \mathbf{v}_2, \ldots, \mathbf{v}_r\}$ or that $\{\mathbf{v}_1, \mathbf{v}_2, \ldots, \mathbf{v}_r\}$ is a **spanning** (or **generating**) **set** for S. ●

On many occasions we shall take S to be the whole space V. Here are some examples.

Examples 4

(i) $\mathbb{R}^4$ is (the subspace of $\mathbb{R}^4$) spanned by {(1, 0, 0, 0), (0, 1, 0, 0), (0, 0, 1, 0), (0, 0, 0, 1)}. It is also spanned by infinitely many other sets including, as just one example, {(1, 2, 3, 4), (5, 6, 7, 8), (9, 10, 11, 12), (5, 1, 2, –3), (4, 9, –4, 9)}. (Am I right?)

(ii) The subspace $S = \{(x, y, z): 2x + 4y - 3z = 0\}$ of $\mathbb{R}^3$ is spanned by $(1, 0, \frac{2}{3})$ and $(0, 1, \frac{4}{3})$, since, if $(a, b, c) \in S$, then $2a + 4b - 3c = 0$ so that $c = (2a + 4b)/3$ and hence $(a, b, c) = (a, b, (2a + 4b)/3) = a(1, 0, \frac{2}{3}) + b(0, 1, \frac{4}{3})$.

(iii) The set of all solutions of the system of (homogeneous) equations

$$\begin{aligned} x + 2y + 3z + 4t &= 0 \\ 5x + 6y + 7z + 8t &= 0, \end{aligned}$$

namely $\{(a + 2b, -2a - 3b, a, b): a, b \in \mathbb{R}\}$, is spanned by the two vectors (1, –2, 1, 0) and (2, –3, 0, 1). Notice (I) how describing the infinitely many solutions in terms of only two of them gives a clearer idea of the nature of this set of solutions; (II) that this set is equally well described as the null space of the matrix

$$\begin{bmatrix} 1 & 2 & 3 & 4 \\ 5 & 6 & 7 & 8 \end{bmatrix}.]$$

(iv) Although Definition 3 deals with finite (spanning) sets it seems reasonable to say that the space of all polynomials, $\mathbb{R}[x]$, is spanned by the infinite set $\{1, x, x^2, x^3 \ldots\}$ of polynomials. [Frequently in problems in mathematics or mathematical physics it is desirable to find other spanning sets $\{p_0(x), p_1(x), \ldots\}$ of $\mathbb{R}[x]$ perhaps with the property that

$$\int_{-1}^{1} p_i(x)p_j(x)\mathrm{d}x = \begin{cases} 0 & \text{if } i \neq j \\ 1 & \text{if } i = j \end{cases}.]$$

PROBLEM 7.6

Show that the set $\{(1, 1, 1), (1, 1, 2), (1, 2, 2), (1, 2, 3)\}$ spans $\mathbb{R}^3$.

PROBLEM 7.7

Find polynomials $p_0(x) = a_0$, $p_1(x) = b_0 + b_1x$, $p_2(x) = c_0 + c_1x + c_2x^2$ such that the property of Example 4(iv) holds.

TUTORIAL PROBLEM 7.1

No finite set of polynomials can span $\mathbb{R}[x]$. Discuss.

Example 4 (cont.)

(v) $\mathbb{C}$ is spanned, as a vector space over $\mathbb{R}$, by $\{1,\mathrm{i}\}$ – also by $\{\mathrm{e} + \mathrm{i}\pi, \mathrm{e} - \mathrm{i}\pi\}$, for example, and by infinitely many other pairs of complex numbers.

(vi) The set (space) of all solutions of $\mathrm{d}^2y/\mathrm{d}x^2 + 4\mathrm{d}y/\mathrm{d}x + 3y = 0$ is spanned by the functions e^{-3x} and e^{-x}. (This is 'because $m^2 + 4m + 3$ has roots -3 and -1'. See Example 2 of Chapter 6.)

(vii) $M_{2\times2}(\mathbb{R})$ is spanned by

$$\begin{bmatrix} 1 & 0 \\ 0 & 0 \end{bmatrix}, \begin{bmatrix} 0 & 1 \\ 0 & 0 \end{bmatrix}, \begin{bmatrix} 0 & 0 \\ 1 & 0 \end{bmatrix}, \begin{bmatrix} 0 & 0 \\ 0 & 1 \end{bmatrix}.$$

Examples (i)–(iii) and (v)–(vii) exhibit finite spanning sets – (iv) shows an infinite spanning set (where no finite spanning set is possible, as you should have discovered in answering Tutorial Problem 7.1.)

TUTORIAL PROBLEM 7.2

(a) Do you think that $\mathbb{R}^4$ could be spanned by just three (carefully chosen) vectors?
(b) What set spans the trivial subspace $\{(0, 0, 0, 0)\}$ of $\mathbb{R}^4$?

The above comments lead, naturally, to the following definition.

Definition 4

If a vector space V is spanned by a finite set of vectors we say that V is **finite dimensional**.

[Many important vector spaces are not finite dimensional; see, for example, Exercise 35 below.]

Examples 5

(i) Each $\mathbb{R}^n$ is finite dimensional (since the n n-vectors $(1, 0, 0, \ldots, 0)$, $(0, 1, 0, \ldots, 0)$, $\ldots (0, 0, 0, \ldots, 1)$ surely span $\mathbb{R}^n$? We leave the verification to you.)

(ii) Let A be an $m \times n$ matrix. The space of all solutions to $A\mathbf{x} = \mathbf{0}$ is finite dimensional. (This generalises Example 4(iii).)

(iii) The space of all solutions of the differential equation

$$a_n \frac{d^n y}{dx^n} + a_{n-1} \frac{d^{n-1} y}{dx^{n-1}} + \ldots + a_1 \frac{dy}{dx} + a_0 y = 0$$

is finite dimensional. (Once again this is a fact: you are not supposed to be able to see why! For example, the space of all solutions of the differential equation

$$\frac{d^3 y}{dx^3} + 4\frac{d^2 y}{dx^2} + \frac{dy}{dx} - 6y = 0$$

is spanned by the functions e^x, e^{-2x} and e^{-3x} 'because' $m^3 + 4m^2 + m - 6 = (m-1)(m+2)(m+3)$ – cf. Example 4(vi). [If there were repeated factors here this approach would need modification.]

(iv) The space $\mathbb{R}[x]$ of all polynomials with real coefficients is not finite dimensional. (For, if it were spanned by polynomials $p_1(x), p_2(x), \ldots, p_k(x)$, and if the greatest degree of these $p_i(x)$ were, say, t, then all polynomials in $\mathbb{R}[x]$ would have to be of degree at most t, a blatant impossibility!)

TUTORIAL PROBLEM 7.3

Is the trivial subspace $\{(0, 0, \ldots, 0)\}$ of $\mathbb{R}^n$ finite dimensional?

One final remark. We have not yet defined the dimension of a finite dimensional vector space! What should the dimension of, say, $\mathbb{R}^3$ be? And, surely, subspaces of finite dimensional vector spaces are again finite dimensional? We shall answer these questions in Chapter 9.

In many circumstances in mathematics it is natural to consider what amount to linear combinations. As a first example, note that if $S_1(x, y) = 0$ and $S_2(x, y) = 0$ are two curves in the x–y plane then $\lambda_1 S_1(x, y) + \lambda_2 S_2(x, y) = 0$ is another curve passing through their points of intersection.

As a second example, recall that any linear combination of solutions of a given homogeneous linear differential equation is again a solution. Somewhat deeper is the following sort of 'converse' statement.

In 1739, in a letter to John Bernoulli, Leonhard Euler explained his method for solving equations of the form of Example 5(iii). This method gave him, in the case of equations of order n, n particular functions. Euler then asserted that, since the most general solution of the given equation must contain n arbitrary constants, each solution is a sum of these n functions, each multiplied by a constant. This would seem to be the first published instance of linear combinations of functions being used, although Bernoulli replied that he had obtained similar results in 1700!

Later (1822) Fourier had occasion to represent certain functions by means of infinte linear combinations when researching into heat flow in solid bodies. [Such infinite sums require discussions of convergence and so are not dealt with in detail here.]

The application presented below began a heated dispute![d] In 1806 Adrien-Marie Lengendre gave the statement and the first published application of the method of least squares. At once Gauss declared that he had been using this method since 1795. This outraged Legendre who did not see why Gauss, the 'Prince of Mathematicians', should (also) want the glory due to others. [As usual the cause was one of communication or, rather, Gauss's lack of it! His motto 'Few, but ripe' summed up his attitude to publishing his work.]

Adrien-Marie Legendre (18 September 1752–9 January 1833) was born in Paris, and in his early life was sufficiently well off financially that he did not need to work. Nevertheless he taught mathematics at the Ecole Militaire from 1775 to 1780. Whilst there he wrote an essay on ballistics which won the 1782 prize offered by the Berlin Academy. His research fell into three areas: celestial mechanics, elliptic functions and number theory. In the study of elliptic functions he was, for 40 years, the sole important researcher; in number theory his attitude to proof again brought him into conflict with Gauss. Briefly, Gauss claimed for himself the first proof of the famous law of quadratic reciprocity, asserting that all earlier proofs were incomplete. Legendre claimed it on the grounds that he had indicated, in detail, how it could be proved. Legendre was co-opted onto many committees, one being charged with standardising the metre. His *Elements of Geometry* (1784), later copies of which contained a proof of the irrationality of π^2 (and, hence, of π) remained a standard substitute for Euclid for 100 years. Aged 40, and with his 'small fortune' gone, he married a 19-year-old girl who, he later stated, helped him put his affairs in order and brought a tranquillity to his life which greatly helped him in his research. (I am tempted to add '!!')

Application

There is a very nice application of the ideas of this chapter – in particular the concept of column space of a matrix – to a problem of obvious practical importance.

[d]Sorry! Accidental pun!

An experiment is carried out to determine the relationship, assumed linear, between two measurable quantities x and y. Suppose that measurements (x_i, y_i) are as obtained below and that we are seeking constants a and b such that $y = ax + b$ is satisfied by all pairs x_i, y_i.

i	1	2	3	4	5	6	7
x_i	3.2	4.1	5.3	6.7	7.1	8.3	9.1
y_i	7.2	8.4	9.9	11.3	12.1	13.3	14.2

Plotting the points (x_i, y_i) shows that no such a, b can exist! But, as the x, y are themselves measured only approximately, we can try to find a straight line which 'most nearly' passes through these points. How does one find such a line and, more important, what criterion does one use to measure 'nearness'? [Before reading on try to find this 'best' line by eye. Then compare your choice with that determined below.]

For each i let us write $y_i = ax_i + b + r_i$ so that r_i measures the error in assuming that $y_i = ax_i + b$ exactly. In matrix notation we can write this as $Y = XU + R$ where

$$Y = \begin{bmatrix} y_1 \\ y_2 \\ \vdots \\ y_7 \end{bmatrix}, \quad X = \begin{bmatrix} x_1 & 1 \\ x_2 & 1 \\ \vdots & \vdots \\ x_7 & 1 \end{bmatrix}, \quad U = \begin{bmatrix} a \\ b \end{bmatrix} \quad \text{and} \quad R = \begin{bmatrix} r_1 \\ r_2 \\ \vdots \\ r_7 \end{bmatrix}.$$

Our aim is to choose the vector R $(= Y - XU)$ as small as possible. Note that

$$X\begin{bmatrix} a \\ b \end{bmatrix} = a\begin{bmatrix} x_1 \\ x_2 \\ \vdots \\ x_7 \end{bmatrix} + b\begin{bmatrix} 1 \\ 1 \\ \vdots \\ 1 \end{bmatrix},$$

so that, if we let a, b range over all of $\mathbb{R}$, we generate a (plane) subspace (S, say) of $\mathbb{R}^7$. It is clear from Fig. 7.1 that R is as small as possible when it is perpendicular to S, that is, to each vector

$$X\begin{bmatrix} r \\ s \end{bmatrix}.$$

This implies that

$$\begin{bmatrix} r & s \end{bmatrix} X^T (Y - X\begin{bmatrix} a \\ b \end{bmatrix}) = 0,$$

the zero scalar. Since this holds for each

$$\begin{bmatrix} r \\ s \end{bmatrix}, \quad \text{we have} \quad X^T (Y - X\begin{bmatrix} a \\ b \end{bmatrix}) = 0,$$

the zero (2×1) matrix. It follows that

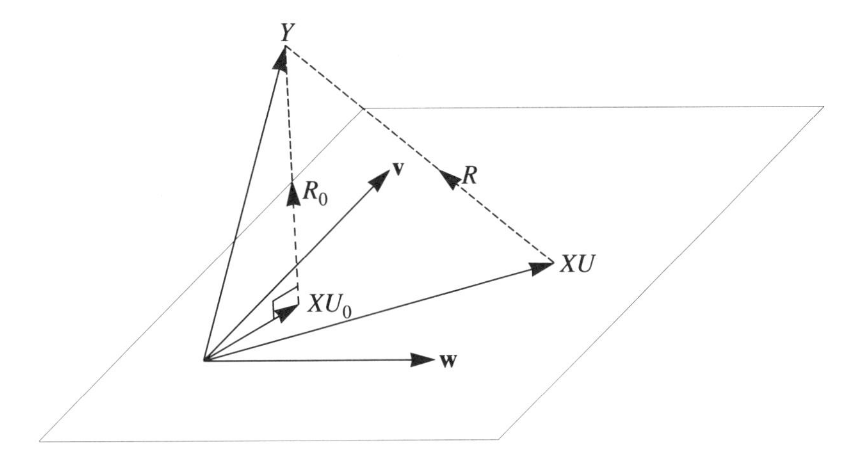

Fig 7.1 $\mathbf{v} = \begin{bmatrix} x_1 \\ \vdots \\ x_7 \end{bmatrix}$ and $\mathbf{w} = \begin{bmatrix} 1 \\ \vdots \\ 1 \end{bmatrix}$. R is shortest (R_0) when U_0 is such that R_0 is perpendicular to the plane

$$\begin{bmatrix} a \\ b \end{bmatrix} = (X^T X)^{-1} X^T Y.$$

In the present case we find that

$$X^T X = \begin{bmatrix} 302.14 & 43.8 \\ 43.8 & 7 \end{bmatrix} \quad \text{and} \quad (X^T X)^{-1} X^T Y = \begin{bmatrix} 1.1801 & \dots \\ 3.5301 & \dots \end{bmatrix}.$$

Thus, to place the most 'accurate' line through the points (x_i, y_i) we should take $a = 1.18\dots$ and $b = 3.53\dots$.

TUTORIAL PROBLEM 7.4

> In the above problem we needed the matrix $X^T X$ to be invertible. Have we any right to expect it will always be?

Since, above, we have minimised the length $\sqrt{(r_1^2 + r_2^2 + \dots + r_7^2)}$ of the vector R, the above method is called the **method of least squares**. Looking at Fig. 7.1, one could argue that other methods of minimising error might be employed, for example why not minimise $|r_1| + |r_2| + \dots + |r_7|$? The answer is merely one of ease of calculation.

The method can easily be adapted to finding best fitting planes, etc. and even to best fitting curves like $y = x^2$ by plotting log y against log x.

Another nice application of subspaces is concerned with coding theory aspects of data communications but we can fairly leave our short discussion until Chapter 9.

Summary

A **subspace** of a vector space V is just a non-empty subset S of V which forms a vector space in its own right (with respect to the same set $\mathbb{F}$ of scalars and the same $\oplus$ and $\circ$ as in V). By Theorem 1, S will be a subspace of V if and only if S is **closed** under $\oplus$ and $\circ$. V itself and $\{\mathbf{0}_v\}$ are the two extreme examples in V. If $\mathbf{v}_1, \mathbf{v}_2, \ldots, \mathbf{v}_r$ are vectors in V the set L of all **linear combinations** $\alpha_1\mathbf{v}_1 + \alpha_2\mathbf{v}_2 + \ldots + \alpha_r\mathbf{v}_r$ (where the $\alpha_i \in \mathbb{F}$) is a subspace of V – in fact the smallest subspace of V containing all of $\mathbf{v}_1, \mathbf{v}_2, \ldots, \mathbf{v}_r$. We then say that L is **spanned** by $\mathbf{v}_1, \mathbf{v}_2, \ldots, \mathbf{v}_r$ and write $L = \text{sp}\{\mathbf{v}_1, \mathbf{v}_2, \ldots, \mathbf{v}_r\}$. The **row space (column space)** of the $m \times n$ matrix A is the subspace of $\mathbb{R}^n$ ($\mathbb{R}^m$) spanned by the rows (columns) of A regarded as vectors in $\mathbb{R}^n$ ($\mathbb{R}^m$). The column space of A may also be thought of as the **image** or **range** $\{A\mathbf{x}:\mathbf{x} \in \mathbb{R}^n\}$ of A. Likewise, the set $\{A\mathbf{x}:\mathbf{x} \in \mathbb{R}^n \text{ and } A\mathbf{x} = 0\}$ is a subspace of $\mathbb{R}^n$ called the **null space** or **kernel** of A.

If V is itself spanned by a finite set of vectors from V we say that V is **finite dimensional**. Most vector spaces considered in this book are finite dimensional. $\mathbb{R}[x]$ is just one example of many which is not finite dimensional.

EXERCISES ON CHAPTER 7

1. (a) Which of the following subsets of $\mathbb{R}^3$ are subspaces of $\mathbb{R}^3$?
   (i) $\{(x, y, z): x + 2y + 3z = 0\}$; (ii) $\{(x, y, z): x + 2y + 3z = 1\}$;
   (iii) $\{(x, y, z): x^2 + y^2 = z^2\}$; (iv) $\{(x, y, 1): x, y \in \mathbb{R}\}$; (v) $\{(x, y, z): x + 2y \in \mathbb{Q}\}$;
   (vi) $\{(x, x, x): 0 \leq x \leq 1\}$; (vii) all of $\mathbb{R}^3$ except for the single point $(1, 1, 1)$.
   (b) Is $\{(x, y, z, t): x, y, z, t \in \mathbb{R} \text{ and } x = 2y, x + y = z + t\}$ a subspace of $\mathbb{R}^4$?
2. Which of the following sets of polynomials is a subspace of $\mathbb{R}[x]$? (Include the zero polynomial in each case.)
   (i) The set of polynomials of degree ≤ 10; (ii) the set of polynomials of degree ≥ 10;
   (iii) $\{a + bx + cx^2 + dx^3 : a + b + c + d = 0\}$; (iv) $\{p(x): p(1) = p(5)\}$;
   (v) $\{p(x): p(1) = \mathrm{d}p/\mathrm{d}x(5)\}$ [Note: $\mathrm{d}p/\mathrm{d}x\,(s)$ is the value of $\mathrm{d}p/\mathrm{d}x$ when $x = s$.];
   (vi) $\{p(x): p(n) = 0 \text{ for each } n \in \mathbb{Z}\}$. (Can you say which polynomials belong to this set?);
   (vii) $\{p(x): p(x) = -p(x) \text{ for all } -1 < x < 1\}$; (viii) $\{p(x): \int_0^1 p(x)\mathrm{d}x = 0\}$;
   (ix) $\{p(x)$: the only real root of $p(x)$ is $1\}$.
3. (a) Which of the following subsets of the vector space $M_3(\mathbb{R})$ of 3×3 matrices with real entries are subspaces of $M_3(\mathbb{R})$?
   (i) $\{A: \det A = 0\}$; (ii) $\{A: \det A$ is an integer$\}$; (iii) $\{A: A$ is invertible$\}$;
   (iv) $\{A: A$ is upper triangular$\}$; (v) $\{A: A$ is symmetric$\}$;
   (vi) $\{A: A$ is skew symmetric$\}$.
   (b) Let B and C be fixed matrices in $M_3(\mathbb{R})$. Let S be the set of all those 3×3 matrices A for which $BAC = 0_{3\times3}$. Is S a subspace of $M_3(\mathbb{R})$?
4. (a) Which of the following subsets are subspaces of the given spaces?
   (i) The first quadrant in the x–y plane; (ii) the line $y = 2x$ in the x–y plane; (iii) the line $y = 2x + 1$ in the x–y plane; (iv) the plane $23x + 42y - 633z = 0$ in x–y–z space; (v) the plane $23x + 42y - 633z = 0.000002$ in x–y–z space.
   (b) Is $\mathbb{R}^2$ a subspace of $\mathbb{R}^3$? [See Chapter 10.]

5. Let $\mathfrak{F}$ denote, as usual, the vector space of all functions $f: \mathbb{R} \to \mathbb{R}$. Let S_1 and S_2 be, respectively, the sets of solutions of the differential equations

$$(1)\ \frac{d^2 f}{dx^2} + \cos x \frac{df}{dx} + e^x f = 0 \quad \text{and} \quad (2)\ \frac{d^2 g}{dx^2} + \cos x \frac{dg}{dx} + e^x = 0.$$

Which, if either, of S_1 and S_2 is a subspace of $\mathfrak{F}$?

6. Show, generalising Example 1(vi), that the solution space of a set of non-homogeneous equations is not a subspace of the appropriate $\mathbb{R}^n$.

7. (a) Confirm that Example 1(iv) is not a subspace of R^4.
(b) Show that the subset $\{(x, y): xy = 0\}$ of $\mathbb{R}^2$ is not a subspace of $\mathbb{R}^2$.
(c) Is the set of all vectors in $\mathbb{R}^3$ which are orthogonal to (1, 2, 3) a subspace of $\mathbb{R}^3$? [Hint: first get a picture in your mind.]
(d) Let S be a (non-empty) subset of the vector space V. Show that S is a subspace of V if and only if, for all $\mathbf{u}, \mathbf{v} \in S$ and for all scalars α, β we have $\alpha\mathbf{u} + \beta\mathbf{v} \in S$. Must S necessarily be a subspace of V if we are only told that, for all $\mathbf{u}, \mathbf{v} \in S$ and scalars α, we have $\alpha\mathbf{u} + \mathbf{v} \in S$?

8. Generalise the column space part of Example 3(ii) as follows. Let S be a subspace of $\mathbb{R}^n$ and A be an $m \times n$ matrix. Prove that the subset $T = \{A\mathbf{x}: \mathbf{x} \in S\}$ of $\mathbb{R}^m$ is a subspace of $\mathbb{R}^m$. [Hint: let $\mathbf{y}_1$ and $\mathbf{y}_2 \in T$. Then there are $\mathbf{x}_1$ and $\mathbf{x}_2 \in S$ such that $\mathbf{y}_1 = A\mathbf{x}_1$ and $\mathbf{y}_2 = A\mathbf{x}_2$. But then $\mathbf{y}_1 \oplus \mathbf{y}_2 = A(\mathbf{x}_1 \oplus \mathbf{x}_2)$. Now $\mathbf{x}_1 \oplus \mathbf{x}_2 \in S$ (why?) Hence $\mathbf{y}_1 \oplus \mathbf{y}_2 \in T$. A similar argument shows that $\alpha\mathbf{x} \in T$ for each real number α.]

9. (a) Let S be the subset $\{(x, 0): x \in \mathbb{R}$ and $x > 0\}$ of $\mathbb{R}^2$. Define $\boxplus$ and $\boxdot$ on S by: $(u, 0) \boxplus (v, 0) = (uv, 0)$ and $\alpha \boxdot (u, 0) = (u^\alpha, 0)$ for $\alpha \in \mathbb{R}$. Show that S is a vector space with respect to $\boxplus$ and $\boxdot$.
[Note, however, that we cannot accept S as a subspace of $\mathbb{R}^2$ since, for example, the 'sum' $(uv, 0)$ of $(u, 0)$ and $(v, 0)$ regarded as elements of S is not the same as their sum, $(u + v, 0)$, when they are regarded as elements of $\mathbb{R}^2$. [Similar objections apply to the two distinct scalar multiplications.]
(b) $\mathbb{R}[x]$, the (real) vector space of polynomials in x with real coefficients, is quite naturally regarded as a subset of $\mathbb{C}[x]$, the vector space of polynomials in x with complex coefficients. But we may not think of $\mathbb{R}[x]$ as a subspace of $\mathbb{C}[x]$ – unless $\mathbb{C}[x]$ is being thought of (unnaturally?) as a vector space over $\mathbb{R}$ rather than over $\mathbb{C}$. Why not?

10. Is $\begin{bmatrix} 1 & -1 & -2 \\ 4 & 6 & 0 \\ 3 & 5 & 2 \end{bmatrix}$ a linear combination of

$$\begin{bmatrix} 1 & 1 & 0 \\ 1 & 1 & 0 \\ 0 & 0 & 0 \end{bmatrix}, \quad \begin{bmatrix} 0 & 1 & 1 \\ 0 & 1 & 1 \\ 0 & 0 & 0 \end{bmatrix}, \quad \begin{bmatrix} 0 & 0 & 0 \\ 1 & 1 & 0 \\ 1 & 1 & 0 \end{bmatrix} \quad \text{and} \quad \begin{bmatrix} 0 & 0 & 0 \\ 0 & 1 & 1 \\ 0 & 1 & 1 \end{bmatrix}?$$

11. Is either of (i) $9x^3 - x^2 - 6$ or (ii) $-4x + 6x^3$ a linear combination of $1 + x$, $1 + 3x^2$ and $1 + x^2 + x^3$?

12. Complete Example 3(i).

13. Show that the vector space of arithmetic progressions is spanned by the two APs (0, 1, 2, 3, . . .) and (1, 1, 1, . . .). Is it also spanned by the pair (1, 2, 3, 4, . . .) and (0, 1, 2, 3, . . .)?

14. (a) Let $\mathbf{v}_1$ and $\mathbf{v}_2$ be vectors in 3-dimensional space. Describe geometrically the conditions under which $\mathbf{v}_1$ and $\mathbf{v}_2$ (i) span a plane, (ii) span just a line.
(b) Explain geometrically why one needs at least three vectors to span $\mathbb{R}^3$.

15. Which of the following sets span (i) $\mathbb{R}^3$; (ii) a plane in $\mathbb{R}^3$? If the former, find as small a subset as possible which spans $\mathbb{R}^3$; if the latter write the equation of the plane in the form $ax + by + cz = 0$.
(i) {(1, 2, 3), (4, 5, 6), (7, 8, 9)}; (ii) {(1, 2, 3), (4, 5, 6), (7, 8, 10)}; (iii) {(1, 1, 4), (2, 1, 5), (0, 1, 3), (3, 2, 9), (1, 1, 1)}; (iv) {(1, 2, 1), (8, 7, −1), (2, 1, −1), (7, 8, 1)}.

16. Do the vectors (1, −1, 0, 0), (0, 1, −1, 0), (0, 0, 1, −1), (−1, 0, 0, 1) span $\mathbb{R}^4$?

17. Let $a, b, c \in \mathbb{R}$. When does $(a, b, c) \in \text{sp}\{(b, c, a), (c, a, b)\}$?

18. (a) Describe in the form $\{(x, y, z): ax + by + cz = 0\}$ the subspace of $\mathbb{R}^3$ spanned by (1, −2, 3) and (4, −5, 6).
(b) Describe in similar form the subspace of $\mathbb{R}^3$ which is the intersection of the planes determined by the equations (i) $x + 2y + 3z = 0$; (ii) $2x + 3y + z = 0$.

19. Exhibit a spanning set for $\mathbb{R}^3$ which contains 20 vectors.

20. Find a set of vectors in $\mathbb{R}^4$ which span the solution space of

$$\begin{aligned} -x_1 - x_2 + x_3 + 2x_4 &= 0 \\ 2x_1 + x_2 + x_3 - x_4 &= 0. \end{aligned}$$

Does the pair {(−16, 33, 5, 6), (−17, 33, 6, 5)} also span it?

21. Find a system of homogeneous linear equations in four variables for which the solution space is spanned by the set {(1, 2, 1, 3), (1, 3, 2, 4)}.

22. Let A be the matrix

$$\begin{bmatrix} 1 & 2 & -1 & 1 \\ 1 & 1 & 2 & -1 \\ 3 & 4 & 3 & -1 \\ 1 & 4 & -7 & 5 \\ 2 & 3 & 1 & 1 \end{bmatrix}:$$

(i) find a spanning set for ker(A); (ii) find as small a set of (row) vectors as possible which spans the row space of A. Repeat (ii) for the column space of A. Compare the sizes of these spanning sets.

23. Show that $\begin{bmatrix} a \\ b \\ c \end{bmatrix}$ is in the column space of $A = \begin{bmatrix} 1 & 3 & 1 & -1 \\ 2 & -1 & 9 & 12 \\ 1 & 2 & 2 & 1 \end{bmatrix}$ if and only if $7c - 5a - b = 0$. Deduce that the column space of A is spanned by

$$\begin{bmatrix} 0 \\ 7 \\ 1 \end{bmatrix} \text{ and } \begin{bmatrix} 1 \\ -5 \\ 0 \end{bmatrix}.$$

24. For $m \times n$ and $n \times p$ matrices A and B show that column space $(AB) \subseteq$ column space A.

25. Show that $\text{sp}\{1+x, 2+2x+x^2, x^2+x^3, 1-x^3\} = \mathbb{R}_3[x]$.

26. If $\mathbf{u}_1, \mathbf{u}_2, \mathbf{u}_3, \mathbf{u}_4 \in \mathbb{R}^3$ are such that $(1, 0, 0), (0, 1, 0), (0, 0, 1) \in \text{sp}\{\mathbf{u}_1, \mathbf{u}_2, \mathbf{u}_3, \mathbf{u}_4\}$ does this mean that $\text{sp}\{\mathbf{u}_1, \mathbf{u}_2, \mathbf{u}_3, \mathbf{u}_4\} = \mathbb{R}^3$?

27. (a) Can you find four invertible matrices which together span $M_2(\mathbb{R})$?
(b) Let A be a 2×2 matrix. Given that no set of three (2×2) matrices can span $M_2(\mathbb{R})$, explain why the set $\{A, A^2, A^3, A^4\}$ cannot span $M_2(\mathbb{R})$. [Hint: Cayley – . . .]

28. Do $S = \{(3, 2, 0, 2), (7, 2, -2, 0), (4, 5, -2, 6)\}$ and

$$T = \{(1,2,1,3),(2,0,-1,-1),(-1,1,0,2),(5,4,3,5),(1,5,7,9)\}$$

span the same subspace of $\mathbb{R}^4$? [Hint: if each element of S lies in $\text{sp}\{T\}$ then $\text{sp}\{S\} \subseteq \text{sp}\{T\}$.]

29. Show that $\text{sp}\{(1, 2, 4), (2, 5, 8), (-1, 5, -4), (3, 7, 1)\} = \mathbb{R}^3$. Find all subsets of three of these vectors which also span $\mathbb{R}^3$.

30. Given that $\mathbf{u}, \mathbf{v}, \mathbf{w}$ span a certain subspace S of the vector space V, show that $\text{sp}\{\mathbf{u}, \mathbf{u}+\mathbf{v}, \mathbf{u}+\mathbf{v}+\mathbf{w}\} = S$.

31. (a) Let $\mathbf{u}, \mathbf{v}, \mathbf{w}$ be vectors in a vector space V. Show that $\text{sp}\{\mathbf{u}, \mathbf{v}, \mathbf{w}\}$ is the (unique) smallest subspace of V containing $\mathbf{u}$, $\mathbf{v}$ and $\mathbf{w}$. [The corresponding result holds, of course, for any number of given vectors.]
(b) Show that, if S_1 and S_2 are subspaces of the vector space V, then the set $\{\mathbf{s}_1 + \mathbf{s}_2 : \mathbf{s}_1 \in S_1 \text{ and } \mathbf{s}_2 \in S_2\}$ – which we naturally denote by $S_1 + S_2$ – is a subspace of V and is, in fact, the smallest subspace of V which contains both S_1 and S_2.

32. Find two vectors in $\mathbb{R}^4$ which span $S_1 \cap S_2$ where $S_1 = \{(x, y, z, t): x+y+z+t=0\}$ and $S_2 = \{(x, y, z, t): x+2y+3z+4t=0\}$.

33. (a) Let $S_1, S_2, \ldots, S_n$ be a finite collection of subspaces of the vector space V. Prove that their set-theoretic intersection $\bigcap_{i=1}^{n} S_i$ is a subspace of V.
(b) Show that if $S_1 = \{(x, 0): x \in \mathbb{R}\}$ and $S_2 = \{(0, y): y \in \mathbb{R}\}$ then $S_1 \cup S_2$ is not a subspace of $\mathbb{R}^2$. [This set has been mentioned in an earlier question. Which?]

34. Prove the result in Example 5(ii).

35. Show that the space of Exercise 15 of Chapter 6 is not finite dimensional.

36. Find two functions which span the solution space of $\dfrac{d^2y}{dx^2} - 6\dfrac{dy}{dx} + 5y = 0$.

COMPUTER PACKAGE PROBLEMS

1. (a) Let $\mathbf{v}_1 = (535, -294, 237, 167)$, $\mathbf{v}_2 = (628, -362, 258, -135)$, $\mathbf{v}_3 = (-357, 177, -163, -462)$, $\mathbf{v}_4 = (-445, 242, 487, 774)$. Determine whether or not $\mathbf{v}_4$ is a linear combination of $\mathbf{v}_1$, $\mathbf{v}_2$ and $\mathbf{v}_3$. [As an alternative to solving, if possible, the equation

$\mathbf{v}_4 = \alpha\mathbf{v}_1 + \beta\mathbf{v}_2 + \gamma\mathbf{v}_3$, the Maple command 'basis' throws out, from a given list, all those vectors which are linear combinations of their predecessors.]

(b) Which of (267, –333, 368, 299, –213), (270, –177, 277, 379, –113), (235, 101, 100, 469, 80), (–205, –392, 449, –4, 224), (230, 50, 108, 429, 41), (259, –290, –417, 483, –37) are linear combinations of their predecessors?

(c) Show that (1009, 1013, 1019, 1021), (2003, 2011, 2017, 2027), (3001, 3011, 3019, 3023), (4001, 4003, 4007, 4013) span $\mathbb{R}^4$. This can be done in 'Maple' by applying the 'basis' option to the above set augmented by the four vectors (1, 0, 0, 0), (0, 1, 0, 0), (0, 0, 1, 0), (0, 0, 0, 1) – and by more straightforward ways later!

(d) Is $\begin{bmatrix} -162 \\ 214 \\ 311 \\ 405 \end{bmatrix}$ in the range of the matrix $\begin{bmatrix} 117 & 1266 & 142 & 100 & 920 & 105 \\ 410 & 1137 & -931 & 400 & -589 & 433 \\ 314 & 615 & -805 & 310 & -781 & 337 \\ 220 & 303 & -709 & 220 & -531 & 237 \end{bmatrix}$?

2. Let $\mathbf{v}_1 = (21, -41, -103, -164)$ and $\mathbf{v}_2 = (93, 79, 65, 316)$. Which pairs of vectors from the set {(1, 0, 0, 0), (0, 1, 0, 0), (0, 0, 1, 0), (0, 0, 0, 1)}, together with $\mathbf{v}_1$ and $\mathbf{v}_2$, form a spanning set for $\mathbb{R}^4$?

3. Use your computer package to generate several 4×6 matrices. The 'basis' option referred to above gives a spanning set of smallest possible size. Find the number of elements in such a (minimal) spanning set for the row space and likewise for the column space of each matrix. Do the results surprise you?

4. Let A be the matrix

$$\begin{bmatrix} 535 & -294 & 237 & 167 \\ 628 & -362 & 258 & -135 \end{bmatrix}$$

and $\mathbf{x}$ be the vector $(x, y, z, t)^T$. Find a pair of vectors in $\mathbb{R}^4$ which span the solution space of $A\mathbf{x} = \mathbf{0}$.

5. Do the sets of vectors (i) {(499, 937, –246, 1107), (452, 1053, –373, 1054), (365, 82, 179, 729), (477, –1520, 465, –821)} and (ii) {(134, 855, –425, 378), (87, 971, –512, 325), (318, 198, 92, 676), (564, –549, –47, –496), (231, –773, 604, 351)} span the same subspace of $\mathbb{R}^4$? [Use the 'basis' option on 'Maple' – see Problem 1 above – or its equivalent. Enter the nine vectors first by entering set (i) before set (ii) and then set (ii) before set (i).]

6. The Fibonacci sequence is the sequence 1, 1, 2, 3, 5, 8, 13, 21, 34, 55, . . . where each term from the third onwards is the sum of the preceding two. Plot the points (1, 1), (2, 3), (5, 8), (13, 21), . . . etc. on a graph. Then use the 'least squares' option to find the straight line which 'best fits': (i) these four points; (ii) the first n such points where $n = 5, 6, 7, 8, 9, 10$. What about the 'best plane' fitting the points (1, 1, 2), (3, 5, 8), (13, 21, 33), . . . etc.?

8 • Linear Dependence and Independence

Chapter 7 showed the usefulness of being able to express each of an infinite set of 'vectors' linearly in terms of a finite number of them. Here we aim to eliminate duplication of expression via the concept of independence. (We simultaneously introduce the 'dual' notion of dependence.) We then tie this up to Chapter 7 by showing (Theorem 1(a)) that, in every linearly dependent set of vectors there exists at least one which is a linear combination of the others. An easy test for linear (in)dependence follows from Theorem 2 which shows that elementary row operations performed on the rows of a matrix may change the set of vectors but does not change their (in)dependence. Theorem 4, which shows that any set of vectors may be replaced by a linearly independent subset spanning the same subspace, prepares the way for the introduction of 'basis' in Chapter 9. A nice application concerns the deciphering of (partly) garbled messages received from outer space.

In order to put the idea of dimension (see Chapter 9) on a firm footing – in particular to help us understand how many 'really different' solutions there are to the system of equations $A\mathbf{x} = \mathbf{0}$ or to a linear differential equation – we need to introduce the concepts of linear dependence and linear independence of vectors.

In the particular case of vectors in 3-dimensional space these concepts arise as follows. Suppose the vectors $\mathbf{v}_1$, $\mathbf{v}_2$, $\mathbf{v}_3$ in $\mathbb{R}^3$ are such that $\mathbf{v}_1$ and $\mathbf{v}_2$ span a plane (necessarily through the origin). If $\mathbf{v}_3$ lies in this plane then $\mathbf{v}_3$ is expressible as a linear combination of $\mathbf{v}_1$ and $\mathbf{v}_2$: $\mathbf{v}_3 = \alpha_1\mathbf{v}_1 + \alpha_2\mathbf{v}_2$, for suitable reals α_1 and α_2. (In the terminology of Chapter 7 $\mathbf{v}_3$ lies in $\text{sp}\{\mathbf{v}_1, \mathbf{v}_2\}$, the span of $\mathbf{v}_1$ and $\mathbf{v}_2$.) If not, then no such expression is possible. In the former case $\mathbf{v}_3$ is, in some sense, *dependent* on $\mathbf{v}_1$ and $\mathbf{v}_2$ and we may write $\alpha_1\mathbf{v}_1 + \alpha_2\mathbf{v}_2 + (-1)\mathbf{v}_3 = \mathbf{0}$. In the latter case the equality $\alpha_1\mathbf{v}_1 + \alpha_2\mathbf{v}_2 + \alpha_3\mathbf{v}_3 = \mathbf{0}$ is seen to be possible only if $\alpha_1 = \alpha_2 = \alpha_3 = 0$ (and then $\mathbf{v}_1$, $\mathbf{v}_2$ and $\mathbf{v}_3$ are, in the same sense, *independent*).

PROBLEM 8.1

Justify the assertion just made, namely that if $\mathbf{v}_1$, $\mathbf{v}_2$ and $\mathbf{v}_3$ are not coplanar and $\alpha_1\mathbf{v}_1 + \alpha_2\mathbf{v}_2 + \alpha_3\mathbf{v}_3 = \mathbf{0}$, then $\alpha_1 = \alpha_2 = \alpha_3 = 0$. ●

As a particular example we have the following.

Example 1

(i) In $\mathbb{R}^3$: if $\mathbf{v}_1 = (1, 3, 2)$, $\mathbf{v}_2 = (2, 0, -2)$ and $\mathbf{v}_3 = (5, 6, 1)$ then $\mathbf{v}_3$ lies in the plane $x - y + z = 0$ spanned by $\mathbf{v}_1$ and $\mathbf{v}_2$ (so that $\mathbf{v}_3$ is, in the above sense, dependent on the vectors $\mathbf{v}_1$ and $\mathbf{v}_2$). On the other hand, $\mathbf{v}_4 = (5, 6, 2) \notin \text{sp}\{\mathbf{v}_1, \mathbf{v}_2\}$. In fact $\mathbf{v}_1$, $\mathbf{v}_2$ and $\mathbf{v}_3$ satisfy the equality $4\mathbf{v}_1 + 3\mathbf{v}_2 - 2\mathbf{v}_3 = \mathbf{0}$ (so that $\mathbf{v}_3 = 2\mathbf{v}_1 + \frac{3}{2}\mathbf{v}_2 \in \text{sp}\{\mathbf{v}_1, \mathbf{v}_2\}$), whereas the equation $\alpha_1\mathbf{v}_1 + \alpha_2\mathbf{v}_2 + \alpha_4\mathbf{v}_4 = \mathbf{0}$ is satisfied only by choosing $\alpha_1 = \alpha_2 = \alpha_4 = 0$. [This can be checked by setting

$$\alpha_1\begin{bmatrix}1\\3\\2\end{bmatrix}+\alpha_2\begin{bmatrix}2\\0\\-2\end{bmatrix}+\alpha_4\begin{bmatrix}5\\6\\2\end{bmatrix}=\begin{bmatrix}0\\0\\0\end{bmatrix}$$

from which we obtain

$$\begin{array}{rcrcrcl} 1\alpha_1 & + & 2\alpha_2 & + & 5\alpha_3 & = & 0\\ 3\alpha_1 & & & + & 6\alpha_3 & = & 0\\ 2\alpha_1 & - & 2\alpha_2 & + & 2\alpha_3 & = & 0 \end{array}$$

the only solution of which is $\alpha_1 = \alpha_2 = \alpha_3 = 0$. Other methods will emerge later.]

(ii) The 'vector' (i.e. polynomial) $1 + x + x^2 + x^3$ is dependent upon the 'vectors' (i.e. polynomials) 1, $2 + 3x + 4x^2 + 5x^3$, $3 - 4x - 7x^2 + 2x^3$ and $x + x^3$ in $\mathbb{R}[x]$ because, for example,

$$16(1) + 3(2 + 3x + 4x^2 + 5x^3) - 1(3 - 4x - 7x^2 + 2x^3) + 6(x + x^3) - 19(1 + x + x^2 + x^3) = Z$$

(the zero polynomial).

PROBLEM 8.2

Let $\mathbf{v}_1 = (3, 1, 2)$, $\mathbf{v}_2 = (4, 2, 3)$, $\mathbf{v}_3 = (0, -2, -1)$, $\mathbf{v}_4 = (1, 1, 3) \in \mathbb{R}^3$. Show that $\mathbf{v}_3 \in \text{sp}\{\mathbf{v}_1, \mathbf{v}_2, \mathbf{v}_4\}$ but that $\mathbf{v}_4 \notin \text{sp}\{\mathbf{v}_1, \mathbf{v}_2, \mathbf{v}_3\}$. Describe this situation geometrically.

From the above ideas we formulate a general definition.

Definition 1

Let V be a vector space over a given set of scalars $\mathbb{F}$. A non-empty set $\{\mathbf{v}_1, \mathbf{v}_2, \ldots, \mathbf{v}_s\}$ of vectors of V is said to be **linearly dependent** (over $\mathbb{F}$) if (and only if) there are, in $\mathbb{F}$, scalars $\alpha_1, \alpha_2, \ldots, \alpha_s$, *not all zero*, such that $\alpha_1\mathbf{v}_1 + \alpha_2\mathbf{v}_2 + \ldots + \alpha_s\mathbf{v}_s = \mathbf{0}$. Otherwise (that is, if the equation $\alpha_1\mathbf{v}_1 + \alpha_2\mathbf{v}_2 + \ldots + \alpha_s\mathbf{v}_s = \mathbf{0}$ has, as its only solution, $\alpha_1 = \alpha_2 = \ldots = \alpha_s = 0$) the set $\{\mathbf{v}_1, \mathbf{v}_2, \ldots \mathbf{v}_s\}$ is said to be **linearly independent** (over $\mathbb{F}$).

For convenience we shall write LI and LD as abbreviations for these two conditions. Let us look immediately at some more examples.

Examples 2

(i) In $\mathbb{R}^4$ let $\mathbf{u}_1 = (-1, 1, 1, 1)$, $\mathbf{u}_2 = (1, -1, 1, 1)$, $\mathbf{u}_3 = (1, 1, -1, 1)$ and $\mathbf{u}_4 = (1, 1, 1, -1)$. Then $\{\mathbf{u}_1, \mathbf{u}_2, \mathbf{u}_3, \mathbf{u}_4\}$ is an LI set in $\mathbb{R}^4$ since, as you can easily check, from $\alpha_1\mathbf{u}_1 + \alpha_2\mathbf{u}_2 + \alpha_3\mathbf{u}_3 + \alpha_4\mathbf{u}_4 = (0, 0, 0, 0)$ we are forced to conclude that $\alpha_1 = \alpha_2 = \alpha_3 = \alpha_4 = 0$. On the other hand, if $\mathbf{v}_1 = (-1, -1, 1, 1)$, $\mathbf{v}_2 = (1, -1, -1, 1)$, $\mathbf{v}_3 = (1, 1, -1, -1)$ and $\mathbf{v}_4 = (-1, 1, 1, -1)$ then $\{\mathbf{v}_1, \mathbf{v}_2, \mathbf{v}_3, \mathbf{v}_4\}$ is an LD set since, for example, $1\mathbf{v}_1 + 0\mathbf{v}_2 + 1\mathbf{v}_3 + 0\mathbf{v}_4 = \mathbf{0}$.

(ii) The matrices

$$\begin{bmatrix} -1 & 1 \\ 1 & 1 \end{bmatrix}, \begin{bmatrix} 1 & -1 \\ 1 & 1 \end{bmatrix}, \begin{bmatrix} 1 & 1 \\ -1 & 1 \end{bmatrix}, \begin{bmatrix} 1 & 1 \\ 1 & -1 \end{bmatrix}$$

form an LI subset of $M_2(\mathbb{R})$. On the other hand

$$\begin{bmatrix} 1 & -1 \\ -1 & 1 \end{bmatrix}, \begin{bmatrix} 1 & 1 \\ -1 & -1 \end{bmatrix}, \begin{bmatrix} -1 & 1 \\ 1 & -1 \end{bmatrix}, \begin{bmatrix} -1 & -1 \\ 1 & 1 \end{bmatrix}$$

form an LD set. (Am I right? You should be able to check almost immediately!)

(iii) The general solution of the linear system

$$\begin{aligned} x+2y-z+4t&=0 \\ 3x+5y+2z+8t&=0 \\ 2x+5y-7z+12t&=0 \end{aligned}$$

is of the form $(-9a+4b, 5a-4b, a, b)$. Consequently the (vector) space of all solutions is spanned by $\mathbf{v}_1 = (-9, 5, 1, 0)$ and $\mathbf{v}_2 = (4, -4, 0, 1)$. Clearly $\{\mathbf{v}_1, \mathbf{v}_2\}$ is an LI subset of $\mathbb{R}^4$. (Clearly? Well, from $\alpha_1\mathbf{v}_1 + \alpha_2\mathbf{v}_2 = \mathbf{0}$ $\{= (0, 0, 0, 0)\}$ you should be able to deduce that $\alpha_1 = \alpha_2 = 0$.)

(iv) In the real vector space $\mathbb{C}$ of complex numbers (see Example 11 of Chapter 6), the set (of 'vectors') $\{1, \mathrm{i}\}$ is an LI set over $\mathbb{R}$, since the equation $r_1 1 + r_2\mathrm{i} = 0$ (where $r_1, r_2 \in \mathbb{R}$) forces $r_1 = r_2 = 0$.

(v) The solution pair $\{\mathrm{e}^{-x}, \mathrm{e}^{-3x}\}$ of the differential equation

$$\frac{\mathrm{d}^2y}{\mathrm{d}x^2} + 4\frac{\mathrm{d}y}{\mathrm{d}x} + 3y = 0$$

forms an LI set. For, if α_1 and α_2 are real numbers such that $\alpha_1\mathrm{e}^{-x} + \alpha_2\mathrm{e}^{-3x} = Z$, the zero function (so that $Z(x) = 0$ for all x), then, on putting, say, $x = 0$ and then $x = 1$, we obtain $\alpha_1 1 + \alpha_2 1 = 0$ and $\alpha_1\mathrm{e}^{-1} + \alpha_2\mathrm{e}^{-3} = 0$, from which $\alpha_1 = \alpha_2 = 0$ immediately. Similarly (I) the pair $\{\mathrm{e}^{-x}, 2\,\mathrm{e}^{-x} + 71\,\mathrm{e}^{-3x}\}$ of functions forms an LI set but (II) the triple $\{\mathrm{e}^{-x}, \mathrm{e}^{-3x}, 2\,\mathrm{e}^{-x} + 71\,\mathrm{e}^{-3x}\}$ of functions does not. [(I) is most easily checked using Theorem 2 below; (II) is immediate.]

PROBLEM 8.3

Show, similarly, that the set of functions $\{\mathrm{e}^{x}, \mathrm{e}^{2x}, \mathrm{e}^{3x}\}$ is an LI set.

Examples 2 (cont.)

(iv) (a) The single element subset $\{\mathbf{v}\}$ of the vector space V is LI if and only if $\mathbf{v}$ is not the zero vector of V; (b) the set $\{\mathbf{v}_1, \mathbf{v}_2\}$ comprising two non-zero vectors is LI iff neither $\mathbf{v}_1$ nor $\mathbf{v}_2$ is a scalar multiple of the other. (See Exercise 18(a) and (b).)

As above, we may always determine the dependence or independence of a given set of vectors by applying Definition 1 directly. For example, the coefficients used in Example 1(ii) can be obtained by rewriting

$$a(1)+b(2+3x+4x^2+5x^3)+c(3-4x-7x^2+2x^3)+d(x+x^3)+e(1+x+x^2+x^3)=Z$$

as

$$(a+2b+3c+e)+(3b-4c+d+e)x+\ldots+(5b+2c+d+e)x^3=Z$$

so that

$$a+2b+3c+e=3b-4c+d+e=\ldots=5b+2c+d+e=0,$$

and finding a particular solution for the resulting system of equations.

A useful variation of Definition 1 (cf. our opening paragraph) is given by the following theorem.

• Theorem 1

(a) The set $\{\mathbf{v}_1, \mathbf{v}_2, \ldots, \mathbf{v}_s\}$ of vectors in the vector space V is LD if and only if some vector $\mathbf{v}_i$, say, is a linear combination $\beta_1\mathbf{v}_1 + \beta_2\mathbf{v}_2 + \ldots + \beta_{i-1}\mathbf{v}_{i-1}$ of its predecessors (and then $\mathrm{sp}\{\mathbf{v}_1, \ldots, \mathbf{v}_i, \ldots, \mathbf{v}_s\} = \mathrm{sp}\{\mathbf{v}_1, \ldots, \hat{\mathbf{v}}_i, \ldots, \mathbf{v}_s\}$).
[^ is the standard symbol used to indicate that what is placed below it has been removed from the list.]

(b) The set $\{\mathbf{v}_1, \mathbf{v}_2, \ldots, \mathbf{v}_s\}$ of vectors is LI if and only if each vector in $\mathrm{sp}\{\mathbf{v}_1, \mathbf{v}_2, \ldots, \mathbf{v}_s\}$ is expressible *in a unique way* as a linear combination of $\mathbf{v}_1, \mathbf{v}_2, \ldots, \mathbf{v}_s$. ●

PROBLEM 8.4

Let $\mathbf{v}_1 = (1, 3, 5)$, $\mathbf{v}_2 = (2, 4, 6)$ and $\mathbf{v}_3 = (9, 9, 9)$. Show that $(1, 1, 1) \in \mathrm{sp}\{\mathbf{v}_1, \mathbf{v}_2\}$ and deduce that the set $\{\mathbf{v}_1, \mathbf{v}_2, \mathbf{v}_3\}$ is LD. ●

We offer you a 'proof' of part (a) of Theorem 1 to show you how such an (obvious?) assertion might formally be confirmed – and also to test your critical faculties. The 'proof' is incomplete. Why?

'PROOF'

(a) (if) Assume $\mathbf{v}_i = \beta_1\mathbf{v}_1 + \ldots + \beta_{i-1}\mathbf{v}_{i-1}$. Then

$$\beta_1\mathbf{v}_1+\ldots+\beta_{i-1}\mathbf{v}_{i-1}+(-1)\mathbf{v}_i+0\mathbf{v}_{i+1}+\ldots+0\mathbf{v}_s=\mathbf{0}.$$

Thus, since $-1 \neq 0$, $\{\mathbf{v}_1, \mathbf{v}_2, \ldots, \mathbf{v}_s\}$ is, by definition, an LD set.
(only if) Assume $\alpha_1\mathbf{v}_1 + \ldots + \alpha_s\mathbf{v}_s = \mathbf{0}$ with not all $\alpha_i = 0$. Let α_k be the 'last' non-zero α (i.e. let k be the largest integer such that $\alpha_k \neq 0$). Then

$$\mathbf{v}_k=-\frac{\alpha_1}{\alpha_k}\mathbf{v}_1-\ldots-\frac{\alpha_{k-1}}{\alpha_k}\mathbf{v}_{k-1}.$$

(b) We leave this to the reader (Exercise 25(a)). ●

TUTORIAL PROBLEM 8.1

The statement in (a), whilst easy to remember as written, is not quite good enough. Can you see why not? [The answer is given in the exercises.]

We make the following immediate observations, all sets mentioned being finite and non-empty:

(i) any set S containing the zero vector is necessarily LD;
(ii) any subset T of an LI set S is LI;
(iii) if a subset T of a set S is LD then so is S.

PROOF

(i) Suppose $S = \{\mathbf{0}, \mathbf{v}_1, \ldots, \mathbf{v}_s\}$. Let α be your favourite (non-zero) number. Then $\alpha\mathbf{0} + 0\mathbf{v}_1 + \ldots + 0\mathbf{v}_s = \mathbf{0}$. So, immediately, we see that S is an LD set.

We leave the proofs of (ii) and (iii) to the exercises.

TUTORIAL PROBLEM 8.2

Ought we to regard the empty set of vectors as LI, LD, both or neither?

A different way of checking some of Examples 1 and 2, without directly introducing systems of equations, is via the following lovely theorem. The theorem shows that linear dependence/independence of a collection of vectors is preserved if (i) the vectors are given in a different order or if (ii) one (and hence any number of the vectors) is modified by multiplication by a non-zero scalar or if (iii) any given vector is modified by adding to it any multiple of any other of the given vectors.

• *Theorem 2*

Let $S_0 = \{\mathbf{v}_1, \ldots, \mathbf{v}_i, \ldots, \mathbf{v}_j, \ldots, \mathbf{v}_s\}$ be a set of vectors in the vector space V. Let

$$\begin{aligned} S_1 &= \{\mathbf{v}_1,\ldots,\mathbf{v}_j,\ldots,\mathbf{v}_i,\ldots,\mathbf{v}_s\} \\ S_2 &= \{\mathbf{v}_1,\ldots,\beta\mathbf{v}_i,\ldots,\mathbf{v}_j,\ldots,\mathbf{v}_s\} \quad (\beta \neq 0, \text{ a scalar}) \\ S_3 &= \{\mathbf{v}_1,\ldots,\mathbf{v}_i+\gamma\mathbf{v}_j,\ldots,\mathbf{v}_j,\ldots,\mathbf{v}_s\} \quad (\gamma \text{ any scalar}). \end{aligned}$$

Then

(i) each of the (four) S_k spans the same subspace of V as any other;
(ii) either all the S_k are LI sets or all are LD sets.

QUESTION Do the changes made in passing from the set S_0 to the sets S_1, S_2 and S_3 above remind you of anything? Answer: Elementary, my dear Watson.[a]

PROOF

(For S_0 and S_3) If $\mathbf{v} = \alpha_1\mathbf{v}_1 + \ldots + \alpha_i\mathbf{v}_i + \ldots + \alpha_j\mathbf{v}_j + \ldots + \alpha_s\mathbf{v}_s \in \mathrm{sp}(S_0)$ then

$$\mathbf{v} = \alpha_1\mathbf{v}_1 + \ldots + \alpha_i(\mathbf{v}_i + \gamma\mathbf{v}_j) + \ldots + (\alpha_j - \alpha_i\gamma)\mathbf{v}_j + \ldots + \alpha_s\mathbf{v}_s \in \mathrm{sp}(S_3).$$

Hence $\mathrm{sp}(S_0) \subseteq \mathrm{sp}(S_3)$.

Conversely, if $\mathbf{w} = k_1\mathbf{v}_1 + \ldots + k_i(\mathbf{v}_i + \gamma\mathbf{v}_j) + \ldots + k_j\mathbf{v}_j + \ldots + k_s\mathbf{v}_s \in \mathrm{sp}(S_3)$ then

$$\mathbf{w} = k_1\mathbf{v}_1 + \ldots + k_i\mathbf{v}_i + \ldots + (k_i\gamma + k_j)\mathbf{v}_j + \ldots + k_s\mathbf{v}_s \in \mathrm{sp}(S_0).$$

[a]Yes, I know, . . . Sherlock Holmes never uttered this particular sequence of words!

Hence $\mathrm{sp}(S_3) \subseteq \mathrm{sp}(S_0)$. We therefore have $\mathrm{sp}(S_0) = \mathrm{sp}(S_3)$, as claimed.

To prove (ii) is somewhat easier. Suppose that S_0 is an LI set. If

$$c_1\mathbf{v}_1 + \ldots + c_i(\mathbf{v}_i + \gamma\mathbf{v}_j) + \ldots + c_j\mathbf{v}_j + \ldots c_s\mathbf{v}_s = 0 \tag{8.1}$$

then

$$c_1\mathbf{v}_1 + \ldots + c_i\mathbf{v}_i + \ldots + (c_i\gamma + c_j)\mathbf{v}_j + \ldots + c_s\mathbf{v}_s = \mathbf{0}.$$

We deduce that

$$c_1 = \ldots = c_i = \ldots = (c_i\gamma + c_j) = \ldots = c_s = 0 \text{ (why?)}$$

and hence that $c_j = 0$ (how?). Therefore, from (8.1), S_3 is an LI set. We leave the other half of (ii) for you to prove. ●

This theorem is quite helpful in practice.

Example 3

(i) Test the set $\{\mathbf{v}_1, \mathbf{v}_2, \mathbf{v}_3\}$ for linear independence in $\mathbb{R}^5$ where

$$\mathbf{v}_1 = (2,0,-2,4,-6), \qquad \mathbf{v}_2 = (1,3,2,-3,4) \quad \text{and} \quad \mathbf{v}_3 = (5,6,1,0,-1).$$

Form successively, by elementary row operations, the matrices

$$\begin{bmatrix} 2 & 0 & -2 & 4 & -6 \\ 1 & 3 & 2 & -3 & 4 \\ 5 & 6 & 1 & 0 & -1 \end{bmatrix} \xrightarrow{\rho_1 \leftrightarrow \rho_2} \begin{bmatrix} 1 & 3 & 2 & -3 & 4 \\ 2 & 0 & -2 & 4 & -6 \\ 5 & 6 & 1 & 0 & -1 \end{bmatrix} \begin{matrix} \rho_2 \to \rho_2 - 2\rho_1 \\ \to \\ \rho_3 \to \rho_3 - 5\rho_1 \end{matrix} \begin{bmatrix} 1 & 3 & 2 & -3 & 4 \\ 0 & -6 & -6 & 10 & -14 \\ 0 & -9 & -9 & 15 & -21 \end{bmatrix}$$

$$\xrightarrow{\rho_2 \to -\frac{1}{6}\rho_2} \begin{bmatrix} 1 & 3 & 2 & -3 & 4 \\ 0 & 1 & 1 & -\frac{5}{3} & \frac{7}{3} \\ 0 & -9 & -9 & 15 & -21 \end{bmatrix} \xrightarrow{\rho_3 \to \rho_3 + 9\rho_2} \begin{bmatrix} 1 & 3 & 2 & -3 & 4 \\ 0 & 1 & 1 & -\frac{5}{3} & \frac{7}{3} \\ 0 & 0 & 0 & 0 & 0 \end{bmatrix}.$$

But $\{(1, 3, 2, -3, 4), (0, 1, 1, -\frac{5}{3}, \frac{7}{3}), (0, 0, 0, 0, 0)\}$ is clearly an LD set in $\mathbb{R}^5$ (why?). Hence the original three vectors form an LD set in $\mathbb{R}^5$. [Note that the equality $3\mathbf{v}_1 + 4\mathbf{v}_2 - 2\mathbf{v}_3 = \mathbf{0}$ is not so immediate by this method.]

(ii) Test[b] $\{(-1, 1, 1, 1), (1, -1, 1, 1), (1, 1, -1, 1), (1, 1, 1, -1)\}$ for linear independence. Here, elementary row operations reduce the given matrix to

$$\begin{bmatrix} -1 & 1 & 1 & 1 \\ 0 & 1 & 1 & 0 \\ 0 & 0 & -1 & 1 \\ 0 & 0 & 0 & 1 \end{bmatrix}.$$

(Remember that echelon form is not (usually) unique so you may obtain a different echelon form from that given here. You should, of course reach the same conclusion regarding LI/LD as I have.)

The claim now is that these vectors form an LI set. Indeed I can 'see' immediately that they do since it is clearly impossible for any row of the matrix to be a linear

[b]Cf. the first four matrices in Example 2(ii) – it is essentially the same example.

combination of any of the rows below it. Theorem 1(a) is then immediately applicable. Alternatively suppose that

$$\alpha(-1,1,1,1)+\beta(0,1,1,0)+\gamma(0,0,-1,1)+\delta(0,0,0,1)=(0,0,0,0). \qquad (8.2)$$

By comparing first components on each side of the equality we see that $\alpha(-1)+\beta 0+\gamma 0+\delta 0=0$. Thus $\alpha=0$. Feeding this information into (8.2) and looking at second components in (8.2) gives $0(1)+\beta 1+\gamma 0+\delta 0=0$. Thus $\beta=0$. Continuing in this way we next find $\gamma=0$ and then $\delta=0$.

And finally:

(iii) Do {(2, 10, 1), (1, 7, 6), (1, 3, 0), (2, 3, −1)} span $\mathbb{R}^3$? Here we first swap the first two vectors, merely for convenience, and then reduce the resulting matrix

$$\begin{bmatrix}1 & 7 & 6\\ 2 & 10 & 1\\ 1 & 3 & 0\\ 2 & 3 & -1\end{bmatrix} \text{ to echelon form obtaining } \begin{bmatrix}1 & 7 & 6\\ 0 & -4 & -11\\ 0 & 0 & 5\\ 0 & 0 & 0\end{bmatrix}.$$

But the first three rows of this matrix already span $\mathbb{R}^3$ (Exercise 14(a)). Consequently so do the given vectors.

PROBLEM 8.5

Test, by the method of (ii) above, the set {(1, 1, 1, 1), (1, 2, 2, 2), (1, 2, 3, 3), (1, 2, 3, 4)} for LI/LD.

We isolate here the property displayed in Examples 3(ii) and make a nice deduction from it.

• *Theorem 3*

(a) Let B be any $m \times n$ matrix in echelon form. Then the non-zero rows of B form an LI set.

(b) The $n \times n$ matrix A is invertible if and only if its rows, regarded as vectors in $\mathbb{R}^n$, form an LI set.

We ask you to prove Theorem 3 as Exercise 27.

Examples 3 show how, given any set of vectors in $\mathbb{R}^3$ (or, more generally, $\mathbb{R}^n$) one may find a linearly independent set of vectors which span the same subspace: we form a matrix with the given vectors as its rows, and then reduce this matrix to echelon form. By Theorem 2(i), the non-zero rows of the reduced matrix have the same span as the given set of vectors. And the echelon form itself ensures, as in Theorem 3(a), that the non-zero rows are LI. Notice, however, that, in general, most of the original vectors are replaced by new ones. For instance, in Example 3(ii) the given vectors (−1, 1, 1, 1), (1, −1, 1, 1), (1, 1, −1, 1), (1, 1, 1, −1) are replaced by the vectors (−1, 1, 1, 1), (0, 1, 1, 0) (0, 0, −1, 1) and (0, 0, 0, 1). In fact we can do better than this as the next result shows.

• *Theorem 4*

Given vectors $\{\mathbf{v}_1, \mathbf{v}_2, \ldots, \mathbf{v}_s\}$ in a vector space V, there exists an LI subset of them which spans the same subspace of V.

[Informally the proof says: go through the given list of vectors and kick out (i) every zero vector and then (ii) every one which is a linear combination of its predecessors. What is left over is what is required!]

PROOF

If $\{\mathbf{v}_1\}$ is an LD subset [in other words if $\mathbf{v}_1 = \mathbf{0}$] throw $\mathbf{v}_1$ away. Otherwise keep it. In general: suppose we already know that $\mathrm{sp}\{\mathbf{v}_1, \ldots, \mathbf{v}_t\} = \mathrm{sp}\{\mathbf{v}_{i_1}, \ldots, \mathbf{v}_{i_k}\}$ where $\{\mathbf{v}_{i_1}, \ldots, \mathbf{v}_{i_k}\}$ is an LI subset of $\{\mathbf{v}_1, \ldots, \mathbf{v}_t\}$. Then:

(i) if $\mathbf{v}_{t+1} \in \mathrm{sp}\{\mathbf{v}_1, \ldots, \mathbf{v}_t\} = \mathrm{sp}\{\mathbf{v}_{i_1}, \ldots, \mathbf{v}_{i_k}\}$ throw away $\mathbf{v}_{t+1}$;

(ii) if $\mathbf{v}_{t+1} \notin \mathrm{sp}\{\mathbf{v}_1, \ldots, \mathbf{v}_t\} = \mathrm{sp}\{\mathbf{v}_{i_1}, \ldots, \mathbf{v}_{i_k}\}$ retain $\mathbf{v}_{t+1}$.

In case (ii) $\mathrm{sp}\{\mathbf{v}_1, \mathbf{v}_2, \ldots \mathbf{v}_t, \mathbf{v}_{t+1}\} = \mathrm{sp}\{\mathbf{v}_{i_1}, \mathbf{v}_{i_2}, \ldots, \mathbf{v}_{i_k}, \mathbf{v}_{t+1}\}$ and the set $\{\mathbf{v}_{i_1}, \mathbf{v}_{i_2}, \ldots, \mathbf{v}_{i_k}, \mathbf{v}_{t+1}\}$ is LI. [For, if it were not, then, by Theorem 1(a), one of the $\mathbf{v}_{i_1}, \mathbf{v}_{i_2}, \ldots, \mathbf{v}_{i_k}, \mathbf{v}_{t+1}$ would be a linear combination of its predecessors. But none of the $\mathbf{v}_{i_r}$ $(1 \le r \le k)$ is such a linear combination (why not?) and nor, by choice, is $\mathbf{v}_{t+1}$ a linear combination of the $\mathbf{v}_{i_1}, \mathbf{v}_{i_2}, \ldots, \mathbf{v}_{i_k}$.] ●

TUTORIAL PROBLEM 8.3

If all the $\mathbf{v}_i = \mathbf{0}$ it appears, from Theorem 4, that the zero subspace is spanned by . . . the empty set! Is this reasonable?

The following example shows how Theorem 4 works in practice.

Example 4

Find a subset of the vectors (1, –2, 3, 1), (2, 2, –2, 1), (5, 2, –1, 3), (11, 2, 1, 7), (2, 8, 2, 3) which is LI and *spans the same subspace of* $\mathbb{R}^4$.

Starting with

$$\begin{bmatrix} 1 & -2 & 3 & 1 \\ 2 & 2 & -2 & 1 \\ 5 & 2 & -1 & 3 \\ 11 & 2 & 1 & 7 \\ 2 & 8 & 2 & 3 \end{bmatrix},$$

use row operations to obtain successively

$$\begin{bmatrix} 1 & -2 & 3 & 1 \\ 0 & 6 & -8 & -1 \\ 0 & 12 & -16 & -2 \\ 0 & 24 & -32 & -4 \\ 0 & 12 & -4 & 1 \end{bmatrix} \text{ and } \begin{bmatrix} 1 & -2 & 3 & 1 \\ 0 & -6 & -8 & -1 \\ 0 & 0 & 0 & 0 \\ 0 & 0 & 0 & 0 \\ 0 & 0 & 12 & 3 \end{bmatrix}.$$

Clearly rows 1, 2 and 5 are LI here. Thus the same is true of the original vectors. That is, {(1, –2, 3, 1), (2, 2, –2, 1), (2, 8, 2, 3)} is an LI subset of the original set of vectors.

[Note that, for this method to succeed, each row operation of type S_3 (in Theorem 2) must only modify a row by adding a multiple of an earlier one. Can you see why?]

An alternative procedure to that given above is as follows:

(i) write the given vectors as the columns of a matrix:

$$\begin{bmatrix} 1 & 2 & 5 & 11 & 2 \\ -2 & 2 & 2 & 2 & 8 \\ 3 & -2 & -1 & 1 & 2 \\ 1 & 1 & 3 & 7 & 3 \end{bmatrix};$$

(ii) apply row operations to (row) echelon this matrix, to get

$$\begin{bmatrix} 1 & 2 & 5 & 11 & 2 \\ 0 & 6 & 12 & 24 & 12 \\ 0 & 0 & 0 & 0 & 12 \\ 0 & 0 & 0 & 0 & 0 \end{bmatrix};$$

(iii) observe that the non-zero rows 'begin' in columns 1, 2 and 5. 'Therefore', of the original vectors, those required are the first, second and fifth!

TUTORIAL PROBLEM 8.4

Why does this work? (The answer is in Exercise 31.)

It is not particularly easy to separate comments on (in)dependence from remarks on spanning (Chapter 7), especially when one knows that a minimal spanning set must also be linearly independent (see Exercise 25(b)) so perhaps it is no surprise that when Euler gave his general solution to a homogeneous linear differential equation of order n as a linear combination of n particular functions, he forgot to require explicitly that the n particular functions be linearly independent.

Early writers on the theory of linear systems often considered $(n \times n)$ systems where there are equal numbers of equations and unknowns. Because they had no notion of the (in)dependence of equations they were often reduced to saying, in cases where no unique solution existed, that 'the problem set is ill-posed'!

The concept of linear independence is so clear in Hamilton's quaternions (see p. 200: surely no non-trivial linear combination of **I**, **i**, **j** and **k** can be zero?) that the concept does not need emphasising. The notion is, however, present in Grassmann's (1862) *Ausdehnungslehre* and explicitly defined in Peano's work of 1888. The idea of linear (in)dependence is also very much present or explicitly defined in the number-theoretic works of Weierstrass and Dedekind towards the end of the 19th century.

An intriguing problem, clearly concerned in some way with linear dependence, is what is called Cramer's paradox. (It was known earlier to Euler.) In general, two plane curves $S_1(x, y) = 0$ and $S_2(x, y) = 0$ each of the degree n will meet in n^2 points. But we noted in Chapter 5 that an nth degree curve can be specified by any $[n(n + 3)]/2$ points on it. Now, for $n > 3$, $n^2 > [n(n + 3)]/2$ so there will be 'too many' equations to be satisfied if we try to find other curves passing through the n^2 points of intersection.

Application

Coded messages sent back from space probes are conveniently transmitted as sequences or 'words' of 0s and 1s, for example 0100 1110 . . . . In transmission, 'noise' may interfere with the signal so that an 'incorrect' codeword is received. To counter this one may send each codeword in triplicate (so that, if 010001100100 is received, one might reasonably infer that 0100 was sent). But multiple repeating is expensive and one wishes to find a better way of detecting (and correcting) errors than by mere brute force.

One way to do this is to use a **generator matrix**, for example

$$G = \begin{bmatrix} I_4 \\ A \end{bmatrix},$$

where, say, A is the 3×4 matrix

$$\begin{bmatrix} 1 & 1 & 0 & 1 \\ 1 & 0 & 1 & 1 \\ 0 & 1 & 1 & 1 \end{bmatrix}.$$

We send the message, say, $\mathbf{m} = [0 \quad 1 \quad 1 \quad 0]^T$ as the codeword $G\mathbf{m} = [0 \quad 1 \quad 1 \quad 0 \quad 1 \quad 1 \quad 0]^T$ (the matrix product of G and $\mathbf{m}$) where we treat 0 and 1 as ordinary integers except that we define 1 + 1 to be 0. [That is, we are using the integers 'mod 2', just like the 24-hour clock uses the integers mod 24.] Thus the last entry in $G\mathbf{m}$ is

$$0.0 + 1.1 + 1.1 + 1.0 = 0 + 1 + 1 + 0 = 0.$$

In fact, for each n, the set of all such n-tuples of 0s and 1s turns out to be a vector space using the integers mod 2 as scalars. Thus the whole of vector space theory is available to us.

Suppose we let $R\mathbf{m}$ be the 7-tuple received when $G\mathbf{m}$ above is sent. If $R\mathbf{m} = G\mathbf{m}$ (though, of course, we cannot know if it is or not!) then the first four symbols of $R\mathbf{m}$ are $\mathbf{m}$ itself. If $R\mathbf{m} \neq G\mathbf{m}$ we have to try and recover $\mathbf{m}$ from the garbled received word $R\mathbf{m}$. We do this using the **parity check** matrix $P = [A \quad \mathbf{I}_3]$. If $R\mathbf{m} = G\mathbf{m}$ we see that

$$P(R\mathbf{m}) = P(G\mathbf{m}) = (PG)\mathbf{m} = (A + A)\mathbf{m} = 0_{3\times 4}\,\mathbf{m} = \begin{bmatrix} 0 \\ 0 \\ 0 \\ 0 \end{bmatrix}.$$

Thus the parity check matrix tells us whether or not the received word $R\mathbf{m}$ is a legitimate codeword. If it is – and if the entries in every two distinct codewords differ in at least k places (so that every two $G\mathbf{m}$ are *distance* at least k apart) – we may reasonably infer, if k is large, that the received word $R\mathbf{m}$ is identical to the codeword sent, namely $G\mathbf{m}$. If $P(R\mathbf{m}) \neq 0$ we know that $R\mathbf{m}$ is not an official codeword. Nevertheless, by studying the (so-called syndrome) vector $P(R\mathbf{m})$ we may – if k is large enough – conclude which codeword $G\mathbf{m}$ is the most likely to have been sent.

Unfortunately there is no room to develop this here except to say that of all possible codes these **linear** (or **group**) codes in which the n-symbol codewords form a subspace of the vector space of all n-tuples of 0s and 1s are very agreeable to work with. Nevertheless there is, even here, an unsolved problem. It can be shown that the minimum distance k between codewords in a linear code is equal to the least number of columns of the parity check matrix which can form an LD set of vectors. Apparently no uniform way is known of determining this number efficiently in all cases.

Incidentally the code determined by G above is the **Hamming (7,4) code**. It is single error correcting and considerably more efficient than the triply repeated code of the first paragraph.

Summary

A non-empty set $C = \{\mathbf{v}_1, \mathbf{v}_2, \ldots, \mathbf{v}_s\}$ of vectors in a vector space V is **linearly independent** (LI) if and only if the only scalars α_i for which $\alpha_1\mathbf{v}_1 + \alpha_2\mathbf{v}_2 + \ldots + \alpha_s\mathbf{v}_s = \mathbf{0}$ are all 0. C is **linearly dependent** (LD) if and only if C is not LI. One can show that C is LD if and only if either C contains the zero vector $\mathbf{0}_v$ or (at least) one of its members is a linear combination of its predecessors.

The set R, say, of row vectors of the $m \times n$ matrix A can be tested for linear (in)dependence by reducing the matrix to echelon form. If one (and hence all – see Chapter 9) of the echelon forms of A has (at least) one row of zeros then R is LD. Otherwise it is LI. Even if R *is* LD one can extract from R an LI subset of R which spans the same subspace of $\mathbb{R}^n$. [This result is provable just as easily for any vector space.]

EXERCISES ON CHAPTER 8

1. Are the following sets of vectors LD or are they LI?
In $\mathbb{R}^3$: (a) $\{(0, -1, 1), (1, 0, -1), (-1, 1, 0)\}$; (b) $\{(0, 1, 1), (1, 0, 1), (1, 1, 0)\}$;
(c) $\{(1, 2, 3), (4, 5, 6), (7, 8, 9)\}$; (d) $\{(-1, 2, -3), (4, -5, 6), (-7, 8, -9)\}$;
(e) $\{(1, 2, 3), (4, 5, 6), (9, 8, 7)\}$; (f) $\{(\sqrt{1}, \sqrt{2}, \sqrt{3}), (\sqrt{4}, \sqrt{5}, \sqrt{6}), (\sqrt{7}, \sqrt{8}, \sqrt{9})\}$.
In $\mathbb{R}^5$: (g) $\{(1, 1, 1, 1, 1), (1, 1, 1, 1, 0), (1, 1, 1, 0, 0), (1, 1, 0, 0, 0), (1, 0, 0, 0, 0)\}$.

2. (a) For which scalars k is $\{(1, -1, 2), (2, -3, k), (4, 2, 1)\}$ an LI set?
(b) For which scalars k is $\{(k, 1, 0), (1, 0, k), (0, k, 1)\}$ an LI set?

3. Are the following sets of vectors (polynomials) in $\mathbb{R}[x]$ LD or are they LI?
(a) $\{-x + x^2, 1 - x^2, -1 + x\}$; (b) $\{x + x^2, 1 + x^2, 1 + x\}$; (c) $\{x\}$;
(d) $\{3 + 8x + 2x^2 + 2x^3, 1 + 2x - 2x^2 + 2x^3, 1 - 2x + 3x^2 + x^3\}$.

4. Is the subset $\left\{\begin{bmatrix}1&1&0\\1&1&0\\0&0&0\end{bmatrix}, \begin{bmatrix}0&1&1\\0&1&1\\0&0&0\end{bmatrix}, \begin{bmatrix}0&0&0\\1&1&0\\1&1&0\end{bmatrix}, \begin{bmatrix}0&0&0\\0&1&1\\0&1&1\end{bmatrix}, \begin{bmatrix}1&0&0\\0&1&0\\0&0&1\end{bmatrix}\right\}$ of $M_3(\mathbb{R})$ LI?

5. Write all solutions of

$$\begin{aligned}2x - y + 2z - 6t &= 0\\ 7x + 4y - 2z - 9t &= 0\\ x + 2y - 2z + t &= 0\end{aligned}$$

as a linear combination of two solutions $\mathbf{v}_1$, $\mathbf{v}_2$ where $\{\mathbf{v}_1, \mathbf{v}_2\}$ is an LI set.

6. (a) Show that the vectors $\mathbf{v}_1 = (2, 3, 1, 4)$, $\mathbf{v}_2 = (1, -2, 0, 2)$ and $\mathbf{v}_3 = (3, 0, 1, 5)$ form an LI set in $\mathbb{R}^4$. Find, if possible, a fourth vector $\mathbf{v}_4 = (a, b, c, d)$ in $\mathbb{R}^4$ such that $\{\mathbf{v}_1, \mathbf{v}_2, \mathbf{v}_3, \mathbf{v}_4\}$ is an LI set in $\mathbb{R}^4$ and such that (i) three of a, b, c, d are zero; (ii) none of a, b, c, d is zero; (iii) $a = c = 0$ and $b = d$.

(b) Can you find vectors $\mathbf{v}_4$ and $\mathbf{v}_5$ in $\mathbb{R}^4$ which, together with $\mathbf{v}_1$, $\mathbf{v}_2$ and $\mathbf{v}_3$ from part (a), form an LI set in $\mathbb{R}^4$?

7. Let $\mathbf{v}_1 = (1, -2, 4, 1)$, $\mathbf{v}_2 = (-2, 1, -3, -1)$ and $\mathbf{v}_3 = (6, 3, -1, 1)$. Show that each pair of these vectors forms an LI set in $\mathbb{R}^4$ but that the set $\{\mathbf{v}_1, \mathbf{v}_2, \mathbf{v}_3\}$ is an LD set. Find two vectors $\mathbf{v}_4$ and $\mathbf{v}_5$ such that $\{\mathbf{v}_1, \mathbf{v}_2, \mathbf{v}_4, \mathbf{v}_5\}$ is an LI set. Will either of $\{\mathbf{v}_1, \mathbf{v}_3, \mathbf{v}_4, \mathbf{v}_5\}$ and $\{\mathbf{v}_2, \mathbf{v}_3, \mathbf{v}_4, \mathbf{v}_5\}$ then also be an LI set? [Hint: think geometrically of $\mathrm{sp}\{\mathbf{v}_1, \mathbf{v}_2\}$, $\mathrm{sp}\{\mathbf{v}_1, \mathbf{v}_3\}$ and $\mathrm{sp}\{\mathbf{v}_2, \mathbf{v}_3\}$.]

8. (a) Show that the set of vectors $\{(\alpha, \beta, \gamma), (a, b, c), (A, B, C)\}$ in $\mathbb{R}^3$ are LI if and only if the determinant $\begin{vmatrix} \alpha & \beta & \gamma \\ a & b & c \\ A & B & C \end{vmatrix} \neq 0$. Deduce that $\begin{bmatrix} \alpha & \beta & \gamma \\ a & b & c \\ A & B & C \end{bmatrix}$ is invertible if and only if $\{(\alpha, \beta, \gamma), (a, b, c), (A, B, C)\}$ is an LI set.

(b) Let $A = \begin{bmatrix} 1 & 1 & -1 \\ 1 & -1 & 1 \\ -1 & 1 & 1 \end{bmatrix}$. Are the rows of A^{58} independent (in $\mathbb{R}^3$)?

9. Test, by the methods of the preceding example, the subset $\{(1, -2, 4, 3), (-2, 1, 2, -1), (-4, -2, 10, -4), (-1, 1, 14, 0)\}$ for linear (in)dependence.

10. In Exercise 8 above show that the set $\{(\alpha, \beta, \gamma), (a, b, c), (A, B, C)\}$ of vectors is LI if and only if the set $\{(\alpha, a, A), (\beta, b, B), (\gamma, c, C)\}$ is LI.

11. Show that the set $\{(1, a, a^2), (1, b, b^2), (1, c, c^2)\}$ is LI if and only if a, b and c are all distinct.

12. Test for LD/LI by the methods of Example 2, the sets:

(i) $\{(-1, 2, 1, 2), (2, -3, 2, 3), (3, 4, -3, 4), (4, 5, 4, -5)\}$;

(ii) $\{(-1, 2, -1, 2, 1), (2, -3, 2, -3, 2), (3, 4, -3, 4, -3)\}$.

13. Repeat Exercise 12 on the sets:

(i) $\{1 + 2x + 2x^2 + 3x^3, 3 + 3x + 4x^2 + 4x^3, 4 + 4x + 5x^2 + 5x^3, 5 + 5x + 5x^2 + 6x^3\}$;

(ii) $\left\{\begin{bmatrix} -1 & 2 \\ 1 & 2 \end{bmatrix}, \begin{bmatrix} 2 & -3 \\ 2 & 3 \end{bmatrix}, \begin{bmatrix} 3 & 4 \\ -3 & 4 \end{bmatrix}, \begin{bmatrix} 4 & 5 \\ 4 & -5 \end{bmatrix}\right\}$.

14. (a) Complete the proof of Example 3(iii);

(b) apply the method of Example 3(iii) to determine whether or not the following set spans $\mathbb{R}^3$: $\{(1, 3, -3), (4, 9, -3), (5, 8, 6), (3, 4, 6), (5, 7, 9)\}$.

15. Let $\mathbf{v}_1, \mathbf{v}_2, \mathbf{v}_3$ be a set of non-zero vectors in a vector space V such that $\mathbf{v}_1 . \mathbf{v}_2 = 0$, $\mathbf{v}_2 . \mathbf{v}_3 = 0$, $\mathbf{v}_3 . \mathbf{v}_1 = 0$. Prove that the set $\{\mathbf{v}_1, \mathbf{v}_2, \mathbf{v}_3\}$ is LI. [Hint: assuming that $\alpha_1 \mathbf{v}_1 + \alpha_2 \mathbf{v}_2 + \alpha_3 \mathbf{v}_3 = \mathbf{0}$ look at $\mathbf{v}_1 . (\alpha_1 \mathbf{v}_1 + \alpha_2 \mathbf{v}_2 + \alpha_3 \mathbf{v}_3) = \mathbf{v}_1 . \mathbf{0} = 0 = \alpha_1 \mathbf{v}_1 . \mathbf{v}_1$.]

16. Give a geometric argument to show that any set of three vectors in $\mathbb{R}^2$ (any set of four vectors in $\mathbb{R}^3$) is LD.

17. Show, by converting the problem into one of solving a system of m homogeneous equations in n unknowns, that, if $m < n$, then any set of n vectors in $\mathbb{R}^m$ must be LD. Deduce that if, in $\mathbb{R}^t$, one can find a set of s linearly independent vectors, then $t \geq s$.

18. (a) Prove that any set $\{\mathbf{v}\}$ containing one non-zero vector is LI. [Hint: use Exercise 25 of Chapter 6.]

(b) Prove that a set containing precisely two non-zero vectors is LD if and only if each vector is a scalar multiple of the other.

(c) Show that Theorem 1(a) fails if $s = 2$, $\mathbf{v}_1 = \mathbf{0}$ and $\mathbf{v}_2 \neq \mathbf{0}$. And what if $s = 1$? Rewrite Theorem 1(a) and its 'proof' to incorporate these extra cases.

19. (a) Let $\{\mathbf{v}_1, \mathbf{v}_2, \mathbf{v}_3\}$ be an LI set of vectors in the vector space V. Are the following subsets of V also necessarily linearly independent? (i) $\{\mathbf{v}_1 + \mathbf{v}_2, \mathbf{v}_2 + \mathbf{v}_3, \mathbf{v}_3 + \mathbf{v}_1\}$; (ii) $\{\mathbf{v}_1 - \mathbf{v}_2, \mathbf{v}_2 - \mathbf{v}_3, \mathbf{v}_3 - \mathbf{v}_1\}$. [Hint for (i): does $\alpha_1(\mathbf{v}_1 + \mathbf{v}_2) + \alpha_2(\mathbf{v}_2 + \mathbf{v}_3) + \alpha_3(\mathbf{v}_3 + \mathbf{v}_1) = 0$ necessarily imply $\alpha_1 = \alpha_2 = \alpha_3 = 0$?]

(b) Let $\{\mathbf{v}_1, \mathbf{v}_2, \mathbf{v}_3, \mathbf{v}_4\}$ be an LI subset in the vector space V. Are the following subsets of V also necessarily linearly independent? (i) $\{\mathbf{v}_1 + \mathbf{v}_2, \mathbf{v}_2 + \mathbf{v}_3, \mathbf{v}_3 + \mathbf{v}_4, \mathbf{v}_4 + \mathbf{v}_1\}$; (ii) $\{\mathbf{v}_1 - \mathbf{v}_2, \mathbf{v}_2 - \mathbf{v}_3, \mathbf{v}_3 - \mathbf{v}_4, \mathbf{v}_4 - \mathbf{v}_1\}$.

20. Give proofs of the following two statements, made earlier (p. 131), concerning non-empty subsets S and T of a vector space V:

(i) any subset T of an LI set S is LI;

(ii) if a subset T of a set S is LD then so is S.

[Hint for (i): if $T = \{\mathbf{v}_1, \mathbf{v}_2, \ldots, \mathbf{v}_t\} \subseteq \{\mathbf{v}_1, \mathbf{v}_2, \ldots, \mathbf{v}_t, \ldots, \mathbf{v}_s\} = S$ and if $\alpha_1\mathbf{v}_1 + \alpha_2\mathbf{v}_2 + \ldots + \alpha_t\mathbf{v}_t = \mathbf{0}$ then $\alpha_1\mathbf{v}_1 + \alpha_2\mathbf{v}_2 + \ldots + \alpha_t\mathbf{v}_t + \ldots + 0\mathbf{v}_s = \mathbf{0}$. But S is LI. . . .]

21. Explain why it is not sensible to talk – as one sometimes might be tempted to – of 'a set of linearly (in)dependent vectors'.

22. To prove a set of functions $\{f_1, f_2, \ldots, f_r\}$ (in the vector space $\mathfrak{F}$, say) is LI we must show that, if $\alpha_1 f_1 + \alpha_2 f_2 + \ldots + \alpha_r f_r = Z$ (the zero function) then $\alpha_1 = \alpha_2 = \ldots = \alpha_r = 0$. By taking, as in Example 2(v), several specific values for x, show that the following sets of vectors in $\mathfrak{F}$ are LI: (i) $\{\sin x, \sin 2x, \sin 3x\}$; (ii) $\{\sin x, \cos x, \sin 2x, \cos 2x\}$; (iii) $\{x, \cos x\}$; (iv) $\{e^x, xe^x, x^2e^x\}$ [Notes: (i), (ii) extend to any number of sines – and cosines – (iv) exhibits independent solutions to

$$\frac{d^3y}{dx^3} - 3\frac{d^2y}{dx^2} + 3\frac{dy}{dx} - y = 0.]$$

23. Let each of f, g, h be twice differentiable functions from $\mathbb{R}$ to $\mathbb{R}$ and let f' and f'' (etc.) denote the first and second derivatives of f (etc.). It can be shown that if the determinant

$$w = \begin{vmatrix} f & g & h \\ f' & g' & h' \\ f'' & g'' & h'' \end{vmatrix} \neq Z \quad \text{(the zero function)}$$

then $\{f, g, h\}$ is an LI set. This determinant is the **Wronskian**[c] of f, g and h. Now show that the sets of functions in Exercise 22(i), (iii) and (iv) are LI.

24. Let A be an $n \times n$ matrix and $\mathbf{v}$ an n-vector such that $A^k\mathbf{v} = \mathbf{0}$ whilst $A^{k-1}\mathbf{v} \neq \mathbf{0}$. Show that $\{\mathbf{v}, A\mathbf{v}, A^2\mathbf{v}, \ldots, A^{k-1}\mathbf{v}\}$ is an LI set. [Hint: from

[c]Josef Maria Hoene-Wronski, 23 August 1776–8 August 1853.

$$\mathbf{u} = \alpha_0\mathbf{v} + \alpha_1 A\mathbf{v} + \alpha_2 A^2\mathbf{v} + \ldots + \alpha_{k-1}A^{k-1}\mathbf{v} = \mathbf{0}$$

deduce $A^{k-1}\mathbf{u} = a_0 A^{k-1}\mathbf{v} = \mathbf{0}$.]

25. (a) Let $\mathbf{v}_1, \mathbf{v}_2, \ldots, \mathbf{v}_r$ be vectors in the vector space V. Show that $\{\mathbf{v}_1, \mathbf{v}_2, \ldots, \mathbf{v}_r\}$ is an LI set if and only if each element $\mathbf{v}$ in $\text{sp}\{\mathbf{v}_1, \mathbf{v}_2, \ldots, \mathbf{v}_r\}$ is expressible uniquely as a linear combination of $\mathbf{v}_1, \mathbf{v}_2, \ldots, \mathbf{v}_r$. [Hint: the difference of two linear combinations for $\mathbf{v}$ gives a linear combination for $\mathbf{0}$.]

(b) Prove that if $S = \text{sp}\{\mathbf{v}_1, \mathbf{v}_2, \ldots, \mathbf{v}_n\}$ and if no proper subset of $\{\mathbf{v}_1, \mathbf{v}_2, \ldots, \mathbf{v}_s\}$ also spans S, then $\{\mathbf{v}_1, \mathbf{v}_2, \ldots, \mathbf{v}_s\}$ is an LI set. [Hint: assume $\{\mathbf{v}_1, \mathbf{v}_2, \ldots, \mathbf{v}_s\}$ is LD and obtain a contradiction.]

26. Prove that if S_1 and S_2 are two subspaces of a vector space V such that $S_1 \cap S_2 = \{\mathbf{0}\}$, and if $\{\mathbf{u}_1, \mathbf{u}_2, \ldots, \mathbf{u}_r\}$ and $\{\mathbf{v}_1, \mathbf{v}_2, \ldots, \mathbf{v}_s\}$ are LI sets in S_1 and S_2 respectively, then $\{\mathbf{u}_1, \mathbf{u}_2, \ldots, \mathbf{u}_r, \mathbf{v}_1, \mathbf{v}_2, \ldots, \mathbf{v}_s\}$ is an LI set in V.

27. Prove Theorem 3. [For (b) see Exercise 8 above.]

28. (a) Let S be the subspace of $\mathbb{R}^4$ spanned by the set $T = \{(1, 1, 2, 1), (2, -1, 1, 2), (4, -5, -1, 4), (5, -1, 4, 5)$ and $(8, -7, 1, 7)\}$. Find an LI subset of T spanning the same subspace. Is $S = \mathbb{R}^4$?

(b) Do the following polynomials span $\mathbb{R}_3[x]$: $2 + x + 3x^3$, $3 - x + 2x^2 + 5x^3$, $3 - 4x + 4x^2 + 8x^3$, $3 + x^2 + 6x^3$, $4 - x + 2x^2$?

29. Let A be an $m \times n$ matrix in echelon form with k non-zero rows where k is the minimum of m and n. Show that (i) if $m \le n$ then the rows of A form an LI set; (ii) if $n \le m$ then the rows of A span $\mathbb{R}^n$.

30. Find an LI subset of the following set of vectors which spans the same subspace of $\mathbb{R}^4$: $\{(-1, 2, 1, 2), (2, -3, 2, 1), (4, -3, 16, 17), (2, -1, 10, 11), (-1, 3, 5, 5)\}$.

31. (a) Let $\mathbf{c}_{i_1}, \mathbf{c}_{i_2}, \ldots, \mathbf{c}_{i_k}$ be columns of the $m \times n$ matrix A. Show that if *row* operations are applied to A thereby producing new columns $\mathbf{d}_{i_1}, \mathbf{d}_{i_2}, \ldots, \mathbf{d}_{i_k}$ then $\{\mathbf{c}_{i_1}, \mathbf{c}_{i_2}, \ldots, \mathbf{c}_{i_k}\}$ is an LD set if and only if $\{\mathbf{d}_{i_1}, \mathbf{d}_{i_2}, \ldots, \mathbf{d}_{i_k}\}$ is an LD set. [Hint: this is part of Theorem 3 of Chapter 9.]

(b) Use (a) to explain the alternative method which was given for dealing with Example 4.

COMPUTER PACKAGE PROBLEMS

1. (a) Show that the set[d] of vectors

$$E = (364, 336, 333, 325, 310), \quad A = (334, 311, 307, 304, 299),$$
$$W = (375, 365, 302, 291, 270), \quad P = (337, 280, 274, 271, 260)$$

form an LI subset of $\mathbb{R}^5$.

(b) Show that the set $\{(103, 117, 255, 500, 123), (-212, 235, 212, -600, 345), (-420, 436, -146, -105, 999), (-65, 177, 233, -101, 234), (123, 231, 312, 588, 153)\}$ of vectors is an LD subset of $\mathbb{R}^5$.

(c) Find an LI subset of this set which contains the final three vectors.

[d] The components of these vectors will be recognised instantly by followers of English, Australian, West Indian and Pakistani test match cricket!

2. Is $\begin{bmatrix} 1 & 1 \\ 1 & 1 \end{bmatrix}, \begin{bmatrix} 2 & 4 \\ 8 & 16 \end{bmatrix}, \begin{bmatrix} 3 & 9 \\ 27 & 81 \end{bmatrix}, \begin{bmatrix} 4 & 16 \\ 64 & 256 \end{bmatrix}$ an LI subset of $M_2(\mathbb{R})$?
3. Can you find nine successive primes $p_1, p_2, \ldots, p_9$ such that (p_1, p_2, p_3), (p_4, p_5, p_6), (p_7, p_8, p_9) is an LD subset of $\mathbb{R}^3$?
4. Determine whether or not there are positive integers (primes if possible) a, b such that $\{(2, 3,5, 7, 11), (13, 17, 19, 23, 29), (31, 37, 41, 43, 47), (53, 59, 61, 67, 71), (a, 79, 83, 89, b)\}$ is an LD set of vectors in $\mathbb{R}^5$.
5. Find integers α, β, γ, δ (all positive, if possible) such that

$$\alpha(123+312x+231x^2)+\beta(132+213x+321x^2)=\gamma(231+123x+312x^2)+ \delta(321+132x+213x^2)$$

in $\mathbb{R}_2[x]$.

6. Repeat Problem 7.5 of Chapter 7 by finding the reduced (row) echelon form of the appropriate matrices. Is the ordinary (row) echelon form of any use here?

9 • Bases and Dimension

This chapter combines the ideas of spanning and independence introduced in Chapters 7 and 8 by defining the concept of 'basis'. By means of a basis B each vector in a vector space V can be expressed – and uniquely so – as a (finite) linear combination of the elements of B. Because of Theorem 2, the basis concept also permits us to give a clear definition of 'dimension'. This allows us to define both the row rank and the column rank of any given $m \times n$ matrix A, say. These ranks turn out to be equal and intimately connected with the number of 'really different' solutions of the linear system $Ax = 0$, as described in the famous formula in Corollary 1.

The concept of (finite) basis is important in linear programming problems (which are often concerned with maximising production and minimising cost).

All vector spaces can be shown to have a basis. We shall deal mainly with finite bases but many infinite dimensional vector spaces are of fundamental importance in mathematics and mathematical physics.

When studying problems involving points lying or moving in a plane or in space it is useful to be able to identify such points by coordinates. In the plane we may choose coordinates in several ways, the most familiar being cartesian coordinates x, y and polar coordinates r, ϑ as in Figs 9.1 and 9.2. Sometimes it is convenient to use oblique coordinates, as in Fig. 9.3. There are occasions when each is to be preferred to the others, but the key thing to note is that, in each system, we need *exactly two measurements* (namely a and b or r and ϑ or $(a - b/2)$ and $b/2$) to determine a point precisely.

Similar remarks apply to problems in 3-dimensional space where it may be convenient to specify the position of a point by giving its cartesian coordinates (a, b, c), it cylindrical polar coordinates (r, ϑ, z) or its spherical polar coordinates (r, ϑ, ϕ). Here we need *precisely three measurements* to tie down exactly the position of a point.

How many 'measurements' does it take to determine uniquely an AP (arithmetic

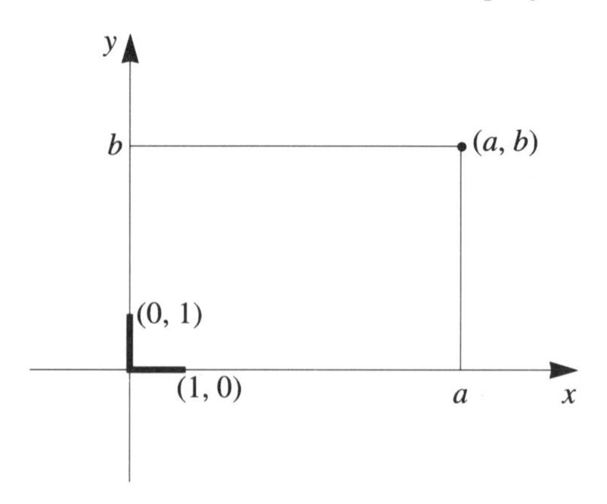

Fig 9.1 $(a, b) = a\,(1, 0) + b\,(0, 1)$

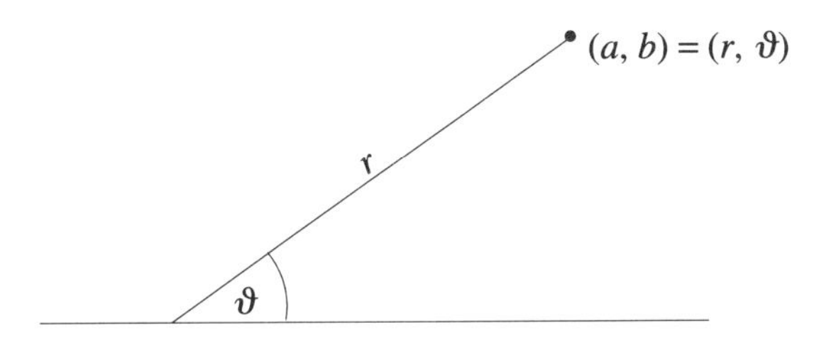

Fig 9.2

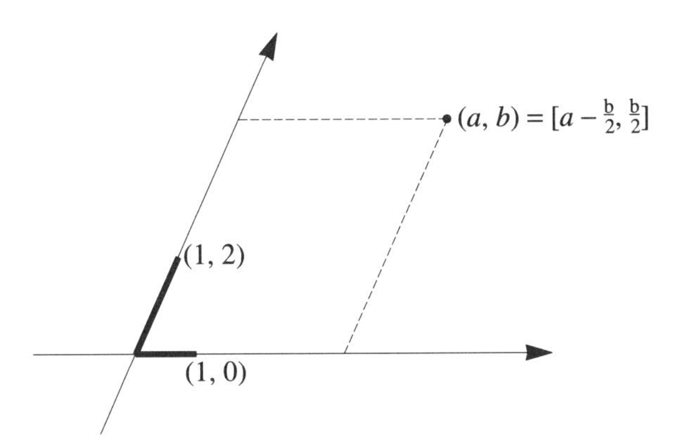

Fig 9.3 $(a, b) = (a - b/2)(1, 0) + b/2(1, 2) = [a - b/2, b/2]$

progression)? The answer is 'two'. For let $A = (a, a + d, a + 2d, a + 3d, \ldots)$ be any AP. It is clear that A is completely determined by the two numbers a and d.

And now for a harder problem! How many of the numbers $a, b, c, \ldots, h, j$ must one specify in order to determine (uniquely) the magic square

$$\begin{matrix} a & b & c \\ d & e & f \\ g & h & j \end{matrix} \quad ?$$

Clearly the answer is not 9. (Why not?) Is it 2? (!?) [After all we must have

$$a+b+c=d+e+f=\ldots=c+f+j=a+e+j=c+e+g$$

which seems to be equivalent to seven homogeneous equations in nine unknowns.] So, what is the answer? We shall see later.

We can explain why the number of 'measurements' are 2, 3 and 2 in the examples given above in another way. Consider, in Fig. 9.1, the point (a, b). If we identify (a, b) with the vector $\mathbf{v}$ which joins $(0, 0)$ to (a, b) and if we let $\mathbf{x}$ denote the vector joining $(0, 0)$ to $(1, 0)$ and $\mathbf{y}$ the vector joining $(0, 0)$ to $(0, 1)$, then we may write $\mathbf{v} = a\mathbf{x} + b\mathbf{y}$.

Likewise, in Fig. 9.3, it seems that, if we let $\mathbf{s}$ and $\mathbf{t}$ denote the vectors joining $(0, 0)$ to $(1, 0)$ and $(0, 0)$ to $(1, 2)$ respectively, then $\mathbf{v} = (a, b) = (a - b/2)\mathbf{s} + b/2\mathbf{t}$. Thus, if we think of $\mathbf{x}$ and $\mathbf{y}$ (in the first case) and $\mathbf{s}$ and $\mathbf{t}$ (in the second case) as 'basic measuring units', we see that the pairs of measurements required to specify **v**, namely a, b and $(a - b/2)$, $b/2$ respectively, describe the amounts of basic units we need to add together to obtain **v**. It then seems reasonable to describe the pairs (a, b) and $[(a - b/2), b/2]$ as the 'coordinates' of **v** with respect to the pairs of basic units **x**, **y** and **s**, **t** respectively.

What basic measuring units can we take in the case of the vector space of all APs? Since

$$A = (a, a+d, a+2d, a+3d, \ldots) = a(1,1,1,1,\ldots) + d(0,1,2,3,\ldots)$$

we see that $(1, 1, 1, 1, \ldots)$ and $(0, 1, 2, 3, \ldots)$ may be taken as basic units, the 'coordinates' of A being, then, a and d.

PROBLEM 9.1

Show that $(-17, -13, -9, -5, \ldots)$ and $(\frac{1}{2}, \frac{3}{2}, \frac{5}{2}, \frac{7}{2}, \frac{9}{2}, \ldots)$ are APs which may be taken as a pair of basic measuring units. Find the coordinates of $(a, a + d, a + 2d, a + 3d, \ldots)$ with respect to these units. ●

PROBLEM 9.2

Write $(1, 1, 1, \ldots)$ in at least three ways as a linear combination of the APs $(0, 1, 2, \ldots)$, $(1, 2, 3, \ldots)$ and $(101, 201, 301, \ldots)$. ●

What the above examples, namely 2-dimensional space and the vector space of APs, appear to suggest is that in any vector space V it might be possible to express each element uniquely in terms of the members of a basic set and that each such basic set for V contains the same number of vectors.

We shall show that this is indeed the case. We begin with an appropriate definition.

• Definition I

Let $B = \{\mathbf{v}_1, \mathbf{v}_2, \ldots, \mathbf{v}_s\}$ be a set of vectors in the vector space V. B is called a **basis** of V if and only if both of the following conditions hold:

(i) $\mathrm{sp}\{\mathbf{B}\} = V$ (i.e. every vector can be expressed as a linear combination of vectors from $\{\mathbf{v}_1, \mathbf{v}_2, \ldots, \mathbf{v}_s\}$) and

(ii) the set B is a linearly independent set. ●

Note that this definition does exactly what we want because (i) says that all elements of V are expressible in terms of $\mathbf{v}_1, \mathbf{v}_2, \ldots, \mathbf{v}_s$ and (ii) tells us that such an expression is unique (by Theorem 1(b) of Chapter 8). [If $\mathbf{v} \in V$ is expressible as $\mathbf{v} = \alpha_1\mathbf{v}_1 + \alpha_2\mathbf{v}_2 + \ldots + \alpha_s\mathbf{v}_s$ then the $\alpha_1, \alpha_2, \ldots, \alpha_s$ which are uniquely determined by **v** are called the *coordinates* of **v** with respect to the basis B. Of course the coordinates of **v** can be expected to change from one basis to another. (See Problem 9.1 and Exercise 9.) Let us give some examples.

Examples I

(i) In $\mathbb{R}^3$ the set $\{(1, 0, 0), (0, 1, 0), (0, 0, 1)\}$ constitutes a basis. So do $\{(3, -1, 6), (2, 7, 0), (1, 2, 1)\}$, $\{(e, \pi, 1), (\pi, 1, e), (1, e, \pi)\}$ (are you sure?), and $\{(-\frac{15}{38}, \frac{7}{57}, \frac{69}{22}), (-\frac{23}{7}, \frac{17}{32}, \frac{81}{97}), (-\frac{41}{77}, \frac{1}{34}, -\frac{11}{12})\}$ (surely?) [Furthermore I am prepared to believe (though I have not checked it!) that any three of these last nine vectors forms a basis for $\mathbb{R}^3$.]

How might one prove these assertions? We resort to Theorem 2 of Chapter 8. For example, for the set $\{(3, -1, 6), (2, 7, 0), (1, 2, 1)\}$ we could form the matrix

$$\begin{bmatrix} 3 & -1 & 6 \\ 2 & 7 & 0 \\ 1 & 2 & 1 \end{bmatrix}$$

and obtain (by devious means – to preserve integer entries!) the echelon form

$$\begin{bmatrix} 1 & 2 & 1 \\ 0 & 21 & -14 \\ 0 & 0 & -5 \end{bmatrix}.$$

One sees immediately, as in Example 3(i) and (iii) of Chapter 8, that these three row vectors (a) form an LI set and (b) span $\mathbb{R}^3$. Hence, by Theorem 2 of Chapter 8, the given set of vectors is both LI and spans $\mathbb{R}^3$. That is, the set $\{(3, -1, 6), (2, 7, 0), (1, 2, 1)\}$ is a basis for $\mathbb{R}^3$.

(ii) Likewise $\{(1, 0, 0, \ldots, 0), (0, 1, 0, \ldots, 0), \ldots, (0, 0, 0, \ldots, 1)\}$ where the ith vector in this list is an n-tuple with a '1' in the ith position and 0s everywhere else, forms a basis for $\mathbb{R}^n$. It is called the **standard basis** for $\mathbb{R}^n$.

(iii) The set $\{e^{-x}, e^{-3x}\}$ containing *two* functions is a basis for the solution space of the *second-order* differential equation

$$\frac{d^2y}{dx^2} + 4\frac{dy}{dx} + 3y = 0.$$

And so is $\{e^{-x} + e^{-3x}, 2e^{-x} - 3e^{-3x}\}$ – and (infinitely) many others! (Cf. Example 2(v) of Chapter 8.)

(iv) The subspace $S = \{(x, y, z): 3x + y + 5z = 0\}$ of $\mathbb{R}^3$ has basis $\{(1, -3, 0), (0, -5, 1)\}$ since $(x, y, z) \in S$ if and only if

$$(x, y, z) = (x, -3x - 5z, z) = x(1, -3, 0) + z(0, -5, 1)$$

where x and z can take any values in $\mathbb{R}$. Thus $B = \{(1, -3, 0), (0, -5, 1)\}$ certainly spans S. And since, from $\alpha(1, -3, 0) + \beta(0, -5, 1) = (0, 0, 0)$, we are forced to conclude that $\alpha = \beta = 0$, we see that B is also an LI set, hence a basis.

(v) The subspace $T = \{(x, y, z, t): x + 2y + 3z = y + 2z + 3t = 0\}$ likewise has basis $\{(1, -2, 1, 0), (6, -3, 0, 1)\}$ since the equations $x + 2y + 3z = y + 2z + 3t = 0$ imply that $y = -2z - 3t$ and $x = z + 6t$ so that $(x, y, z, t) = (z + 6t, -2z - 3t, z, t)$. Of course (of course?) $\{(1, -2, 1, 0), (6, -3, 0, 1)\}$ is an LI set.

(vi) The subspace U of $\mathbb{R}_3[x]$ comprising polynomials p for which $p(-1) = 0$ has basis $\{1 + x^3, x - x^3, x^2 + x^3\}$ since, if $p(x) = a + bx + cx^2 + dx^3$, then $p(-1) = a - b + c - d = 0$. Consequently $p \in U$ if and only if

$$p(x) = a + bx + cx^2 + (a - b + c)x^3 = a(1 + x^3) + b(x - x^3) + c(x^2 + x^3).$$

Thus $1+x^3, x-x^3, x^2+x^3$ certainly span U. We leave it to you to prove that if $a(1+x^3)+b(x-x^3)+c(x^2+x^3)=Z$ (the zero polynomial) then $a=b=c=0$; that is $\{(1+x^3), (x-x^3), (x^2+x^3)\}$ is an LI set. Consequently these three polynomials together form a basis for U.

(vii) $\{(1, 2, 3), (1, 3, 5), (2, 4, 8), (11, 11, 100)\}$ is not a basis for $\mathbb{R}^3$. (Do you have any feeling why it cannot be? In due course, see Corollary 2.) Nor (for a totally different reason, namely . . .? – but see the same corollary!) is $\{(1, 2, 3), (2, 4, 8)\}$.

(viii) $\{1, \mathrm{i}\}$ is a basis for the set $\mathbb{C}$ of complex numbers when it is regarded as a vector space over the field of real numbers $\mathbb{R}$ – since (I) each $c \in \mathbb{C}$ is expressible in the form $c=\alpha 1+\beta\mathrm{i}$ and (II) $c=0$ implies that $\alpha=\beta=0$. $\{\pi-\frac{22}{7}\mathrm{i}, \mathrm{e}-\frac{27}{10}\mathrm{i}\}$ is also a basis for $\mathbb{C}$. (Prove it!) Indeed $\{a+b\mathrm{i}, c+d\mathrm{i}\}$ is a basis for $\mathbb{C}$ over $\mathbb{R}$ if and only if . . .? (See Exercise 6.)

PROBLEM 9.3

Find the coordinates of $1+0\mathrm{i}$ and $0+1\mathrm{i}$ in terms of the basis $\{\pi-\frac{22}{7}\mathrm{i}, \mathrm{e}-\frac{27}{10}\mathrm{i}\}$.

Examples 1 (cont.)

(ix) Theorem 4 of Chapter 8 implies that each finite dimensional vector space V must have a (finite) basis. [For, since V is finite dimensional, it has, by Definition 4 of Chapter 7, a finite spanning set G, say. But then Theorem 4 tells us that we can find a finite subset of G which both spans V and is LI.]

Of course, $\mathbb{R}[x]$ cannot have a finite basis. Why not? (See Exercise 8.) On the other hand, if anything deserves to be called a basis for $\mathbb{R}[x]$ it is surely the set $\{1, x, x^2, x^3, \ldots\}$. [But then what about $\{1, 1+x, x+x^2, x^2+x^3, x^3+x^4, \ldots\}$?

In fact *every vector space V has a basis* – that is, a subset $\mathfrak{B}$ of elements such that (α) each element of V is a linear combination of (finitely many – why?) element from $\mathfrak{B}$ and (β) each (finite – why?) subset of $\mathfrak{B}$ is LI. (See, for example, Grossmann, S. I. *Elementary Linear Algebra*.)

TUTORIAL PROBLEM 9.1

For each real number r define a function f_r: $\mathbb{R}\to\mathbb{R}$ by: $f_r(x)=1$ if $x=r$ and $f_r(x)=0$ if $x\neq r$. The set $\mathfrak{B}=\{f_r: r\in\mathbb{R}\}$ looks like an infinite analogue of the standard bases for $\mathbb{R}^n$. Is it a basis for the vector space $\mathfrak{F}$ of all functions from $\mathbb{R}$ to $\mathbb{R}$?

TUTORIAL PROBLEM 9.2

(Very tricky) What might constitute a basis for the vector space of infinite sequences in Exercise 15 of Chapter 6? And what about a basis for $\mathbb{R}$, thought of as a vector space over the rational numbers $\mathbb{Q}$?

To establish the truth of the suggestion about 'fixed basis number' we call upon a result which, together with Example 1(ii), confirms the easily believed fact that, in $\mathbb{R}^n$, no set of $n + 1$ (or more) distinct vectors can be LI nor can any set of $n - 1$ (or fewer) vectors span $\mathbb{R}^n$.

• *Theorem 1*

Let V be a vector space. Suppose that $\{\mathbf{v}_1, \mathbf{v}_2, \ldots, \mathbf{v}_s\}$ spans V and $\{\mathbf{w}_1, \mathbf{w}_2, \ldots, \mathbf{w}_t\}$ is LI in V. Then $s \geq t$.

PROOF

Clearly (are you sure it is clear?) we may assume that none of $\mathbf{v}_1, \mathbf{v}_2, \ldots, \mathbf{v}_s$ is the zero vector. First note that

$$\mathrm{sp}\{\mathbf{w}_1, \mathbf{v}_1, \mathbf{v}_2, \ldots, \mathbf{v}_s\} \supseteq \mathrm{sp}\{\mathbf{v}_1, \mathbf{v}_2, \ldots, \mathbf{v}_s\} = V.$$

Now $\{\mathbf{w}_1, \mathbf{v}_1, \mathbf{v}_2, \ldots, \mathbf{v}_s\}$ is LD (since $\mathbf{w}_1$ is a linear combination of $\mathbf{v}_1, \mathbf{v}_2, \ldots, \mathbf{v}_s$ – because $\mathbf{v}_1, \mathbf{v}_2, \ldots, \mathbf{v}_s$ span all of V). Hence, by Theorem 1(a) of Chapter 8, there exists a $\mathbf{v}_1$ which is a linear combination of its predecessors. (Or maybe $\mathbf{w}_1$ is $\mathbf{0}$? Oh, no, it cannot be. Why not?) But this means that $\{\mathbf{w}_1, \mathbf{v}_1, \ldots, \hat{\mathbf{v}}_i, \ldots, \mathbf{v}_s\}$ spans V. Consequently, adjoining $\mathbf{w}_2$ to this list, $\{\mathbf{w}_2, \mathbf{w}_1, \mathbf{v}_1, \ldots, \hat{\mathbf{v}}_i, \ldots, \mathbf{v}_s\}$ is LD. (Why?) Using Theorem 1(a) of Chapter 8 again, there exists some vector $\mathbf{v}_j$, say, which is a linear combination of its predecessors and so can be omitted. (Why cannot it be $\mathbf{w}_1$ which is omitted this time? Answer: $\{\mathbf{w}_2, \mathbf{w}_1\}$ is a . . . set.)

Thus we have $\mathrm{sp}\{\mathbf{w}_2, \mathbf{w}_1, \mathbf{v}_1, \ldots, \hat{\mathbf{v}}_i, \ldots, \hat{\mathbf{v}}_j, \ldots, \mathbf{v}_s\} = V$. Continuing in this manner (note the neat sidestepping of a formal proof using mathematical induction), we may, at each step, introduce a $\mathbf{w}_k$ and kick out a $\mathbf{v}_l$ – *and preserve the spanning property*. Now, if $s < t$, then, at step s, we would find that $\{\mathbf{w}_s, \mathbf{w}_{s-1}, \ldots, \mathbf{w}_1\}$ spans V, so that $\mathbf{w}_{s+1}$ is a linear combination of $\mathbf{w}_s, \mathbf{w}_{s-1}, \ldots, \mathbf{w}_1$. But this is a blatant contradiction (to what?). Hence s cannot be less than t, so that $s \geq t$ follows. ●

TUTORIAL PROBLEM 9.3

Why are we entitled to say 'clearly' at the start of Theorem 1?

• *Example 2*

The set $\{(1, 2, 1, 1, 5), (-2, 3, 1, 2, -4), (3, 3, 0, 3, 4), (2, 6, 0, 0, 3)\}$ cannot possibly span $\mathbb{R}^5$; the set $\{(1, 0, 6, 6), (1, 4, 9, 2), (1, 7, 2, 9), (1, 8, 1, 5), (1, 9, 4, 2)\}$ cannot possibly be an LI subset of $\mathbb{R}^4$.

And now for a beautiful[a] theorem.

• *Theorem 2*

Let V be a finite dimensional vector space. Then *every* two bases have the same number of elements.

[a]Because it is clearly an important statement with a compact and easily understood proof.

PROOF

Let $U = \{\mathbf{u}_1, \mathbf{u}_2, \ldots, \mathbf{u}_s\}$ and $W = \{\mathbf{w}_1, \mathbf{w}_2, \ldots, \mathbf{w}_t\}$ be bases for V. Since U spans V and W is LI we deduce, from Theorem 1, that $s \geq t$. But . . . (can you see how we are going to argue?) . . . since W spans V and U is LI we also know (Theorem 1) that $t \geq s$. Consequently, $s = t$, as claimed. ●

Applying this result to Examples 1(i) and (ii) we see that every basis of $\mathbb{R}^3$ ($\mathbb{R}^n$) must have exactly three (exactly n) elements. As we naturally think of $\mathbb{R}^3$ ($\mathbb{R}^n$) as being 3-dimensional (n-dimensional) it seems natural to make the following definition.

• Definition 2

Let V be a finite dimensional vector space. The (fixed) number of elements in each basis of V is called the **dimension of V**. We denote this number by dim V. ●

• Examples 3

(i) $\mathbb{R}^n$ has dimension n – according to Definition 2.

(ii) The subspace $S = \text{sp}\{(1, -2, 3, 1), (2, 2, -2, 1), (5, 2, -1, 3), (11, 2, 1, 7), (2, 8, 2, 3)\}$ of $\mathbb{R}^4$ – see Example 4 of Chapter 8 – has dimension 3.

[Please, please, note a common misunderstanding: S comprises 4-tuples but this does NOT mean that S necessarily has dimension 4. (!) Dimension is not necessarily equal to the number of coordinates! However, the fact that S is a subspace of $\mathbb{R}^4$ DOES imply that the dimension of S CANNOT EXCEED 4 (see Theorem 5′, below).]

(iii) The solution space of $\mathrm{d}^2y/\mathrm{d}x^2 + 4\mathrm{d}y/\mathrm{d}x + 3y = 0$ has dimension 2, $\{e^{-x}, e^{-3x}\}$ being one basis for it. (See Example 1(iii).) Likewise, the solution space of each third-order differential equation with constant coefficients has dimension 3.

(iv) The vector space of 3×3 magic squares actually has dimension 3. •

PROBLEM 9.4

(i) Prove the statement just made by showing that

$$\mathbf{v}_1 = \begin{matrix} 1 & 1 & 1 \\ 1 & 1 & 1 \\ 1 & 1 & 1 \end{matrix}, \quad \mathbf{v}_2 = \begin{matrix} 1 & -1 & 0 \\ -1 & 0 & 1 \\ 0 & 1 & -1 \end{matrix} \quad \text{and} \quad \mathbf{v}_3 = \begin{matrix} 0 & 1 & -1 \\ -1 & 0 & 1 \\ 1 & -1 & 0 \end{matrix}$$

form an LI set in the space of 3×3 magic squares and that the magic square

$$\begin{matrix} a & b & c \\ d & e & f \\ g & h & j \end{matrix}$$

is $e\mathbf{v}_1 + (a - e)\mathbf{v}_2 + (g - e)\mathbf{v}_3$.

[Perhaps you might even try to show how these particular magic squares arise by solving the system of equations $a + b + c = d + e + f = \ldots = c + e + g$ mentioned earlier.]

(ii) What is the dimension of the subspace (is it one?) of magic squares with centre component 0? ●

Examples 3 (cont.)

(v) Let A be an $m \times n$ matrix and let Ech(A) be an echelon form of A. Now although Ech(A) is not (for most matrices) uniquely determined by A (see Exercise 3 of Chapter 2; that is why we say 'an' rather than 'the') the *number of rows* in each Ech(A) *is the same*. For the rows of each Ech(A) (regarded as vectors in $\mathbb{R}^n$) span the same subspace of $\mathbb{R}^n$ as the rows of A itself (by Theorem 2 of Chapter 8). But the non-zero rows of each Ech(A) form an LI set (see Theorem 3(a) of Chapter 8) spanning the row space of A. Hence each Ech(A) contains the same number of non-zero rows. This number is called the **row rank** of A.

We can, of course, define the **column rank** of A in an analogous manner.

PROBLEM 9.5

Find the row rank and the column rank of each of the following matrices:

$$\begin{bmatrix} 3 & 3 & 0 & 1 \\ 2 & 1 & -1 & -3 \\ 5 & 7 & 2 & 9 \end{bmatrix}, \quad \begin{bmatrix} 3 & 7 & 5 \\ 0 & 4 & 2 \\ 1 & -3 & -1 \\ 2 & 4 & 3 \end{bmatrix}, \quad \begin{bmatrix} 1 & 1 & 1 & 1 & 1 \\ 1 & 2 & 2 & 2 & 2 \\ 1 & 2 & 3 & 3 & 3 \\ 1 & 2 & 3 & 4 & 5 \end{bmatrix}, \quad \begin{bmatrix} 1 & 2 & 0 & 3 & 0 & 0 & -6 \\ 0 & 0 & 1 & -1 & 0 & 0 & 2 \\ 0 & 0 & 0 & 0 & 1 & 0 & 1 \\ 0 & 0 & 0 & 0 & 0 & 0 & 0 \end{bmatrix}.$$

Do the results you obtained surprise you? In fact they will not once you have seen the following theorem.

Theorem 3

Let A be any $m \times n$ matrix. Then the row rank of A is equal to the column rank of A.

PROOF

We have just observed that any echelon matrix obtained from A will have the same row space and hence the same row rank as A. Now elementary *row* operations applied to A will yield a matrix C, say, whose column *space* will, in general, be different from that of A. Nevertheless the column *rank* of C can be shown to equal that of A. For suppose $C = EA$ where E is an elementary matrix and assume that columns $i_1, i_2, \ldots, i_r$ of A give rise to an LD set of vectors $\mathbf{c}_{i_1}, \mathbf{c}_{i_2}, \ldots, \mathbf{c}_{i_r}$ so that $\alpha_1\mathbf{c}_{i_1} + \alpha_2\mathbf{c}_{i_2}, + \ldots + \alpha_r\mathbf{c}_{i_r} = \mathbf{0}$ for suitable scalars α_k. Immediately, we deduce that

$$\alpha_1 E\mathbf{c}_{i_1} + \alpha_2 E\mathbf{c}_{i_2} + \ldots + \alpha_r E\mathbf{c}_{i_r} = E(\alpha_1\mathbf{c}_{i_1} + \alpha_2\mathbf{c}_{i_2} + \ldots + \alpha_r\mathbf{c}_{i_r}) = E\mathbf{0} = \mathbf{0},$$

that is, the vectors $E\mathbf{c}_{i_1}, E\mathbf{c}_{i_2}, \ldots, E\mathbf{c}_{i_r}$ (in columns $i_1, i_2, \ldots, i_r$) of C (= EA, see Exercise 9(a) of Chapter 3) also form an LD set. Conversely, since we can write $A = E^{-1}C$, we may equally well deduce that if the vectors in columns $j_1, j_2, \ldots, j_s$ of C form an LD set then so do the vectors in the corresponding columns of A. We have therefore shown that if a set of columns of A is LD then the corresponding set of columns of C is LD and vice versa. Finally choose as large an LI set as possible from the set of columns of A and take the corresponding set of columns for C. By Exercise 17(b), these sets are bases for their respective column spaces. Thus C (= EA) and A have the same column rank. Likewise, if $B = E_t \ldots E_2E_1A$ where $E_t, \ldots, E_2, E_1$ is any succession of ($m \times m$) elementary matrices, we find column rank B = column rank A and, of course, row rank B = row rank A. Now, by correct choice of $E_t, \ldots, E_2, E_1$ we may take B to be the (unique[b]) reduced (row) echelon form of A. But, just as in the final matrix of Problem 5, if B has k non-zero rows it is fairly easy to see that the row rank and column rank of B are both equal to k. Hence row rank A and column rank A are equal (to k). This completes the proof. ●

PROBLEM 9.6

Change the matrix

$$\begin{bmatrix} 3 & 3 & 0 & 1 \\ 2 & 1 & -1 & -3 \\ 5 & 7 & 2 & 9 \end{bmatrix}$$

of Problem 9.5 to a reduced (row) echelon matrix B and verify that the row and column ranks of B are identical. ●

This common value of the row and column ranks of A is termed simply the **rank** of A.

Because of the uniqueness of the number of the non-zero rows in any echelon form of A we see (answering a question posed at the end of Chapter 2) that:

the dimension of the solution space for the system of equations $A\mathbf{x} = \mathbf{0}$ is the number of arbitrary constants needed in the full solution of $A\mathbf{x} = \mathbf{0}$

$$\begin{aligned}
&= (\text{number of unknowns}) - (\text{number of non-zero rows in any echelon form for } A) \\
&\{= (\text{number of unknowns}) - (\text{maximum number of 'independent' equations in } A\mathbf{x} = \mathbf{0})\} \\
&= (\text{number of cols of } A) - (\text{row rank of } A) \\
&= (\text{number of cols of } A) - (\text{column rank of } A).
\end{aligned}$$

[Notice that we get this without calling on the uniqueness of the reduced (row) echelon form of A.]

By defining the **nullity** of A to be the dimension of the null space of A (see Examples 1(vi) of Chapter 7) we obtain a result for matrices which we rederive later in a more general setting. [See Theorem 2 of Chapter 10.]

• *Corollary 1*

Let A be an $m \times n$ matrix. Then nullity $A = n -$ rank A, in brief: rank + nullity = n. ●

PROBLEM 9.7

Check this result on the matrices of Problem 9.5. ●

From Theorem 1 we can immediately deduce two simple criteria which generalise results we have already noted for $\mathbb{R}^n$ (just prior to Theorem 1).

• *Corollary 2*

Let V be a vector space of finite dimension n. Then:

(i) each set C of more than n vectors in V must be LD.

[b]We now admit that this result is too long to prove here. See, for example, Finkbeiner, D. T. *Matrices and Linear Transformations*.

(ii) no set D of fewer than n vectors in V can span V.
(In particular neither type of set can be a basis for V.)

PROOF
Let B be a basis for V. If C were LI Theorem 1 would give $|C| \leq |B|$ (since B spans V), a contradiction. If D were to span V then Theorem 1 would show $|D| \geq |B|$ (since B is an LI set). ●

Examples 4

(i) $\{(-43, -26, -9, 8, 25, \ldots), (71, 61, 51, 41, \ldots), (4, 3, 2, 1, 0, -1, -2, \ldots)\}$ is not an LI subset of the vector space of arithmetic progressions. (In brief: there are too *many* vectors to be independent.)
(ii) $3 + x + 4x^2 + x^3, 5 + 9x + 2x^2 + 6x^3, 5 + 3x + 5x^2 + 8x^3$ cannot span $\mathbb{R}_3[x]$. (In brief: there are too *few* vectors to span.)

Surely each **proper subspace** S of a finite dimensional vector space V (i.e. $S \subset V$) is also finite dimensional (as a vector space in its own right) and with smaller dimension? Is this not obvious? Well, not without proof! For the simple proof we need the following theorem.

• Theorem 4

Any LI set of vectors in the finite dimensional vector space V can be extended to a basis of V.

PROOF
Suppose that $W_k = \{\mathbf{v}_1, \mathbf{v}_2, \ldots, \mathbf{v}_k\}$ is an LI subset of V. If $\text{sp}\{W_k\} \subset V$ then there exists $\mathbf{v}_{k+1}$ which is not a linear combination of $\mathbf{v}_1, \mathbf{v}_2, \ldots, \mathbf{v}_k$. Thus $W_{k+1} = \{\mathbf{v}_1, \mathbf{v}_2, \ldots, \mathbf{v}_k, \mathbf{v}_{k+1}\}$ is LI. (It could not be LD. Why not?) If $\text{sp}\{W_{k+1}\} \subset V$ repeat this step. Assuming V has (finite) dimension n this process must stop (after $n - k$ steps! – exactly?) at which point we shall have $\text{sp}\{W_n\} = V$. ●

An easy example is the following.

Example 5

$\{(3, 1, 2), (0, 2, 4)\}$ is an LI set in $\mathbb{R}^3$. Then $\{(3, 1, 2), (0, 2, 4), (0, 0, 1)\}$ is a basis for $\mathbb{R}^3$. (Is it? And could we not take $(1, 0, 0)$ or $(0, 1, 0)$ or $(617, -e^2, 1212)$ as the 'third' element instead? Please check!)

So now for the result mentioned above concerning proper subspaces.

• Theorem 5

Let S be a subspace of the finite dimensional vector space V. Then $\dim S \leq \dim V$.

PROOF
Let $\{\mathbf{v}_1, \mathbf{v}_2, \ldots, \mathbf{v}_k\}$ be a basis for S. This LI set of elements of V can, by Theorem 4, be extended to a basis of V. Hence, immediately, $\dim S \leq \dim V$. ●

Did you like that proof? *You shouldn't have done! It's invalid!!* Why? (Read the first sentence again . . . and again. . . .)

Here is a better proof – and a better stated theorem!

• Theorem 5′

Let S be a subspace of the finite dimensional vector space V. Then S *is also finite dimensional* and $\dim S \leq \dim V$.

PROOF

Let $\dim V = n$ and let $T = \{\mathbf{u}_1, \mathbf{u}_2, \ldots, \mathbf{u}_k\}$ be any LI set of vectors of V all lying in S. Then $k \leq n$ (by Theorem 1 or Corollary 2(i)). If $\mathrm{sp}\{T\} = S$ then T is a finite basis for S. Extending this set to a basis of V (by Theorem 4) we see that the desired result holds. If T does not span S choose $\mathbf{u}_{k+1} \in S \setminus \mathrm{sp}\{T\}$. Then $\{\mathbf{u}_1, \mathbf{u}_2, \ldots, \mathbf{u}_k, \mathbf{u}_{k+1}\}$ is an LI subset of S (and of V (?)) as in Theorem 4. Note that, then, $k + 1 \leq n$ by Theorem 1 (or Corollary 2(i)). When this process stops (after $\leq n - k$ steps) we have an LI spanning set (i.e. a basis) B_S, say, for S. [In particular we have *proved* that S is finite dimensional. We have not assumed it.] Since B_S is an LI subset of V we may extend it (by Theorem 4) to a basis of V. If $S = V$ no actual extension takes place; if $S \subset V$ then $\dim S < \dim V$. ●

Theorem 4 also leads quickly to Grassmann's theorem, which follows.

• Theorem 6

Let S and T be subspaces of the finite dimensional vector space V. Then

$$\dim S + \dim T = \dim(S + T) + \dim(S \cap T).$$

[Recall: $S + T = \{\mathbf{s} + \mathbf{t} : \mathbf{s} \in S, \mathbf{t} \in T\}$ is the smallest subspace of V which contains both S and T as subspaces. See Exercise 31(b) of Chapter 7.)

PROOF

Select a basis $L = \{\mathbf{u}_1, \mathbf{u}_2, \ldots, \mathbf{u}_\lambda\}$ of $S \cap T$. Using Theorem 4, extend it to a basis

$$M = \{\mathbf{u}_1, \mathbf{u}_2, \ldots, \mathbf{u}_\lambda, \mathbf{s}_1, \mathbf{s}_2, \ldots, \mathbf{s}_\mu\}$$

of S and to a basis

$$N = \{\mathbf{u}_1, \mathbf{u}_2, \ldots, \mathbf{u}_\lambda, \mathbf{t}_1, \mathbf{t}_2, \ldots, \mathbf{t}_\nu\}$$

of T. We claim that

$$K = \{\mathbf{u}_1, \mathbf{u}_2, \ldots, \mathbf{u}_\lambda, \mathbf{s}_1, \mathbf{s}_2, \ldots, \mathbf{s}_\mu, \mathbf{t}_1, \mathbf{t}_2, \ldots, \mathbf{t}_\nu\}$$

is then a basis of $S + T$.

Since each element of S (respectively, of T) is a linear combination of the $\mathbf{u}_i$ and the $\mathbf{s}_j$ (respectively the $\mathbf{u}_i$ and the $\mathbf{t}_k$) K certainly spans the subspace $S + T$. (All right?) To show K is an LI set assume

$$\alpha_1\mathbf{u}_1 + \alpha_2\mathbf{u}_2 + \ldots + \alpha_\lambda\mathbf{u}_\lambda + \beta_1\mathbf{s}_1 + \beta_2\mathbf{s}_2 + \ldots + \beta_\mu\mathbf{s}_\mu + \gamma_1\mathbf{t}_1 + \gamma_2\mathbf{t}_2 + \ldots + \gamma_\nu\mathbf{t}_\nu = \mathbf{0}. \quad (9.1)$$

Then $\alpha_1\mathbf{u}_1 + \alpha_2\mathbf{u}_2 + \ldots + \alpha_\lambda\mathbf{u}_\lambda + \beta_1\mathbf{s}_1 + \beta_2\mathbf{s}_2 + \ldots + \beta_\mu\mathbf{s}_\mu = -(\gamma_1\mathbf{t}_1 + \gamma_2\mathbf{t}_2 + \ldots + \gamma_\nu\mathbf{t}_\nu)$ which is therefore an element of $S \cap T$ (why?). This implies that there exist scalars δ_i, say, such that

$$\gamma_1\mathbf{t}_1 + \gamma_2\mathbf{t}_2 + \ldots + \gamma_\nu\mathbf{t}_\nu = \delta_1\mathbf{u}_1 + \delta_2\mathbf{u}_2 + \ldots + \delta_\lambda\mathbf{u}_\lambda \quad \text{(why?)}$$

and this tells us that

$$\gamma_1\mathbf{t}_1+\gamma_2\mathbf{t}_2+\ldots+\gamma_\nu\mathbf{t}_\nu-\delta_1\mathbf{u}_1-\delta_2\mathbf{u}_2-\ldots-\delta_\lambda\mathbf{u}_\lambda=\mathbf{0}.$$

We deduce, immediately, that

$$\gamma_1=\gamma_2=\ldots=\gamma_\nu=\delta_1=\delta_2=\ldots=\delta_\lambda=0 \quad \text{(why?)}.$$

In a similar manner we can show that $\beta_1=\beta_2=\ldots=\beta_\mu=0$ and then, from (9.1), that $\alpha_1=\alpha_2=\ldots=\alpha_\lambda=0$. This shows that K is a basis for $S+T$. Now, counting the number of **u**, **s** and **t** finishes the proof. (Isn't that wonderful?) ●

PROBLEM 9.8

Let $S=\text{sp}\{(1,0,0,0),(0,1,0,0),(0,0,1,0)\}$ and

$$T=\text{sp}\{(1,2,3,4),(5,6,7,8),(9,10,11,12)\}\subseteq\mathbb{R}^4.$$

Work out $\dim S$ and $\dim T$ and show that $S+T=\mathbb{R}^4$. Deduce the dimension of $S\cap T$. ●

Theorem 4 also generates the following labour-saving corollary.

● *Corollary 3*

Let V be a finite dimensional vector space with $\dim V=n$ and let $\{\mathbf{v}_1,\mathbf{v}_2,\ldots,\mathbf{v}_n\}$ be a set of n vectors in V. Then:

(i) If $\{\mathbf{v}_1,\mathbf{v}_2,\ldots,\mathbf{v}_n\}$ is an LI set then $\text{sp}\{\mathbf{v}_1,\mathbf{v}_2,\ldots,\mathbf{v}_n\}=V$. Hence $\{\mathbf{v}_1,\mathbf{v}_2,\ldots,\mathbf{v}_n\}$ is a basis for V.

(ii) If $\text{sp}\{\mathbf{v}_1,\mathbf{v}_2,\ldots,\mathbf{v}_n\}=V$ then $\{\mathbf{v}_1,\mathbf{v}_2,\ldots,\mathbf{v}_n\}$ is an LI set. Hence $\{\mathbf{v}_1,\mathbf{v}_2,\ldots,\mathbf{v}_n\}$ is a basis for V.

PROOF

(i) Set $U=\text{sp}\{\mathbf{v}_1,\mathbf{v}_2,\ldots,\mathbf{v}_n\}$. Then $\dim U=n$ ($=\dim V$). Hence $U=V$.

(ii) If $\{\mathbf{v}_1,\mathbf{v}_2,\ldots,\mathbf{v}_n\}$ is LD then (by Theorem 4 of Chapter 8) some LI subset also spans V. Thus V (of dimension n) has a basis of $<n$ elements – a contradiction. ●

Example 6

Do $(1,2,3,1),(1,2,-1,4),(-1,0,3,-3),(2,2,-1,5)$ form a basis for $\mathbb{R}^4$?

SOLUTION

$$\begin{bmatrix}1&2&3&1\\1&2&-1&4\\-1&0&3&-3\\2&2&-1&5\end{bmatrix}\rightarrow\text{(echelon)}\begin{bmatrix}1&2&3&1\\0&2&6&-2\\0&0&-4&3\\0&0&0&1\end{bmatrix}.$$

These four row vectors clearly form an LI set (Theorem 3(a) of Chapter 8). Hence, by Corollary 3 they form a basis for $\mathbb{R}^4$. (Of course, from their form, you can see also immediately (cf. Exercise 29(ii) of Chapter 8) that these final four vectors span $\mathbb{R}^4$.

Applications

There are many applications of the concept of basis. The one we have chosen concerns linear programming, mainly because of its close association with systems of linear equations. We shall give you an example of a special case; for more details see, for example, Gass, S. I. *Linear Programming; Methods and Applications*.

Consider the problem of minimising the quantity $x + y$ given that $x - 2y \le 2$, $3x - y \ge 4$, $2x + y \ge 6$ and that $x \ge 0$ and $y \ge 0$. [I leave you to imagine real-life settings in which $x + y$ might be the cost of manufacturing some item, the inequalities being constraints on the size of an order and the capacity of a factory to produce the goods, etc.] This 2-dimensional problem (it involves only x and y) can be solved graphically by identifying the region of the x–y plane in which the five inequalities hold simultaneously and by finding the value of c for which the line $x + y = c$ just meets this region at a 'corner' point. [Of course real-life problems can come with thousands of variables and constraints and we need a method which works equally well in multidimensional space.]

The above problem can be rewritten: minimise $x + y$ subject to $x - 2y \le 2$, $-3x + y \le -4$, $-2x - y \le -6$; $x \ge 0$, $y \ge 0$; and then as: minimise $x + y$ subject to

$$\begin{aligned} x - 2y + u \qquad\qquad &= 2 \\ -3x + y \quad + v \qquad &= -4 \\ -2x - y \qquad\quad + w &= -6 \end{aligned}$$

with $x, y, u, v, w \ge 0$ which we may again rewrite as

$$x\begin{bmatrix}1\\3\\2\end{bmatrix} + y\begin{bmatrix}-2\\-1\\1\end{bmatrix} + u\begin{bmatrix}1\\0\\0\end{bmatrix} + v\begin{bmatrix}0\\-1\\0\end{bmatrix} + w\begin{bmatrix}0\\0\\-1\end{bmatrix} = \begin{bmatrix}2\\4\\6\end{bmatrix} \qquad (9.2)$$

By inspection, one solution is $x = 2$, $y = 2$, $u = 4$, $v = w = 0$. [Note that, for example, $u = 2$, $v = -4$, $w = -6$, $x = y = 0$ is not!] x, y, u are then called **basic variables**, with v and w being **non-basic**.

For these x, y, u, v, w we have $x + y$ (= z, say) = 4. Is this the required minimum? Note that

$$\begin{bmatrix}1\\3\\2\end{bmatrix}, \begin{bmatrix}-2\\-1\\1\end{bmatrix}, \begin{bmatrix}1\\0\\0\end{bmatrix}$$

form a basis for $\mathbb{R}^3$. Hence we can solve (9.2) for x, y, u. We obtain

$$\begin{aligned} x \qquad - \tfrac{1}{5}v - \tfrac{1}{5}w &= 2 \\ y \quad + \tfrac{2}{5}v - \tfrac{3}{5}w &= 2 \\ u + \ \ v - \ \ w &= 4 \end{aligned} \qquad (9.3)$$

from which we find that $z\ (= x + y) = 4 - \tfrac{1}{5}v + \tfrac{4}{5}w$. Again we see that $z = 4$ corresponds to taking $v = w = 0$. The theory now tells us to reduce the value of z further by increasing the value of v (from 0) whilst letting one of x, y, u decrease to 0. Which one? Certainly not x

since, in (9.3), increasing v *increases* x. In the other equalities of (9.3) v may be increased to 5 before y has to become negative whereas v can only be increased to 4 before u is forced to become negative. Thus we are limited to increasing v up to 4 and this determines that it is u which becomes a non-basic variable. That is we solve for x, y, v in terms of u and w. Doing so we obtain

$$\begin{aligned} x \quad + \tfrac{1}{5}u \qquad - \tfrac{2}{5}w &= \tfrac{14}{5} \\ y - \tfrac{2}{5}u \qquad - \tfrac{1}{5}w &= \tfrac{2}{5} \\ u + v - \quad w &= 4 \end{aligned}$$

and then $z = \frac{16}{5} + \frac{1}{5}u + \frac{3}{5}w$.

Since $u, v \geq 0$, it is clear that the minimum value of z is $\frac{16}{5}$ corresponding to $u = w = 0$, that is, to the basis

$$\left\{ \begin{bmatrix} 1 \\ 3 \\ 2 \end{bmatrix}, \begin{bmatrix} -2 \\ -1 \\ 1 \end{bmatrix}, \begin{bmatrix} 0 \\ -1 \\ 0 \end{bmatrix} \right\}$$

of $\mathbb{R}^3$. [If we had wished to maximise $x + y$ we could have minimised $-x - y$ and taken the negative of the result obtained.]

This demonstration raises many more questions than it answers. Our hopes are that you will be convinced of the non-trivial use of the basis concept in a very important application and that you will be encouraged to read up more on the topic.

Summary

A **basis** for the (finite dimensional) vector space V is a subset $B = \{\mathbf{v}_1, \mathbf{v}_2, \ldots, \mathbf{v}_s\}$ of V which (i) spans V and (ii) is LI. Given B, each element $\mathbf{v}$ of V can be expressed as a linear combination $\alpha_1\mathbf{v}_1 + \alpha_2\mathbf{v}_2 + \ldots + \alpha_s\mathbf{v}_s$ of the $\mathbf{v}_i$ and the α_i are uniquely determined by $\mathbf{v}_i$. [We may therefore think of the α_i as the 'coordinates' of $\mathbf{v}$ with respect to the particular basis B.]

Most of our examples of vector spaces, being finite dimensional, necessarily have a finite basis: one proves that any two bases of a given finite dimensional vector space have the same number of elements. This unique number is the **dimension** of V. All subspaces of V, other than V itself, have smaller dimension than that of V. Furthermore $\mathbb{R}^n$ does indeed have dimension n. More important, the space of all solutions of the differential equation

$$\frac{\mathrm{d}^n y}{\mathrm{d}x^n} + a_{n-1}\frac{\mathrm{d}^{n-1} y}{\mathrm{d}x^{n-1}} + \ldots + a_1\frac{\mathrm{d}y}{\mathrm{d}x} + a_0 y = 0$$

also has dimension n.

For the $m \times n$ matrix A the **row rank** (**column rank**) of A is the dimension of the row space (column space) of A. These two ranks are equal so we call each the **rank** of A. The dimension of the subspace $\{\mathbf{x}\colon \mathbf{x} \in \mathbb{R}^n \text{ and } A\mathbf{x} = \mathbf{0}\}$ (the so-called null space) of $\mathbb{R}^n$ is called the **nullity** of A. We then have: rank A + nullity $A = n$. In particular, the dimension of the solution space of the system of equations $A\mathbf{x} = \mathbf{0}$ is the number of columns of A – the rank of A (the rank – rather than merely the number of rows of A – reflecting the number of 'independent' equations in the system $A\mathbf{x} = \mathbf{0}$).

In any n-dimensional vector space V any (sub)set of more than n vectors must be LD whilst no set of fewer than n vectors can span V. On the other hand, a set of exactly n vectors in V is LI if and only if it spans V. Thus to confirm that a set C, say, of n vectors in an n-dimensional vector space is a basis, one only needs to check either of the two conditions given in Definition 1.

The key result on bases is, of course, that, in any particular vector space, they are all of the same size, thus allowing the idea of dimension to be defined unambiguously. As we have noted, Cayley and Grassmann had no conceptual difficulty in passing beyond three dimensions, though Cayley's ideas were rooted in straightforward n-tuples whereas Grassmann's were more geometrical. Grassmann's 1862 *Ausdehnungslehre* contains the ideas of basis, dimension and subspace and it is here for the first time that the formula

$$\dim S + \dim T = \dim (S + T) + \dim (S \cap T)$$

appears. It is only with the result that a matrix has a well-defined rank (the idea is due to Sylvester (1850), the name to Frobenius (1879)) that a proper treatment of the solution spaces of linear systems can be given. The corresponding notion of nullity (for square matrices only) was introduced by Sylvester in 1884. Four years later Peano explicitly defined the concept of dimension and remarks that polynomials form an infinite dimensional space.

Giuseppe Peano (27 August 1858–20 April 1932) was born in Cuneo, Italy, one of four brothers and a sister. For educational reasons he moved to Turin in his early teens. He was a successful student, winning prizes and honours at school, college and at the University of Turin. In 1880 he joined the staff at the university, becoming a professor in 1890. Peano's name is inextricably linked with symbolic logic and axiomatics, in particular with his list of postulates for the positive integers, but his real concern seems to have been in encouraging a more critical attitude in mathematical discussion. He himself supplied several examples showing that commonly held beliefs were simply wrong. The most famous, wholly counter to everyone's intuition, was the 'space filling' curve whose coordinates $(x(t), y(t))$ change continuously with t $(0 \le t \le 1)$ and yet passes through every point of the square with corners at (0, 0), (1, 0), (1, 1) and (0, 1). Peano was, for 24 years, president of the academy set up to promote the international language Interlingua.

The application given above has a much more recent history. Maximising resources available becomes especially critical in wartime. The mathematical formulation of such problems was developed by George Dantzig who proposed the above **simplex method** of solution in 1947.

EXERCISES ON CHAPTER 9

1. Determine whether or not the following sets are bases for the vector spaces indicated;
   (i) For $\mathbb{R}^3$ (a) $\{(1, 2, 3), (4, 5, 6), (7, 8, 9)\}$; (b) $\{(1, 2, 3), (6, 5, 4), (7, 8, 9)\}$;
   (c) $\{(1, 0, 3), (0, 5, 0), (7, 0, 9), (11, 13, 15)\}$; (d) $\{(13, 12, 11), (10, 9, 8)\}$;
   (e) $\{(\sqrt{1}, \sqrt{2}, \sqrt{3}), (\sqrt{4}, \sqrt{5}, \sqrt{6}), (\sqrt{7}, \sqrt{8}, \sqrt{9})\}$.
   [(c) and (d) can be done much more easily after you have read Corollary 2.]
   (ii) For $\mathbb{R}^4$: (a) $\{(2, 1, 7, 7), (-5, -3, 53, 48), (-1, 0, 2, 3), (1, 2, 10, 15)\}$;
   (b) $\{(1, 1, 1, 1), (1, 1, 1, 0), (1, 1, 0, 0), (1, 0, 0, 0)\}$;
   (c) $\{(3, 1, 4, 1), (5, 9, 2, 6), (5, 3, 5, 8), (9, 7, 9, 3)\}$;
   (d) $\{(1, 1, -1, 2), (2, 1, 1, -1,), (-1, 2, 1, 1), (1, -1, 2, 1)\}$;
   (e) $\{(2, -3, 4), (-5, 6, -7), (1, 1, 1)\}$.
2. (a) Is $\{1 + 2x, 3 + 4x, 1 + x + 2x^2 + 2x^3, 1 - x^2 - x^3\}$ a basis for $\mathbb{R}_3[x]$?
   (b) What about $\{3 + x + 4x^2 + x^3, 5 + 9x + 2x^2 + 6x^3, 5 + 3x + 5x^2 + 8x^3, 9 + 7x + 9x^2 + 3x^3\}$?
3. (a) Show that $\left\{\begin{bmatrix}1 & 0\\0 & 0\end{bmatrix}, \begin{bmatrix}0 & 1\\0 & 0\end{bmatrix}, \begin{bmatrix}0 & 0\\1 & 0\end{bmatrix}, \begin{bmatrix}0 & 0\\0 & 1\end{bmatrix}\right\}$ is a basis of $M_2(\mathbb{R})$.
   (b) Is $\left\{\begin{bmatrix}1 & 1\\0 & 1\end{bmatrix}, \begin{bmatrix}2 & 3\\1 & 2\end{bmatrix}, \begin{bmatrix}5 & 8\\3 & 5\end{bmatrix}, \begin{bmatrix}13 & 21\\8 & 13\end{bmatrix}\right\}$ a basis of $M_2(\mathbb{R})$?
   (c) And what about $\left\{\begin{bmatrix}3 & 1\\4 & 1\end{bmatrix}, \begin{bmatrix}5 & 9\\2 & 6\end{bmatrix}, \begin{bmatrix}5 & 3\\5 & 8\end{bmatrix}, \begin{bmatrix}9 & 7\\9 & 3\end{bmatrix}\right\}$?
   (d) Is $\{I, X, Y, Z\}$, as in Exercise 7 of Chapter 3, a basis for $M_2(\mathbb{C})$ over $\mathbb{C}$?
   What do you notice about Exercises 1(ii)(c), 2(b) and 3(c)?
4. Find bases for the vector spaces in Exercises 3(a)(iv), (v) and (vi) at the end of Chapter 7. Show that any $n \times n$ matrix can be expressed as a sum of a symmetric matrix and a skew-symmetric matrix. [Hint: $A + A^T$ $(A - A^T)$ is symmetric (skew-symmetric).] Hence find a basis for $M_3(\mathbb{R})$ comprising only symmetric and skew-symmetric matrices.
5. Can you find a basis for $\mathbb{R}_3[x]$ which contains $1 + x, -1 + x^2, 1 + x^3$ and whose fourth element is a polynomial with all coefficients equal?
6. Show that $\{a + ib, c + id\}$ form a basis for $\mathbb{C}$ over $\mathbb{R}$ if and only if $\det \begin{bmatrix}a & b\\c & d\end{bmatrix} \neq 0$.

7. Exhibit 10 different bases for $\mathbb{R}^3$.

8. Show that $\mathbb{R}[x]$ cannot have a finite basis. (See Examples 5(iv) in Chapter 7.)

9. Determine the coordinates of $(4, 7, 2, -5)$ with respect to each of the following bases of $\mathbb{R}^4$:

(i) $\{(1, 0, 0, 0), (0, 1, 0, 0), (0, 0, 1, 0), (0, 0, 0, 2)\}$;
(ii) $\{(4, 0, 0, 0), (0, 7, 0, 0), (0, 0, 2, 0), (1, 1, 1, 1)\}$.

10. Find a basis for each of the following subspaces:

(i) $\{(x, y, z): 4x + 5y - z = 0\}$ in $\mathbb{R}^3$; (ii) $\{(x, y, z, t): x + 3y = 5z + 7t\}$ in $\mathbb{R}^4$;
(iii) $\{(x, y, z, w, t): 3x + y + 4z + w + 5t = 9x + 2y + 6z + 5w + 3t = 3x - 2z + 3w - 7t = 0\}$ in $\mathbb{R}^5$; (iv) $\{p(x): p(3) = 0\}$ in $\mathbb{R}_4[x]$; (v) $\{p(x): p(3) = p(1) = 0\}$ in $\mathbb{R}_6[x]$.

11. (a) Find a basis for the solution space of the system of equations:

$$\begin{aligned} 5x-3y-3z-t&=0\\ 4x+6y-3z+2t&=0\\ 7x+y-3z-t&=0\\ 3x-3y+3z-5t&=0 \end{aligned}$$

(b) Let A be the matrix

$$\begin{bmatrix} 2 & 1 & -1 \\ -1 & 2 & 1 \\ 1 & -1 & 2 \end{bmatrix}$$

and let W be the subspace $\{\mathbf{v}: A\mathbf{v} = 2\mathbf{v}\}$ of $\mathbb{R}^3$. Find a basis for W.

12. Find a basis for the subspace $\{a_0 + a_1x + a_2x^2 + a_3x^3: a_0 + a_1 + a_2 + a_3 = 0\}$ of $\mathbb{R}_3[x]$.

13. Write down a basis for $\{(x, y, z, t): ax + by + cz + dt = 0$ (a, b, c, d are all fixed and all are non-zero)$\}$ which contains the vector $(-b - c - d, a, a, a)$.

14. (a) Let $\mathbf{v}_1 = (1, 2, 1, -1)$ and $\mathbf{v}_2 = (2, -1, 1, 1)$. Show that $\mathbf{v}_1.\mathbf{v}_2 = \mathbf{0}$. Find further vectors $\mathbf{v}_3$, $\mathbf{v}_4$ such that $\mathbf{v}_i.\mathbf{v}_j = \mathbf{0}$ for all $i \neq j$ ($1 \leq i \leq j \leq 4$). ($\{\mathbf{v}_1, \mathbf{v}_2, \mathbf{v}_3, \mathbf{v}_4\}$ is then called an **orthogonal** basis for $\mathbb{R}^4$.) [Hint: set $\mathbf{v}_3 = (a, b, c, d)$ and determine a, b, c, d from the orthogonality conditions. Repeat for $\mathbf{v}_4$.]

(b) Find an (orthogonal) basis $\{p_0(x), p_1(x), p_2(x)\}$ for $\mathbb{R}_2[x]$ such that, for $0 \leq i, j \leq 2$, we have $\int_0^1 p_i(x)p_j(x)dx = 1$ (if $i = j$) and 0 (if $i \neq j$). [Hint: take $p_0(x) = a_0$, $p_1(x) = a_1 + b_1x$, etc. and determine $a_0, a_1, b_1, \ldots$ etc. successively.]

15. Let $\{\mathbf{v}_1, \mathbf{v}_2, \ldots, \mathbf{v}_r\}$ be a set of vectors in the vector space V. Show that, if each element of V is expressible uniquely as a linear combination of elements of $\{\mathbf{v}_1, \mathbf{v}_2, \ldots, \mathbf{v}_r\}$ then $\{\mathbf{v}_1, \mathbf{v}_2, \ldots, \mathbf{v}_r\}$ is a basis for V. [Hint: you are given that each $\mathbf{v} \in V$ is expressible as a linear combination and that each such expression is unique.]

16. Let $\{\mathbf{v}_1, \mathbf{v}_2, \ldots, \mathbf{v}_k\}$ span the vector space V and $\{\mathbf{w}_1, \mathbf{w}_2, \ldots, \mathbf{w}_k\}$ (same k) be LI in V. Show that each is a basis for V. [Hint: if $\{\mathbf{v}_1, \mathbf{v}_2, \ldots, \mathbf{v}_k\}$ is LD a proper subset of it will also span V. This contradicts Theorem 1. Now use Corollary 3.]

17. (a) Suppose that $\{\mathbf{v}_1, \mathbf{v}_2, \ldots, \mathbf{v}_k\}$ spans the vector space V but that no proper subset does. Prove that $\{\mathbf{v}_1, \mathbf{v}_2, \ldots, \mathbf{v}_k\}$ is a basis for V.

(b) Suppose that $\{\mathbf{v}_1, \mathbf{v}_2, \ldots, \mathbf{v}_k\}$ spans the vector space V and that the subset $\{\mathbf{v}_{i_1}, \mathbf{v}_{i_2}, \ldots, \mathbf{v}_{i_t}\}$ is an LI set. Prove that, if no larger subset of $\{\mathbf{v}_1, \mathbf{v}_2, \ldots, \mathbf{v}_k\}$ is LI then $\{\mathbf{v}_{i_1}, \mathbf{v}_{i_2}, \ldots, \mathbf{v}_{i_t}\}$ is a basis for V.

18. Show that there are, in $\mathbb{R}^4$, infinitely many different subspaces of dimension 1. Are there, likewise, infinitely many of dimension (a) 2; (b) 3; (c) 4; (d) 0?

19. What is the dimension of the subspace of $M_n(\mathbb{R})$ comprising: (i) all upper triangular matrices; (ii) all skew-symmetric matrices?

20. For which α, if any, is $\{(1, 1, 1, \alpha), (1, 1, \alpha, \alpha), (1, \alpha, \alpha, \alpha), (\alpha, \alpha, \alpha, \alpha)\}$ a (i) 2-dimensional; (ii) 3-dimensional subspace of $\mathbb{R}^4$?

21. Let AP(4,3) be the set of 4×3 matrices of the form

$$\begin{bmatrix} a & a+b & a+2b \\ a+c & a+c+b & a+c+2b \\ a+2c & a+2c+b & a+2c+2b \\ a+3c & a+3c+b & a+3c+2b \end{bmatrix},$$

where $a, b, c \in \mathbb{R}$. Prove that AP(4,3) is a vector space and find a basis for it.

22. (a) For the matrix

$$\begin{bmatrix} 1 & 1 & 1 & 1 & 6 \\ -1 & 1 & 1 & 1 & 1 \\ 1 & -1 & 1 & 1 & 2 \\ 1 & 1 & -1 & -1 & 3 \end{bmatrix}$$

find its rank and nullity. Is their sum what you would expect it to be?

(b) Let A be any matrix. Show that the equation $A\mathbf{x} = \mathbf{b}$ has a solution if and only if rank $A = \text{rank}[A \mid \mathbf{b}]$.

(c) Show that the $n \times n$ matrix A is non-singular if and only if rank $A = n$. [Thus yet another equivalent condition can be added to Theorem 5 of Chapter 4.]

(d) Find a basis for $M_{2\times3}(\mathbb{R})$ comprising matrices all of rank 2.

23. Let A be an $m \times n$ matrix and B an $n \times p$ matrix. Prove:

(α) $\text{rank}(AB) \le \min\{\text{rank } A, \text{rank } B\}$ [Hint: use Exercise 24 of Chapter 7 and row rank = column rank.] This is part of Sylvester's **law of nullity** (1884), the other part being:

(β) $\text{rank}(AB) \ge \text{rank } A + \text{rank } B - n$. Why nullity? Well, the two results quoted are equivalent (for $n \times n$ matrices) to:

(γ) $\text{nullity}(AB) \ge \max\{\text{nullity } A, \text{nullity } B\}$;

(δ) $\text{nullity}(AB) \le \text{nullity } A + \text{nullity } B$.

Can you prove these equivalences?

24. Prove that, for any matrix A, $\text{rank}(A^TA) = \text{rank } A$. [Hint: (i) $Ax = 0 \Rightarrow A^TAx = 0$; (ii) $A^TAx = 0 \Rightarrow x^TA^TAx = 0 \Rightarrow$ (why?) $Ax = 0$.] How does this help answer Tutorial Problem 7.4 of Chapter 7?

25. If V is n-dimensional as a vector space over $\mathbb{C}$ what is its dimension when regarded as a vector space over $\mathbb{R}$?

26. Extend $T = \{(1, 2, 3, 0), (2, 3, 4, 1), (3, 4, 5, 3)\}$ to a basis of $\mathbb{R}^4$. Now find another extension of T not using any multiple of the vector just chosen.

27. Find a basis in $\mathbb{R}^4$ for the intersection of the subspace $\{(x, y, z, t): x + y + z + t = 0\}$ with that of Exercise 10(ii). State the dimension of the sum of these two subspaces.

28. Find the dimension of the subspace $S = \text{sp}\{(1, 3, 1, 4), (1, 1, 0, 1), (4, 6, 1, 7)\}$. Does either of the subspaces $\text{sp}\{(1, 5, 3, 8), (1, 9, -2, 5), (2, 4, 3, 7)\}$ or $\text{sp}\{(3, 5, 1, 6), (5, 7, 1, 8), (-1, 1, 1, 2)\}$ intersect S in more than $\{(0, 0, 0, 0)\}$?

29. By writing down two bases, exhibit two subspaces S_1 and S_2 of R^{10} such that each has dimension 7 and their intersection has dimension 4. What can you say about the subspace $S_1 + S_2$?

30. Show that if S_1, S_2 are subspaces of a vector space V such that $S_1 \cap S_2 = \{\mathbf{0}\}$, then $\dim(S_1 + S_2) = \dim S_1 + \dim S_2$. Show that, if $S_2 = \{\mathbf{0}\}$, then, for the last formula to remain valid, we must define $\dim\{\mathbf{0}\}$ to be 0. Deduce also that, if $\dim S_1 + \dim S_2 > \dim V$ then $S_1 \cap S_2 > \{\mathbf{0}\}$.

31. Let $S_1 = \text{sp}\{\mathbf{c}_1, \mathbf{c}_2, \mathbf{c}_3\}$ and $S_2 = \text{sp}\{\mathbf{d}_1, \mathbf{d}_2, \mathbf{d}_3\}$ where $\mathbf{c}_1 = (2, 3, 5, 4, 7)$, $\mathbf{c}_2 = (1, 2, 4, 4, 6)$, $\mathbf{c}_3 = (1, 0, -2, -4, -4)$, $\mathbf{d}_1 = (3, 4, 7, 5, 5)$, $\mathbf{d}_2 = (2, 4, 6, 4, 5)$, $\mathbf{d}_3 = (1, 2, 2, 0, -1)\}$. Find a basis $\{\mathbf{x}_1, \ldots, \mathbf{x}_k, \mathbf{u}_1, \ldots, \mathbf{u}_l, \mathbf{v}_1, \ldots, \mathbf{v}_m\}$ for $\mathbb{R}^5$ such that $\{\mathbf{x}_1, \ldots, \mathbf{x}_k\}$, $\{\mathbf{x}_1, \ldots, \mathbf{x}_k, \mathbf{u}_1, \ldots, \mathbf{u}_l\}$, $\{\mathbf{x}_1, \ldots, \mathbf{x}_k, \mathbf{v}_1, \ldots, \mathbf{v}_m\}$ are bases for $S_1 \cap S_2$, S_1 and S_2 respectively.

COMPUTER PACKAGE PROBLEMS

1. Find bases for the row space and the column space of the matrix

$$\begin{bmatrix} 88 & 85 & 45 & 71 & 61 & -43 \\ 99 & 89 & -62 & -97 & 79 & 79 \\ 110 & 97 & -30 & -45 & 80 & 40 \\ 99 & 93 & 77 & 123 & 62 & -82 \\ -165 & 87 & 128 & 90 & 101 & 53 \end{bmatrix}.$$

2. Find all x, if any, for which $\{(x, 5, 7, x), (13, 17, 19, 23), (29, 31, 37, 41), (x, 47, 53, x)\}$ is not a basis for $\mathbb{R}^4$.

3. Find a basis B, say, for the solution space of the linear system

$$\begin{aligned} 47x + 57y - 94z + 36t - 36w &= 0 \\ 28x + 47y - 56z + 53t + 18w &= 0 \\ 37x + 84y - 74z + 66t - 81w &= 0. \end{aligned}$$

Which triples of vectors from the standard basis $\{(1, 0, 0, 0, 0), \ldots, (0, 0, 0, 0, 1)\}$ of $\mathbb{R}^5$ can be used to extend B to a basis of $\mathbb{R}^5$?

4. Is $\{(0, 1, 1, 1), (1, 0, 1, 1), (1, 1, 0, 1), (1, 1, 1, 0)\}$ a basis for $\mathbb{R}^4$? Consider the same question for $\mathbb{R}^5$ (i.e. use $(0, 1, 1, 1, 1), \ldots, (1, 1, 1, 1, 0)$). Do you see any pattern? Either: find an example where the pattern breaks down or give a proof that the pattern always persists.

5. By taking the vectors (117, 220, 314, 410), (105, 237, 337, 433), (1266, 303, 615, 1137), (142, –709, –805, –931), (920, –531, –781, –589), (100, 220, 310, 400) as the columns of a 4×6 matrix and by applying row reduction, find a subset of the given set of vectors which is a basis for the space they span.

6. Let S_1, S_2 be the subspaces of $\mathbb{R}^6$ spanned by (i) the first three rows; (ii) the last three rows, of the matrix in Problem 1. Find a basis for the subspace $S_1 \cap S_2$ of $\mathbb{R}^6$. Do the same for the first two and the last two columns of that matrix.

7. Find, by obtaining the general solution the appropriate homogeneous system of 9(?) simultaneous linear equations, the dimension of the vector space of 4×4 magic squares. Try to guess the dimension before finding it.

8. π is (approximately) 3.14159 26535 89793 23846 26433 83279 50288 41971 69399 37510. . . .Look at sets of vectors such as {(3, 1, 4), (1, 5, 9), (2, 6, 5)}, {(2, 6, 5), (3, 5, 8), (9, 7, 9)}, {(3, 1, 4, 1), (5, 9, 2, 6), (5, 3, 5, 8), (9, 7, 9, 3)} to see if you can find any instances where the sets are not bases of $\mathbb{R}^3$, $\mathbb{R}^4$, etc. as appropriate.

9. Let $\mathbf{v}_1$, $\mathbf{v}_2$, $\mathbf{v}_3$, $\mathbf{v}_4$ be as in Problem 1 of Chapter 7 and let $\mathbf{b}_1$, $\mathbf{b}_2$, $\mathbf{b}_3$ be, respectively, the vectors (1, 0, 0, 0), (0, 1, 0, 0), (0, 0, 1, 0). {Using the MAPLE commands 'intbasis' and 'sumbasis' or their equivalents} find a basis for: (i) the subspace which is the intersection of the subspaces $\text{sp}\{\mathbf{v}_1, \mathbf{v}_2, \mathbf{v}_3\}$ and $\text{sp}\{\mathbf{v}_4, \mathbf{b}_1, \mathbf{b}_2\}$ and (ii) the smallest subspace of $\mathbb{R}^4$ containing both of the subspaces $\text{sp}\{\mathbf{v}_1, \mathbf{v}_2\}$ and $\text{sp}\{\mathbf{b}_1, \mathbf{b}_2\}$.

10. Repeat Problem 9 given the subspaces $S_1 = \{(x, y, z, t): x + 2y + 3z + 4t = 0\}$ and $S_2 = \{(x, y, z, t): x + y + z + t = x - y + 2z - 3t = 0\}$ of $\mathbb{R}^4$.

10 · Linear Transformations (and Matrices)

Mathematics is much concerned with functions which map one set into another. When these sets are both vector spaces, the natural maps to consider are those which respect the structure, that is send linear combinations to linear combinations. Such maps, called linear transformations, are studied here. In the exercises we shall confirm that, in finite dimensional vector spaces, linear transformations are nothing more than matrices in disguise and calculations involving linear maps are often carried out using matrices. Nevertheless it is often more useful to deal with transformations rather than with matrix representations. In particular, the matrix versions of a transformation are of little use in Definition 2 which enables us to express precisely what is meant by two vector spaces being 'essentially the same'.

When studying particular instances of vector spaces, certain types of function mapping one vector space to another keep on arising.

We begin with some examples.

Examples 1

(i) Consider the function T, mapping the vector space $\mathbb{R}^2$ to itself, given by $T(x, y) = (x, -y)$. Geometrically T reflects each point of the plane in the x-axis. Note that, if we replace each point (x, y) by the vector (matrix)

$$\begin{bmatrix} x \\ -y \end{bmatrix}$$

then we can equally well describe T by matrix multiplication:

$$T\begin{bmatrix} x \\ y \end{bmatrix} = \begin{bmatrix} x \\ -y \end{bmatrix} = \begin{bmatrix} 1 & 0 \\ 0 & -1 \end{bmatrix}\begin{bmatrix} x \\ y \end{bmatrix}$$

(see Problem 3.5 of Chapter 3).

(ii) Let R be the function from $\mathbb{R}^2$ to $\mathbb{R}^2$ which rotates each point of $\mathbb{R}^2$ (except (0, 0)!) through an angle ϑ anticlockwise about the origin (0, 0). Then the effect of R can also be described by matrix multiplication:

$$R\begin{bmatrix} x \\ y \end{bmatrix} = \begin{bmatrix} x\cos\vartheta - y\sin\vartheta \\ x\sin\vartheta + y\cos\vartheta \end{bmatrix} = \begin{bmatrix} \cos\vartheta & -\sin\vartheta \\ \sin\vartheta & \cos\vartheta \end{bmatrix}\begin{bmatrix} x \\ y \end{bmatrix}$$

(see Exercise 15 in Chapter 3).

PROBLEM 10.1

(i) Find the matrix which corresponds to the function $E:\mathbb{R}^2 \to \mathbb{R}^2$ given by $E(x, y) = (x, 2y)$.

(ii) Is there a matrix corresponding to the function $S:\mathbb{R}^2 \to \mathbb{R}^2$ given by $S(x, y) = (x^3, y^3)$?

(iii) Plot several points in $\mathbb{R}^2$ (including (0, 0), (0, 1), (1, 3), (2, 5), (3, 7), (1, 1) and (1, 0)) and also their images $F(x, y)$ where

$$F\begin{bmatrix} x \\ y \end{bmatrix} \text{ is given by } \quad (\alpha)\ F\begin{bmatrix} x \\ y \end{bmatrix} = \begin{bmatrix} 4 & 2 \\ 2 & 4 \end{bmatrix}\begin{bmatrix} x \\ y \end{bmatrix}; \qquad (\beta)\ F\begin{bmatrix} x \\ y \end{bmatrix} = \begin{bmatrix} 4 & 2 \\ 2 & 1 \end{bmatrix}\begin{bmatrix} x \\ y \end{bmatrix}.$$

Do the same for the functions E and S in parts (i) and (ii). ●

You should find that Problems 10.1(i) and (iii)(α) – as well as Examples 1(i) and (ii) – map $\mathbb{R}^2$ *onto* $\mathbb{R}^2$ (that is, each (u, v) in $\mathbb{R}^2$ is the image of some point (x, y) in $\mathbb{R}^2$) whereas all the points

$$F\begin{bmatrix} x \\ y \end{bmatrix}$$

in Problem 10.1(iii)(β) lie on a straight line in $\mathbb{R}^2$.

Of course there is no reason why we should restrict ourselves to maps from $\mathbb{R}^2$ to $\mathbb{R}^2$; given $\mathbb{R}^n$ and $\mathbb{R}^m$, each $m \times n$ matrix A yields a function M mapping the vector space $\mathbb{R}^n$ to the vector space $\mathbb{R}^m$ defined by $M(\mathbf{u}) = A\mathbf{u}$ (where $\mathbf{u}$ is a $n \times 1$ column matrix or vector).

Now all of the above examples (except Problem 10.1(ii)) have the property that they map straight lines in $\mathbb{R}^n$ to straight lines (or, maybe, single points) in $\mathbb{R}^m$. (This follows from the fact that for each $m \times n$ matrix A, for all $n \times 1$ matrices $\mathbf{u}$ and $\mathbf{v}$ and for all $\alpha \in \mathbb{R}$, we have $A(\mathbf{u} + \mathbf{v}) = A\mathbf{u} + A\mathbf{v}$ and $A(\alpha\mathbf{u}) = \alpha A\mathbf{v}$. See Exercise 15.) They are therefore examples of what are called *linear* maps (or transformations) – see Definition 1 below.

PROBLEM 10.2

Let $P:\mathbb{R}^3 \to \mathbb{R}^2$ given by $P(x, y, z) = (x + y, y + 2z)$. Show that the line $\{\alpha(2, -2, 1);\ \alpha \in \mathbb{R}\}$ in $\mathbb{R}^3$ is mapped by P to a single point in $\mathbb{R}^2$ and that the plane $\{(x, y, z)\colon 4x + 3y + 2z = 0\}$ in $\mathbb{R}^3$ is mapped to a line in $\mathbb{R}^2$. What is the equation of this line? ●

Before we give Definition 1, here is a different type of example – even more familiar, and very important in the theory of differential equations – which cannot be described by (finite sized) matrices.

Example 2

Let $\mathfrak{D}$ be the vector space of all those functions of a real variable which are differentiable for all $x \in \mathbb{R}$ and $\mathfrak{F}$ be the vector space of all functions from $\mathbb{R}$ to $\mathbb{R}$. Let $\mathrm{D} = \mathrm{d}/\mathrm{d}x$ denote the operation of differentiation (so that D is a mapping from the vector space $\mathfrak{D}$ to the vector space $\mathfrak{F}$). Then, for all 'vectors' (i.e. functions) f and g in $\mathfrak{D}$ and for all real numbers α, we have $\mathrm{D}(f + g) = \mathrm{D}f + \mathrm{D}g$ and $\mathrm{D}(\alpha f) = \alpha\mathrm{D}f$. (These are just the well-known results that the derivative of a sum of functions is the sum of their derivatives – and the corresponding result for scalar multiples of functions.)

Very many similar examples could be given (see the exercises). They all suggest the introduction of the following definition.

• Definition 1

Let V, W be any two vector spaces in which addition and scalar multiplication are denoted by $\oplus$, $\circ$ and $+$, $\bullet$ respectively. A function $T:V \to W$ is called a **linear transformation** iff, for all $\mathbf{u}, \mathbf{v} \in V$ and for all scalars α, (i) and (ii) both hold where

(i)[a] $T(\mathbf{u} \oplus \mathbf{v}) = T\mathbf{u} + T\mathbf{v}$ and (ii)[b] $T(\alpha \circ \mathbf{u}) = \alpha \bullet T\mathbf{u}$. •

Note that the summations and scalar multiplications on the two sides of the equalities in (i) and (ii) are taking place inside (possibly) different vector spaces. Hence the (temporary) use of different summation and product signs (to help us avoid possible confusion and the making of unwarranted assumptions at the beginning of our investigations).

Talking of unwarranted assumptions, what about the following?

PROBLEM 10.3

'The map $f:\mathbb{R} \to \mathbb{R}$ given by $f(x) = x^2$ is clearly not a linear transformation (since, for example, $f(2 + 3) = f(5) = 25$ but $f(2) + f(3) = 4 + 9$ ($\neq 25$)): the map $g:\mathbb{R} \to \mathbb{R}$ given by $g(x) = 2x + 1$ just as clearly is.' Am I right? (See Exercise 17 below.) •

We have seen that matrix multiplication and also ordinary differentiation provide us with linear maps. Let us try to get a better feel for what kind of functions define linear transformations by considering some more examples.

• Examples 3

QUESTION Which of the following are linear transformations?

(i) Let $T_1:\mathbb{R}^3 \to \mathbb{R}^2$ be given by[c] $T_1(x, y, z) = (x - 2y, y - 3z)$.

(ii) Let $T_2:\mathbb{R}^3 \to \mathbb{R}^2$ be given by $T_2(x, y, z) = (x + y + z, \{xyz\}^{1/3})$.
[Here $\{xyz\}^{1/3}$ denotes the (unique) real cube root of xyz.]

(iii) Let $T_3:\mathbb{R}^3 \to \mathbb{R}^2$ be given by $T_3(x, y, z) = (x + y, 1)$.

(iv) Let $\mathfrak{F}$ be the vector space of all functions from $\mathbb{R}$ to $\mathbb{R}$. Define $T_4: \mathfrak{F} \to \mathbb{R}$ by $T_4(f) = f(2)$. That is, with each function f in $\mathfrak{F}$ we associate the value of f at $x = 2$.

(v) Let $\mathcal{J}$ be the vector subspace of $\mathfrak{F}$ comprising those functions which can be integrated between 0 and 1. Define $T:\mathcal{J} \to \mathbb{R}$ by $T_5(g) = \int_0^1 g(x)\mathrm{d}x$.

ANSWERS[d]

(i) We use Definition 1 directly to show that T_1 is a linear transformation. To do this we take $\mathbf{u} = (x, y, z)$ and $\mathbf{v} = (x_1, y_1, z_1) \in \mathbb{R}^3$. Then

[a] The map of the sum is the sum of the maps!

[b] The map of the product . . .

[c] Note that we sometimes write $T\mathbf{u}$ or $T(x, y, z)$ rather than $T(\mathbf{u})$ or $T((x, y, z))$.

[d] In (i) only we use $\oplus$ and $\circ$ to help us distinguish vector addition and scalar multiplication from (ordinary) addition and multiplication in $\mathbb{R}$.

$$\begin{aligned}T_1(\mathbf{u}\oplus\mathbf{v}) &= T_1(x+x_1, y+y_1, z+z_1)\\ &= (\{x+x_1\}-2\{y+y_1\}, \{y+y_1\}-3\{z+z_1\})\\ &= (\{x-2y\}+\{x_1-2y_1\}, \{y-3z\}+\{y_1-3z_1\})\\ &= (x-2y, y-3z)\oplus(x_1-2y_1, y_1-3z_1)\\ &= T_1\mathbf{u}\oplus \mathrm{T}_1\mathbf{v}.\end{aligned}$$

Consequently the first part of Definition 1 is satisfied by the function T_1. What about the second part?

To check that, let α be any scalar. Then

$$\begin{aligned}T_1(\alpha\circ\mathbf{u}) &= T_1(\alpha x, \alpha y, \alpha z)\\ &= (\{\alpha x-2\alpha y\}, \{\alpha y-3\alpha z\})\\ &= (\alpha\{x-2y\}, \alpha\{y-3z\})\\ &= \alpha\circ(x-2y, y-3z) = \alpha\circ(T_1\mathbf{u}).\end{aligned}$$

Thus T_1 satisfies both parts of Definition 1 and so is a linear transformation.

Alternatively, we could have proved T_1 to be linear by showing (cf. Examples 1) that T is equally well described by matrix multiplication and by using the (proven) linear properties $A(\mathbf{u}+\mathbf{v}) = A\mathbf{u}+A\mathbf{v}$ and $A(\alpha\mathbf{u}) = \alpha A\mathbf{u}$ of matrix multiplication.

PROBLEM 10.4

Give this alternate proof. (That is, find the matrix describing T_1. Cf. Example 1(i) and Example 5.)

Examples 3 (cont.)

(ii) This time there appears to be no matrix capable of describing T_2 and so one might speculate that T_2 is not a linear transformation. Recall that to confirm this, it suffices to give one single instance of the breakdown of the requirements of Definition 1. Take a couple of vectors, say $\mathbf{u} = (1, 2, 3)$ and $\mathbf{v} = (4, 5, 6)$, essentially at random. Then $\mathbf{u}+\mathbf{v} = (5, 7, 9)$. Hence $T_2(\mathbf{u}+\mathbf{v}) = (5+7+9, \{5.7.9\}^{1/3})$. On the other hand

$$T_2(\mathbf{u})+T_2(\mathbf{v}) = (1+2+3, \{1.2.3\}^{1/3})+(4+5+6, \{4.5.6\}^{1/3})$$

and I can see readily that $T_2(\mathbf{u})+T_2(\mathbf{v}) \neq T_2(\mathbf{u}+\mathbf{v})$ since they have different second components (i.e. $\{5.7.9\}^{1/3} \neq \{1.2.3\}^{1/3}+\{4.5.6\}^{1/3}$). Thus T_2 is proved not to be a linear transformation.

[Observe that the *first* components of $T_2(\mathbf{u})+T_2(\mathbf{v})$ and $T_2(\mathbf{u}+\mathbf{v})$ *are* equal – and, furthermore, $T_2(\alpha\mathbf{u}) = \alpha(T_2\mathbf{u})$ for all (real) scalars α and for all vectors $\mathbf{u}$ in $\mathbb{R}^3$. But this is all irrelevant and need not even be mentioned: the fact that T_2 does not fully satisfy the first condition of Definition 1 is sufficient to disqualify T_2 from being a linear transformation.]

QUESTION Did the actual definition of T_2 lead you to suspect that T_2 would turn out not to be linear? Perhaps it is a bit soon to expect you to 'smell a rat'. But the product term xyz in $T_2(x, y, z)$ might just have inclined you to conjecture nonlinearity here.

Furthermore, when, later, you find it easier to speculate that a given mapping might not be a linear transformation, you might try to think of using **u** and **v** where the checking that $T\mathbf{u} + T\mathbf{v} \neq T(\mathbf{u} + \mathbf{v})$ (or $T(\alpha\mathbf{u}) \neq \alpha T\mathbf{u}$) involves less arithmetic. For example, try $\mathbf{u} = \mathbf{v} = (1, 1, 1)$. [Why will the even 'simpler' $\mathbf{u} = (1, 0, 0)$, $\mathbf{v} = (0, 1, 0)$ not do?]

(iii) Is T_3 linear? What is your gut feeling? Here I notice that $T_3(0, 0, 0) = (0, 0, 1)$. Consequently

$$T_3((0,0,0)+(0,0,0)) = T_3(0,0,0) = (0,0,1)$$

whereas

$$T_3(0,0,0)+T_3(0,0,0) = (0,0,1)+(0,0,1) = (0,0,2) \neq (0,0,1).$$

Hence T_3 is not a linear transformation. [Theorem 1(II) below will allow us to check the nonlinearity here in a different way.]

(iv) Here, although I cannot see a suitable matrix interpretation to help me, I feel that T_4 is a linear transformation. Expecting to have to check both parts of Definition 1, let f, g be two functions from $\mathfrak{F}$ and let α be any scalar. We then have

$$T_4(f \oplus g) = [\text{why?}]\,(f \oplus g)(2) = [\text{why?}]\, f(2) + g(2) = T_4(f) + T_4(g)$$
$$T_4(\alpha f) = (\alpha f)(2) = [\text{why?}]\, \alpha\{f(2)\} = \alpha\{T_4(f)\}.$$

(v) Here $\int_0^1 \{f(x) \oplus g(x)\}\mathrm{d}x = \int_0^1 f(x)\mathrm{d}x + \int_0^1 g(x)\mathrm{d}x$ and $\int_0^1 \{\alpha f(x)\}\mathrm{d}x = \alpha \int_0^1 f(x)\mathrm{d}x$ for all functions $f, g \in \mathcal{I}$ and for all $\alpha \in \mathbb{R}$. Thus T_5 is seen to be linear.

NOTE 1 We saw above that each $m \times n$ matrix gives rise, by multiplication, to a linear transformation from a vector space of n-tuples to one of m-tuples and we used this as a guide in speculating that the mapping T_2 above is not linear. Was this a good ploy? In fact we shall see (cf. Example 5 below and, especially, Exercise 22) that each linear transformation between any two finite dimensional vector spaces (FDVSs) is so expressible.

Now, following the spirit of Chapter 6, we deduce, from Definition 1, some properties which all linear transformations (even those which nobody has yet had occasion to define!) must possess. Since matrix multiplication and linear transformations are closely related, it is no real surprise to find the following theorem.

• Theorem 1

Let $T: V \to W$ be a linear transformation. Then:

(I) For any vectors $\mathbf{v}_1, \mathbf{v}_2, \ldots, \mathbf{v}_n \in V$ and for any real number $\alpha_1, \alpha_2, \ldots, \alpha_n$:

$$T(\alpha_1\mathbf{v}_1 + \alpha_2\mathbf{v}_2 + \ldots + \alpha_n\mathbf{v}_n) = \alpha_1 T\mathbf{v}_1 + \alpha_2 T\mathbf{v}_2 + \ldots + \alpha_n T\mathbf{v}_n;$$

(II) $T(\mathbf{0}_V) = \mathbf{0}_W$; [Here $\mathbf{0}_V$ and $\mathbf{0}_W$ are, respectively, the zero elements of V and W.]

(III) (Cf. Example 1(vi) of Chapter 7) Denoting $\{\mathbf{v}: \mathbf{v} \in V \text{ and } T(\mathbf{v}) = \mathbf{0}_W\}$ by $N_T(V)$ we have: $N_T(V)$ is a subspace of V.

(IV) (Cf. Example 3(ii) and Exercise 8 of Chapter 7) If V_1 is a subspace of V, then $\{T(\mathbf{v}_1): \mathbf{v}_1 \in V_1\}$ is a subspace of W. [We often denote the set $\{T(\mathbf{v}_1): \mathbf{v}_1 \in V_1\}$ by the rather appropriate symbol $T(V_1)$.]

Following the matrix lead (cf. Examples 1(vi) and 3(ii) of Chapter 7) we call the subspace $N_T(V)$ (also written $\ker(T)$) of V the **kernel** (or **null space**) of T and the subspace $T(V)$ of W the **image** (or **range**) of T, usually written $\mathrm{im}(T)$.

We can represent what is going on here rather nicely by a picture (Fig. 10.1).

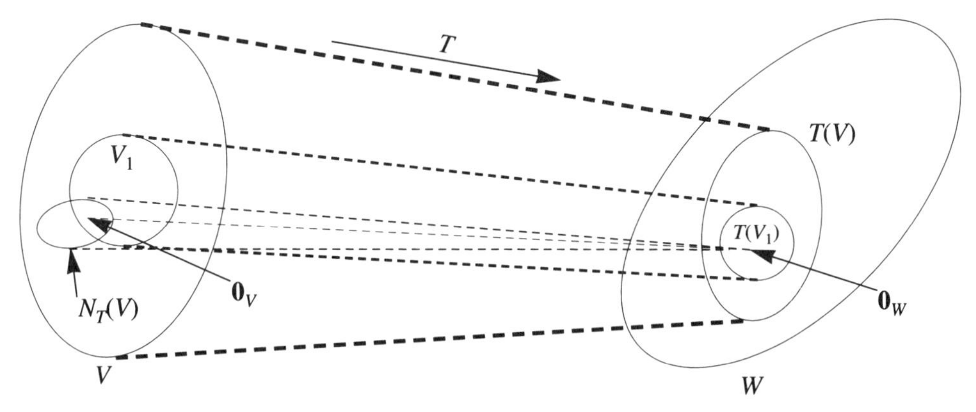

Fig 10.1

PROOF

(I) We leave this to Exercise 14(a).

(II) Let us observe that $\mathbf{0}_V = \mathbf{0}_V + \mathbf{0}_V$ (cf. the proof of Theorem 1 of Chapter 6). Then $T(\mathbf{0}_V) = T(\mathbf{0}_V + \mathbf{0}_V) = T(\mathbf{0}_V) + T(\mathbf{0}_V)$. But $T(\mathbf{0}_V) = T(\mathbf{0}_V) + \mathbf{0}_W$. (Why?) Hence, using Theorem 1 of Chapter 6, we deduce that $T(\mathbf{0}_V) = \mathbf{0}_W$.

(III) Let $\mathbf{u}, \mathbf{v} \in N_T(V)$ and let $\alpha \in \mathbb{R}$. Then

(i) $T(\mathbf{u} + \mathbf{v}) = T\mathbf{u} + T\mathbf{v}$ (why?) $= \mathbf{0}_W + \mathbf{0}_W$ (why?) $= \mathbf{0}_W$. Thus $\mathbf{u} + \mathbf{v} \in N_T(V)$.

(ii) $T(\alpha\mathbf{u}) = \alpha(T\mathbf{u})$ (why?) $= \alpha\mathbf{0}_W$ (why?) $= \mathbf{0}_W$ (why?). Thus $\alpha\mathbf{u} \in N_T(V)$. Thus $N_T(V)$ is a subspace of V. Is it non-empty? Oh yes! Of course! Why?)

(IV) This is proved by a virtual repeat of the argument of Exercise 8 of Chapter 7. We leave it to you (with hints once again) in Exercise 16. ●

As examples involving parts of Theorem 1 we offer the following example.

Example 4

(i) The linear transformation $T:\mathbb{R}^3 \to \mathbb{R}^4$ is defined by $T(1, 1, 2) = (1, 1, 1, 0)$, $T(2, 1, 1) = (0, 1, 1, 1)$, $T(1, 2, 1) = (1, 0, 1, 1)$. What is $T(1, 0, 0)$?

Since $\{(1, 1, 2), (2, 1, 1), (1, 2, 1)\}$ is a basis of $\mathbb{R}^3$ (is it?) we know $(1, 0, 0) = \alpha(1, 1, 2) + \beta(2, 1, 1) + \gamma(1, 2, 1)$ for suitable scalars α, β, γ. Indeed we find $\alpha = \gamma = -\frac{1}{4}$ and $b = \frac{3}{4}$. Hence, by (I),

$$T(1,0,0) = \alpha T(1,1,2) + \beta T(2,1,1) + \gamma T(1,2,1) = \left(-\tfrac{1}{2}, \tfrac{1}{2}, \tfrac{1}{4}, \tfrac{1}{2}\right).$$

(ii) Part (II) confirms easily that Example 3(iii) is not a linear transformation.

(iii) In the case where $V = \mathbb{R}^n$ and $T\mathbf{x} = A\mathbf{x}$, A being an $m \times n$ matrix, (III) reconfirms that the solution space of the system of equations $A\mathbf{x} = 0$ is a subspace of $\mathbb{R}^n$. (Cf. Example 1(vi) of Chapter 7.)

NOTES 2

(i) Theorem 1(I) explains clearly why linear transformations are the most important maps between vector spaces: vector spaces comprise elements which are linear combinations (of basis elements) and (somewhat loosely stated) linear transformations are precisely the maps which send linear combinations to linear combinations. Theorem 1(I) also tells us that the action of T on the infinitely many elements of V is totally determined by how it behaves on the (frequently) finitely many elements of some basis of V.

(ii) It also follows from Theorem 1(I) that, given a vector space V of dimension n, a basis $\mathbf{v}_1, \mathbf{v}_2, \ldots, \mathbf{v}_n$ of V and any n elements $\mathbf{w}_1, \mathbf{w}_2, \ldots, \mathbf{w}_n$ of W (including $\mathbf{0}_W$ and/or repeats, if you like) we can define a linear transformation $T:V \to W$ with the property that $T(\mathbf{v}_i) = \mathbf{w}_i$ for each i. Indeed, each element $\mathbf{v}$ of V can be expressed uniquely (why?) as a linear combination $\mathbf{v} = \alpha_1\mathbf{v}_1 + \alpha_2\mathbf{v}_2 + \ldots + \alpha_n\mathbf{v}_n$, and, then, Theorem 1(I) tells us we *must* define $T(\mathbf{v})$ to be

$$\alpha_1 T(\mathbf{v}_1) + \alpha_2 T(\mathbf{v}_2) + \ldots + \alpha_n T(\mathbf{v}_n) = \alpha_1\mathbf{w}_1 + \alpha_2\mathbf{w}_2 + \ldots + \alpha_n\mathbf{w}_n$$

if we wish to make T into a linear transformation. (In particular T is, thereby, uniquely determined.)

PROBLEM 10.5

Show that there exists a linear transformation $T:\mathbb{R}^3 \to \mathbb{R}^2$ in which $T(1, 2, 3) = (1, 1)$, $T(1, 2, 2) = (2, -1)$ and $T(1, 1, 1) = (1, 2)$, by finding a 3×2 matrix A such that $T\mathbf{x} = A\mathbf{x}$ for all $\mathbf{x} \in \mathbb{R}^3$. [You will need to determine $T(1, 0, 0)$, $T(0, 1, 0)$ and $T(0, 0, 1)$ first.] ●

NOTES 2 (cont.)

(iii) Continuing our extension of the terminology from the matrix case, we call the dimension, dim $T(V)$, of $T(V)$, the **rank** of T and dim $N_T(V)$ the **nullity** of T.

(iv) Let $T:\mathbb{R}^n \to \mathbb{R}^m$ be a linear transformation. As we have already asserted, it can be shown (see Exercise 22) that T can be realised by multiplication of the n-tuples by a suitable (uniquely determined) matrix A, say (as in Example 1(i)). We have seen (Examples 3(ii) and 1(vi) of Chapter 7) that im(T) is precisely the column space of A whilst the kernel of T is the set of solutions of the system of equations $A\mathbf{x} = \mathbf{0}$. Thus, for such $T:\mathbb{R}^n \to \mathbb{R}^m$, the rank and nullity of T coincide with the rank and nullity of the corresponding matrix.

Let us look at an example.

Example 5

Let $T:\mathbb{R}^3 \to \mathbb{R}^2$ be defined by

$$T\begin{bmatrix} x \\ y \\ z \end{bmatrix} = \begin{bmatrix} x + y \\ 3x + y + 2z \end{bmatrix}.$$

Then

$$T\begin{bmatrix}x\\y\\z\end{bmatrix}=A\begin{bmatrix}x\\y\\z\end{bmatrix} \quad \text{where } A=\begin{bmatrix}1&1&0\\3&1&2\end{bmatrix}.$$

Thus

$$\text{im}(T)=\left\{\begin{bmatrix}x+y\\3x+y+2z\end{bmatrix}:x,y,z\in\mathbb{R}\right\}=\left\{x\begin{bmatrix}1\\3\end{bmatrix}+y\begin{bmatrix}1\\1\end{bmatrix}+z\begin{bmatrix}0\\2\end{bmatrix}:x,y,z\in\mathbb{R}\right\}$$

which is clearly the column space of A.

$$\left\{\begin{bmatrix}1\\3\end{bmatrix},\begin{bmatrix}1\\1\end{bmatrix}\right\}$$

is one possible basis for im(T) which therefore has dimension 2. Likewise

$$\ker(T)=\{\mathbf{x}:A\mathbf{x}=\mathbf{0}\}=\left\{\begin{bmatrix}x\\y\\z\end{bmatrix}:x+y=0,3x+y+2z=0\right\}$$

$$=\left\{\begin{bmatrix}x\\-x\\-x\end{bmatrix}:x\in\mathbb{R}\right\}=\left\{x\begin{bmatrix}1\\-1\\-1\end{bmatrix}:x\in\mathbb{R}\right\}.$$

Thus

$$\left\{\begin{bmatrix}1\\-1\\-1\end{bmatrix}\right\}$$

constitutes a basis for ker(T). Consequently, ker(T) has dimension 1.

Note, as one would expect from Corollary 1 of Chapter 9, that rank T + nullity T (= 2 + 1 = 3) = dim $\mathbb{R}^3$.

Because of the connection between matrices and linear transformations on (finite dimensional) vector spaces, you may feel you already believe the following theorem to be true. But, because of the theorem's importance, we take the opportunity to offer you a proof in a more abstract setting. It's really beautiful! So – on first reading – just sit back and enjoy it!

• *Theorem 2*

Let $T: V \to W$ be a linear transformation from the FDVS (see p. 166) V to the FDVS W. Then dim{im(T)} + dim{ker(T)} = dim V, i.e. rank(T) + nullity(T) = dim V.

[Once again a picture (Fig. 10.2) might help keep in your mind exactly where the elements and subspaces mentioned lie.]

PROOF

Choose a basis $\mathbf{u}_1, \mathbf{u}_2, \ldots, \mathbf{u}_s$ of ker(T) and (using Theorem 4 of Chapter 9) extend it to a basis $\mathbf{u}_1, \ldots, \mathbf{u}_s, \mathbf{u}_{s+1}, \ldots, \mathbf{u}_n$ of V. We show that $T\mathbf{u}_{s+1}, \ldots, T\mathbf{u}_n$ is a basis for im(T).

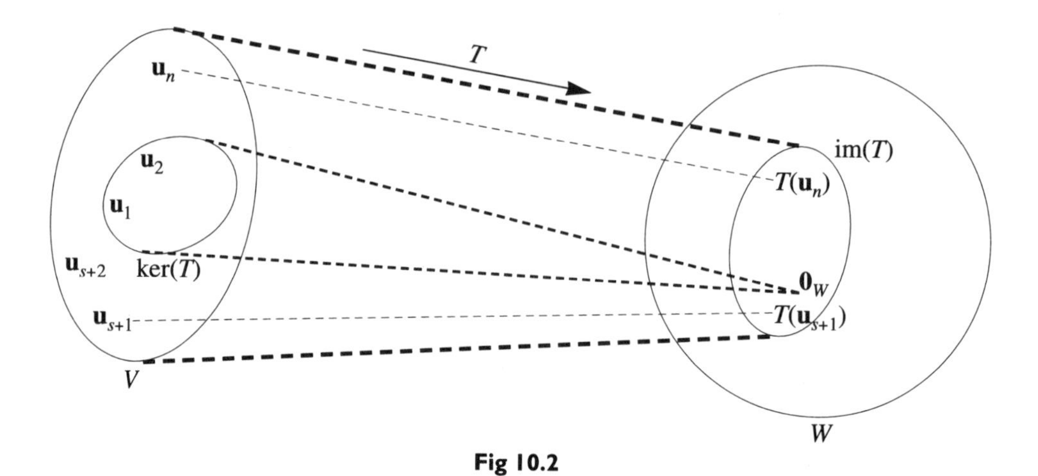

Fig 10.2

(i) *Spanning* Let $\mathbf{a} \in \text{im}(T)$. Then $\mathbf{a} = T\mathbf{v}$ where $\mathbf{v} = \alpha_1\mathbf{u}_1 + \ldots + \alpha_n\mathbf{u}_n$. But

$$T\mathbf{v} = \alpha_1(T\mathbf{u}_1) + \ldots + \alpha_s(T\mathbf{u}_s) + \alpha_{s+1}(T\mathbf{u}_{s+1}) + \ldots + \alpha_n(T\mathbf{u}_n)$$
$$= \mathbf{0}_W + \ldots + \mathbf{0}_W + \alpha_{s+1}(T\mathbf{u}_{s+1}) + \ldots + \alpha_n(T\mathbf{u}_n)$$

so $\{T\mathbf{u}_{s+1}, \ldots, T\mathbf{u}_n\}$ do span im(T).

(ii) *Independence* Suppose that

$$\beta_{s+1}(T\mathbf{u}_{s+1}) + \ldots + \beta_n(T\mathbf{u}_n) = \mathbf{0}_W. \qquad (10.1)$$

Then $T(\beta_{s+1}\mathbf{u}_{s+1} + \ldots + \beta_n\mathbf{u}_n) = \mathbf{0}_W$. But this just says that

$$\beta_{s+1}\mathbf{u}_{s+1} + \ldots + \beta_n\mathbf{u}_n \in \ker(T).$$

Hence we may write

$$\beta_{s+1}\mathbf{u}_{s+1} + \ldots + \beta_n\mathbf{u}_n = \gamma_1\mathbf{u}_1 + \ldots + \gamma_s\mathbf{u}_s$$

for suitable $\gamma_1, \ldots, \gamma_s \in \mathbb{R}$ (why?). Rewriting this we have

$$\gamma_1\mathbf{u}_1 + \ldots + \gamma_s\mathbf{u}_s - \beta_{s+1}\mathbf{u}_{s+1} - \ldots - \beta_n\mathbf{u}_n = \mathbf{0}.$$

But this implies that $\beta_i = 0$ for each i ($s + 1 \le i \le n$) – and, in fact, each $\gamma_i = 0$, though we do not need to know that! Hence, from (10.1) above, we see that $\{T\mathbf{u}_{s+1}, \ldots, T\mathbf{u}_n\}$ is a linearly independent set and hence, by part (i), a basis for im(T).

It follows that $\dim\{\text{im}(T)\} + \dim\{\ker(T)\} = (n - s) + s = n = \dim V$, as claimed. ●

TUTORIAL PROBLEM 10.1

Discuss this proof in the cases $s = 0$ and $s = n$. Do you see any problems with the proof?

We now turn to an important theoretical use of the concept of linear transformation.

A question which is very important in many areas of mathematics, including all of algebra, is: 'When are two mathematical objects essentially indistinguishable?' In linear algebra one such question of obvious important is: When are two finite dimensional vector spaces 'essentially the same'? [One can be fairly confident that the answer will either delight you – or disappoint you dreadfully! But, first, we must explain what we mean by the phrase 'essentially the same'.]

Thinking aloud (or, whatever is the written equivalent!), let us suppose that we are presented with two vector spaces V and V', say. Suppose the elements of V are $\{\mathbf{u}, \mathbf{v}, \mathbf{w}, \dots\}$ and the elements of V' are $\{\mathbf{u}', \mathbf{v}', \mathbf{w}', \dots\}$ so that the pairing of $\mathbf{u}$ with $\mathbf{u}'$, $\mathbf{v}$ with $\mathbf{v}'$, $\mathbf{w}$ with $\mathbf{w}'$, etc. matches all the elements of V with all those of V' in a 1–1 manner. Under this pairing it may (or may not!) be the case that, for all $\mathbf{u}, \mathbf{v} \in V$ and all $\alpha \in \mathbb{R}$, we have both

(i) $\mathbf{u}+\mathbf{v}$ pairs with $\mathbf{u}'+\mathbf{v}'$ (i.e. $(\mathbf{u}+\mathbf{v})' = \mathbf{u}'+\mathbf{v}'$) and

(ii) $\alpha\mathbf{u}$ pairs with $\alpha\mathbf{u}'$ (i.e. $(a\mathbf{u})' = a(\mathbf{u}')$).

If so, then the two vector spaces, though comprising distinct elements, are surely 'structurally identical' in the sense that any relationship existing between the $\mathbf{u}, \mathbf{v}, \mathbf{w}, \dots$ of V is reflected faithfully by an identical relationship between the $\mathbf{u}', \mathbf{v}', \mathbf{w}', \dots$ of V' obtained from the former relationship by 'dashing'. [For example, if $3\mathbf{u} + 2\mathbf{v} - 17\mathbf{w} = \mathbf{0}_V$, we could surely conclude that $3\mathbf{u}' + 2\mathbf{v}' - 17\mathbf{w}' = \mathbf{0}_{V'}$.] This idea seems to convey the correct idea of 'sameness' so let us formalise it and use the formalisation as a definition of this 'sameness'.

Before we do, let us consider the vector spaces $\mathbb{R}^2$ and $\mathbb{R}^3$. Having continually thought of them as (representations of) 2- and 3-dimensional space respectively, it is perhaps a bit of a surprise (Exercise 4(b) of Chapter 7) to find that $\mathbb{R}^2$ is not a subspace of $\mathbb{R}^3$. Indeed $\mathbb{R}^2$ is not even a *subset* of $\mathbb{R}^3$! (For $\mathbb{R}^2$ comprises pairs (a, b) and none of these is a triple in $\mathbb{R}^3$!) However, Definition 2 saves us: it shows that $\mathbb{R}^2$ is 'essentially' a subspace of $\mathbb{R}^3$. (We leave this to Exercise 26.)

• *Definition 2*

Let V and W be vector spaces over the same set (field) of scalars $\mathbb{F}$. Suppose that T is a 1–1 mapping from V *onto* W such that for all $\mathbf{u}, \mathbf{v} \in V$ and for all $\alpha \in \mathbb{F}$:

$$\text{(i) } T(\mathbf{u}+\mathbf{v}) = T(\mathbf{u}) + T(\mathbf{v}) \quad \text{and} \quad \text{(ii) } T(\alpha\mathbf{u}) = \alpha T(\mathbf{u})$$

(i.e. T is a linear transformation which is 1–1 and onto) then T is called an **isomorphism** between V and W and V and W are said to be **isomorphic**. We write $V \cong W$. ●

[The word 'isomorphism' comes from the Greek and means 'same form'.]

Using Definition 2 we can establish the remarkable fact that each (real) FDVS is nothing more than one of the $\mathbb{R}^n$ in disguise! This means that all (real) FDVSs – of any given dimension – are algebraically indistinguishable; in other words the only real difference is in the nature of their elements.

• *Theorem 3*

Let V be any (real) vector space of dimension n. Then $V \cong \mathbb{R}^n$.

PROOF

Let $B = \{\mathbf{v}_1, \mathbf{v}_2, \ldots, \mathbf{v}_n\}$ be a basis of V. Define the map $T: V \to \mathbb{R}^n$ as follows. If

$$\mathbf{v} = \alpha_1\mathbf{v}_1 + \alpha_2\mathbf{v}_2 + \ldots + \alpha_n\mathbf{v}_n \tag{10.2}$$

then $T(\mathbf{v}) = (\alpha_1, \alpha_2, \ldots, \alpha_n) \in \mathbb{R}^n$. Since, in (10.2), $\mathbf{v}$ determines the α_i (the coordinates of $\mathbf{v}$ with respect to the basis B) uniquely (why?), each $\mathbf{v}$ unambiguously determines the n-tuple $T(\mathbf{v})$ in $\mathbb{R}^n$. Conversely, each n-tuple in $\mathbb{R}^n$ arises in this way. The map T is, therefore, 1–1 and onto. We check property (i) of Definition 2, leaving part (ii) to you.

So, let $\mathbf{u} = \beta_1\mathbf{v}_1 + \beta_2\mathbf{v}_2 + \ldots + \beta_n\mathbf{v}_n$ and $\mathbf{v} = \alpha_1\mathbf{v}_1 + \alpha_2\mathbf{v}_2 + \ldots + \alpha_n\mathbf{v}_n$. Then

$$\begin{aligned} T(\mathbf{u}+\mathbf{v}) &= T\{(\beta_1+\alpha_1)\mathbf{v}_1 + (\beta_2+\alpha_2)\mathbf{v}_2 + \ldots + (\beta_n+\alpha_n)\mathbf{v}_n\} \\ &= (\beta_1+\alpha_1, \beta_2+\alpha_2, \ldots \beta_n+\alpha_n) \\ &= (\beta_1, \beta_2, \ldots, \beta_n) + (\alpha_1, \alpha_2, \ldots, \alpha_n) \\ &= T(\mathbf{u}) + T(\mathbf{v}). \end{aligned}$$

This completes part (i). Proving that (ii) holds will show that T is an isomorphism – hence $V \cong \mathbb{R}^n$. ●

As an example we give the following.

Example 6

$\mathbb{R}_2[x] \cong \mathbb{R}^3$ via the (most obvious?) map S defined by

$$S(a_0 + a_1x + a_2x^2) = (a_0, a_1, a_2),$$

but also via the map given by

$$T(a_0 + a_1x + a_2x^2) = (2a_0 + 4a_1, a_0 + 4a_1 + 2a_2, 3a_0 + 2a_1 + 3a_2).$$

[Can you see why?) ●

Theorem 3 generalises, by an almost identical 'component comparing' proof, to show that any two vector spaces of the same dimension n are isomorphic. (See Exercise 28.) And, of course, each is then isomorphic to $\mathbb{R}^n$. As in Example 6 there will, in general, be many maps which will establish an isomorphism, although, often, one of them stands out as being more 'natural' than the rest.

PROBLEM 10.6

To which $\mathbb{R}^n$ is the vector space of arithmetic progressions (see Definition 1 of Chapter 6) isomorphic? Confirm your assertion by finding an appropriate isomorphism. ●

Finally, two problems the first of which stands out like a sore thumb.

TUTORIAL PROBLEM 10.2

(Finite dimensional) vector spaces of the same dimension are isomorphic, but are isomorphic vector spaces necessarily of the same dimension? (You have enough information in previous chapters to answer this!)

TUTORIAL PROBLEM 10.3

Can you find two infinite dimensional vector spaces which are not isomorphic?

POSTSCRIPT The essential interchangeability of the concepts of matrix and linear transformation suggests that, just as $m \times n$ matrices can be added and multiplied by scalars and then turned into a vector space (of dimension mn), so too ought we be able to add and scalarly multiply the set of all linear transformations from an n-dimensional vector space V say, to one of dimension m, W say, and turn the result into a vector space! (Indeed sums and scalar multiples of functions are discussed in the Appendix – and, in fact, the above programme can be carried out.)

If V and W coincide then $m = n$ and, since two $n \times n$ matrices can also be multiplied, one might expect that, on the set (vector space) of all linear transformations from V to itself, *products* of linear transformations could be defined. Indeed they can! That will be in the follow-up book to this one – we just note that the idea is closely connected with the concept of 'change of basis' in vector spaces which lies behind much of the goings-on in Chapter 11.

Applications

Various examples have shown that multiplication by a suitable matrix produces a linear transformation from $\mathbb{R}^n$ to $\mathbb{R}^m$ and, conversely (we claim), all such transformations can be so obtained (Exercise 22). Thus the applications here include all those listed in Chapter 3. Replacing the matrix by a single symbolic letter leads, oftentimes, to more succinct proofs (e.g. Theorem 2) of theorems and also encourages us to consider mappings on infinite dimensional spaces – cf. Examples 2 and 3(iv) and (v).

Summary

Amongst all the functions T which can be defined from one vector space, V, to another, W, the most natural to consider are those which 'preserve' the structure: that is, those which (roughly speaking) map linear combinations to linear combinations. In particular, we have $T(\alpha\mathbf{u} + \beta\mathbf{v}) = \alpha T(\mathbf{u}) + \beta T(\mathbf{v})$. One consequence of this is that the action of T on V is fully determined once its action on a basis of V is known. Another is that subspaces (of V) are mapped by T to subspaces (of W). In particular $T(V)$ is a subspace of W – the **image** (or **range**) of T. The other important subspace associated with T is its **kernel** (or **null space**) – the set of elements of V which T maps to $\mathbf{0}_W$. These subspaces and T are connected by the equality

$$\text{rank } T + \text{nullity } T \ (= \dim(\text{im}T) + \dim(\ker T)) = \dim V.$$

Finally linear transformations which are 1–1 and onto were used to define the concept of **isomorphic vector spaces** and it was found that each (real) n-dimensional vector space is isomorphic to the (real) vector space $\mathbb{R}^n$ of n-tuples of (real) numbers.

Linear transformations (under the name linear substitutions) have been around a long time. They make an appearance in the 17th and 18th centuries in the process of change of coordinates which allowed classification of curves and surfaces according to their degree. They also appear later in the work of Lagrange and Gauss concerning the set of all values that can be taken by a quadratic form (for example $f(x, y) = 2x^2 + 5xy + 4y^2$) when x, y are restricted to being integers. (Provided $\alpha, \beta, \gamma, \delta \in \mathbb{Z}$ and $\alpha\delta - \beta\gamma = \pm 1$ the totality of values taken by $f(x, y)$ and $f(\alpha x + \beta y, \gamma x + \delta y)$ coincide so that $f(x, y)$ and $f(\alpha x + \beta y, \gamma x + \delta y)$ can be regarded as equivalent.) Yet another is in projective geometry. Introduced by Desargues in the 17th century, it was resurrected after a period of stagnation using both purely geometric and purely algebraic methods. In the latter case 'projective transformations' and linear substitutions become linked, different substitutions representing the same transformation when their coefficients are in proportion. And, of course, we must not forget the linear vector functions such as curl and ∇ introduced by Maxwell and others in the late 19th century.

A formal definition of linear transformation (just like the above) was given by Peano in 1888 and by Weyl in his *Space, Time and Matter* of 1918.

EXERCISES ON CHAPTER 10

1. Let A be an $n \times m$ matrix. Regarding the elements of $\mathbb{R}^n$ and $\mathbb{R}^m$ as row vectors define a map $S: \mathbb{R}^n \to \mathbb{R}^m$ as follows. For $\mathbf{u} \in \mathbb{R}^n$ set $S(\mathbf{u}) = \mathbf{u}A$. Show that S is a linear transformation of $\mathbb{R}^n$ into $\mathbb{R}^m$.

2. (a) Describe geometrically the linear transformations of $\mathbb{R}^2$ which the following matrices represent:

$$\text{(i)} \begin{bmatrix} 0 & 1 \\ 1 & 0 \end{bmatrix}; \quad \text{(ii)} \begin{bmatrix} 0 & 1 \\ -1 & 0 \end{bmatrix}; \quad \text{(iii)} \begin{bmatrix} 2 & 0 \\ 0 & -1 \end{bmatrix}; \quad \text{(iv)} \begin{bmatrix} 1 & \frac{1}{3} \\ 0 & 1 \end{bmatrix};$$

$$\text{(v)} \begin{bmatrix} \frac{1}{2} & -\sqrt{\frac{3}{2}} \\ \sqrt{\frac{3}{2}} & \frac{1}{2} \end{bmatrix}; \quad \text{(vi)} \begin{bmatrix} 1 & 0 \\ 0 & r \end{bmatrix}; \quad \text{(vii)} \begin{bmatrix} 0 & 1 \\ r & 0 \end{bmatrix}; \quad \text{(viii)} \begin{bmatrix} \cos\vartheta & \sin\vartheta \\ \sin\vartheta & -\cos\vartheta \end{bmatrix};$$

$$\text{(ix)} \begin{bmatrix} \cos\vartheta & \sin\vartheta \\ -\sin\vartheta & \cos\vartheta \end{bmatrix}; \quad \text{(x)} \begin{bmatrix} \sin\vartheta & -\cos\vartheta \\ \cos\vartheta & \sin\vartheta \end{bmatrix}.$$

(In (vi) and (vii) you may need to consider ranges $0 < r < 1$, $r < -1$, etc. separately.)

(b) Which, if any, of (ii), (iv) and (vi) map the circle $x^2 + y^2 = 1$ to a circle?

3. Write down matrices which in $\mathbb{R}^2$ describe:

(i) reflection in the y-axis;

(ii) reflection in the line $y = 2x$;

(iii) perpendicular projection onto the line $y = x$;

and which, in $\mathbb{R}^3$, describe:

(iv) reflection in the z-plane;

(v) rotation in a right-hand screw direction through an angle ϑ about the y-axis.

4. Let Det be the determinant map and let Tr be the trace map from $M_2(\mathbb{R})$ to $\mathbb{R}$ given by

$$\text{Det}\begin{bmatrix} a & b \\ c & d \end{bmatrix} = ad - bc \quad \text{and} \quad \text{Tr}\begin{bmatrix} a & b \\ c & d \end{bmatrix} = a + d.$$

Determine (!) whether or not Det and Tr are linear maps.

5. (a) Let D_0 be the 'second derivative' map given by $D_0(f) = d^2f/dx^2$. On what space of 'vectors' can D_0 most naturally be defined? Show that D_0 is then a linear map.

(b) Let g be a fixed differentiable function. Define the maps D_1 and D_2 from the space $\mathfrak{D}$ of differentiable functions to the space $\mathfrak{F}$ of all functions by: $D_1(f) = fg' + gf'$; $D_2(f) = ff'$ (f' denoting, as usual, the derivative of f). Decide which, if either, of D_1 and D_2 is/are linear.

6. Let T_β be the mapping of $\mathbb{R}^3$ to itself under which each plane $3x + 2y - 17z = \alpha$ is moved to the plane $3x + 2y - 17z = \alpha + \beta$. Describe this movement geometrically and show that T_β is not a linear transformation unless $\beta = 0$.

7. Let $T: \mathbb{R}^5 \to \mathbb{R}^4$ be the map defined by:

$$T(x, y, z, w, t) = (x + y + 2z, x + 2y + 4z + 2w + t, 2z + 2w + t, 2x + 2y - 4w - 2t).$$

Do you believe that T is linear? Give the reasons for your belief. Now, verify your belief by means of a proof or a counterexample.

8. (a) Let $T:\mathbb{R}^2 \to \mathbb{R}^2$ be defined by $T(x, y) = (2x + 3y + 4, 5x - y)$. Show that T is not linear.

(b) Let $p(x)$ be a fixed polynomial in $\mathbb{R}[x]$. Define $T:\mathbb{R}[x] \to \mathbb{R}[x]$ by: for each polynomial $q(x) \in \mathbb{R}[x]$, $T(q(x)) = p(x)q(x)$. Is T a linear map?

9. Intuitively is there a 'degree changing' linear map $T:\mathbb{R}_3[x] \to \mathbb{R}_3[x]$ satisfying: $T(1) = 1 + x$, $T(x) = 1 + x^2$, $T(x^2) = 1 + x^3$ and $T(x^3) = 1$? Explain why not – or prove that there is.

10. (i) Let $T:\mathbb{R}^3 \to \mathbb{R}^3$ be a linear transformation such that $T(5, 8, 8) = (1, 5, 3)$, $T(3, 3, 5) = (1, 2, 2)$ and $T(a, b, c) = (9, 9, 9)$. Find $T(1, 2, 3)$, given that $a = b = c = 9$.

(ii) In (i) what values must one take for a, b, c in order that $T(1, 2, 3) = (-1, -2, -3)$? Write down the matrices corresponding to T (cf. Example 5) in each case.

11. Is there a linear transformation $T:\mathbb{R}^3 \to \mathbb{R}^4$ such that $T(1, 2, 3) = (0, 1, 0, 1)$, $T(2, 3, 4) = (2, -3, 1, 7)$ and $T(3, 4, 5) = (4, -7, 2, 14)$? If not, why not?

12. Let C be a fixed invertible matrix from $M_2(\mathbb{R})$. Define the map $T:M_2(\mathbb{R}) \to M_2(\mathbb{R})$ by $T(A) = C^{-1}AC$ for all $A \in M_2(\mathbb{R})$. Is T linear?

13. Let V, W be vector spaces. Define the map $Z: V \to W$ by $Z(\mathbf{v}) = \mathbf{0}_W$ for every $\mathbf{v} \in V$. Show that Z is a linear transformation (the **zero** transformation). Do the same for the map $\text{Id}:V \to V$ given by $\text{Id}(\mathbf{v}) = \mathbf{v}$ for all $\mathbf{v} \in V$. (Id is the **identity** transformation.)

14. (a) Show that the map $T: V \to W$ is a linear transformation if and only if, for all $\mathbf{u}, \mathbf{v} \in V$ and for all $\alpha \in \mathbb{R}$, we have $T(\alpha\mathbf{u} + \beta\mathbf{v}) = \alpha T(\mathbf{u}) + \beta T(\mathbf{v})$. [Thus this one requirement can replace the two of Definition 1.] Now use the method of induction to extend this result to that of Theorem 1(I).

(b) Let $T:V \to W$ be a linear map. Prove, formally, that, for $\mathbf{v} \in V$, we have $T(-\mathbf{v}) = -T(\mathbf{v})$. Deduce that, for all $\mathbf{u}, \mathbf{v} \in V$ and all $\alpha, \beta \in \mathbb{R}$, we have $T(\alpha\mathbf{u} - \beta\mathbf{v}) = \alpha T(\mathbf{u}) - \beta T(\mathbf{v})$.

15. (a) Show that the line in $\mathbb{R}^3$ which passes through $\mathbf{a} = (a, b, c)$ and is parallel to the line joining $\mathbf{0} = (0, 0, 0)$ to $\mathbf{u} = (u, v, w)$ comprises all points of the form $\mathbf{a} + \alpha\mathbf{u}$ where $\alpha \in \mathbb{R}$. Hence show that any linear transformation from $\mathbb{R}^3$ to any $\mathbb{R}^n$ maps this line to a line – or just a point.

(b) If l is the line given by the intersection of the planes $x + 2y + 3z = 0$ and $3x + y + 2z = 0$ and if $T: \mathbb{R}^3 \to \mathbb{R}^2$ is the linear transformation given by $T(x, y, z) = (x + y, y + z)$, find the equation of the line to which T maps l.

16. Prove Theorem 1(IV). [Method: let $\mathbf{w}_1, \mathbf{w}_2 \in T(V_1)$. Then $\mathbf{w}_1 = T(\mathbf{v}_1)$, $\mathbf{w}_2 = T(\mathbf{v}_2)$ for suitable $\mathbf{v}_1, \mathbf{v}_2, \in V_1$. Hence $\mathbf{w}_1 + \mathbf{w}_2 = T(\mathbf{v}_1 + \mathbf{v}_2) \in T(V_1)$ (why?)]

17. Show that each linear transformation $T: \mathbb{R} \to \mathbb{R}$ takes the form $T(r) = \alpha r$ where α is a suitable real number. Can you identify geometrically the different types of linear transformations $T: \mathbb{R}^2 \to \mathbb{R}^2$ (e.g. rotation, shear, etc.)? [This might be easier after studying Chapter 11.]

18. (See Exercise 15 of Chapter 6.) Let T be the map $\mathcal{S} \to \mathcal{S}$ given by $T(\{x_i\}) = \{y_i\}$, where $\{x_i\} = (x_1, x_2, \ldots)$ and $y_i = x_{i+1}$. Show that T is linear. What is its kernel?

19. Repeat Exercise 18 in the cases where T is given by:
(i) $T(\{x_i\}) = \{y_i\}$ where $y_1 = 0$ and $y_i = x_{i-1}$;
(ii) $T(\{x_i\}) = \{y_i\}$ where $y_i = x_i + 2x_{i+1}$.

20. Let $T: \mathbb{R}_3[x] \to M(\mathbb{R}_2)$ be given by

$$T(a_0 + a_1x + a_2x^2 + a_3x^3) = \begin{bmatrix} a_0 + a_1 & a_1 + a_2 \\ a_2 + a_3 & a_3 + a_0 \end{bmatrix}.$$

Check that T is linear. What is its kernel? Determine a basis for its image.

21. Let D_3 be the linear map defined on the set of all twice differentiable functions by: $\mathrm{D}_3(f) = f'' + 4f' + 3f$. What is the kernel of this map? (Cf. equation 6.2.)

22. (a) Let $T: \mathbb{R}^n \to \mathbb{R}^m$ be given by $T(\mathbf{v}_i) = \beta_{1i}\mathbf{w}_1 + \beta_{2i}\mathbf{w}_2 + \ldots + \beta_{mi}\mathbf{w}_m$ where, for each i, j $(1 \leq i \leq n, 1 \leq j \leq m)$, $\mathbf{v}_i$ $(\mathbf{w}_j)$ is the $n \times 1$ $(m \times 1)$ column vector with a '1' in its ith place (jth place) and 0s elsewhere. (That is, the $\mathbf{v}$ and $\mathbf{w}$ constitute the usual standard basis – see Example 1(ii) of Chapter 9.) Write down the matrix determined by T and the $\mathbf{v}_i$, $\mathbf{w}_j$.

(b) [Generalising part (a)] Let T be a linear transformation from the vector space V with selected basis $B_V = \{\mathbf{v}_1, \mathbf{v}_2, \ldots, \mathbf{v}_n\}$ to the vector space W with selected basis $B_W = \{\mathbf{w}_1, \mathbf{w}_2, \ldots, \mathbf{w}_m\}$. Suppose that, for each i $(1 \leq i \leq n)$, $T(\mathbf{v}_i) = \beta_{1i}\mathbf{w}_1 + \beta_{2i}\mathbf{w}_2 + \ldots + \beta_{mi}\mathbf{w}_m$ and let M be the matrix $[\beta_{ij}]$. Suppose further that $\mathbf{v} = \alpha_1\mathbf{v}_1 + \alpha_2\mathbf{v}_2 + \ldots + \alpha_n\mathbf{v}_n$, that $\mathbf{w} = \gamma_1\mathbf{w}_1 + \gamma_2\mathbf{w}_2 + \ldots + \gamma_m\mathbf{w}_m$ and that $T(\mathbf{v}) = \mathbf{w}$ (so that

$$\begin{bmatrix} \alpha_1 \\ \vdots \\ \alpha_n \end{bmatrix} \text{ and } \begin{bmatrix} \gamma_1 \\ \vdots \\ \gamma_m \end{bmatrix}$$

may be regarded as the 'coordinate vectors' (see the start of Chapter 9) for $\mathbf{v}$ and $\mathbf{w}$ with respect to the given bases B_V and B_W.) Show that

$$\begin{bmatrix} \gamma_1 \\ \vdots \\ \gamma_m \end{bmatrix} = M \begin{bmatrix} \alpha_1 \\ \vdots \\ \alpha_n \end{bmatrix}.$$

[Hint: express each $T(\mathbf{v})$ in terms of the $T(\mathbf{v}_i)$ and then $T(\mathbf{v}_i)$ in terms of the $\mathbf{w}_j$.]

23. Find bases for the kernel and for the image and, hence, find the rank and nullity of the map of Exercise 7.

24. Find a basis for the kernel and for the image of the linear transformation $T: M_{3\times3}(\mathbb{R}) \to M_{3\times3}(\mathbb{R})$ given by

$$T(M) = \begin{bmatrix} 1 & 2 & 1 \\ 2 & 1 & 3 \\ 1 & -1 & 2 \end{bmatrix} M.$$

Confirm the result of Theorem 2 on this example.

25. Find the rank and nullity of the maps f, g given by:
(i) $f(x, y, z) = (x + y, y + z, z + x)$;
(ii) $g(x, y, z, t) = (x - y, y - z, z - t, t - x)$. [$x, y, z, t \in \mathbb{R}$, of course.]

26. Show that the vector space $\mathbb{R}^2$ is isomorphic to the subspace $\{(x, y, 0): x, y \in \mathbb{R}\}$ of $\mathbb{R}^3$. Show that it is also isomorphic to each of the subspaces:
(i) $\{(x, y, z): x + y + z = 0\}$;
(ii) $\{(x + y, y + z, z - x): x, y, z \in \mathbb{R}\}$.
[Hint: choose a basis in each.]
Although it is unnecessary to give more than one, give two isomorphisms in each case.

27. Extend the first part of Exercise 26 by showing that, for $n > m > 0$, $\mathbb{R}^n$ has a subspace (indeed infinitely many subspaces) isomorphic to $\mathbb{R}^m$.

28. Prove directly that any two vector spaces of the same dimension (and over the same scalars $\mathbb{F}$) are isomorphic. [Hint: choose a basis for each, then pair off $\alpha_1\mathbf{v}_1 + \ldots + \alpha_n\mathbf{v}_n$ with $\alpha_1\mathbf{w}_1 + \ldots + \alpha_n\mathbf{w}_n$.]

29. Are the vector spaces $\{(x, y, z): x + y + z = 0\}$ and $\{(u, v, w): u + 2v + 3w = 0\}$ isomorphic? If so, try to find a specific isomorphism which establishes this.

30. Is there an isomorphism $T: M(\mathbb{R}_2) \to \mathbb{R}_3[x]$ in which

$$T\begin{bmatrix} 1 & 0 \\ 0 & 0 \end{bmatrix} = 1 + x + x^3\,?$$

Either prove there is not or define T completely by writing down

$$T\begin{bmatrix} 0 & 1 \\ 0 & 0 \end{bmatrix}, \quad T\begin{bmatrix} 0 & 0 \\ 1 & 0 \end{bmatrix} \quad \text{and} \quad T\begin{bmatrix} 0 & 0 \\ 0 & 1 \end{bmatrix}.$$

31. (a) Show that if $m \neq n$ then $\mathbb{R}^m \not\cong \mathbb{R}^n$. [How can you be sure that there is not a rather peculiarly defined isomorphism? Indeed there do exist 1–1 onto maps pairing off the elements of $\mathbb{R}^n$ and $\mathbb{R}^m$ – see the historical remarks in Chapter 9. Hint: isomorphisms send bases to . . .?]

(b) Let A be an $m \times n$ matrix with $m \neq n$. Is the row space of (n-tuples of) A isomorphic to the column space of (m-tuples of) A?

32. Place the following vector spaces into distinct categories so that two spaces are in the same category if and only if they are isomorphic:

(i) $M_2(\mathbb{R})$;

(ii) the space of arithmetic progressions;

(iii) the subspace of skew-symmetric matrices in $M_3(\mathbb{R})$;

(iv) the subspace of symmetric matrices in $M_2(\mathbb{R})$;

(v) the solution space of the system of equations

$$\begin{aligned} x-2y-2z+10t&=0\\ x-y-z+2t&=0\\ 2x+y+z&=0. \end{aligned}$$

(vi) $\mathbb{R}_3[x]$;

(vii) the space of 3×3 magic squares.

33. Show that, over $\mathbb{R}$, $\mathbb{C}^2$ is isomorphic to $\mathbb{R}^4$. Is $\mathbb{C}^2$ (as a vector space over $\mathbb{C}$) isomorphic to $\mathbb{R}^2$ as a vector space over $\mathbb{R}$? (Surely each has dimension 2?)

COMPUTER PACKAGE PROBLEMS

1. Are the following subspaces of $\mathbb{R}^5$ isomorphic?

(i) sp{(24, –89, –112, –268, –67), (25, –13, –24, –229, 6), (–62, –26, –4, 58, –20), (64, 72, 56, 40, 56), (–55, 27, 56, –69, 46), (–47, 19, 41, 178, 10)}

(ii) sp{(72, –176, –240, 360, 8), (–100, 95, 158, –414, 3), (81, 216, 216, 135, –27), (22, 66, 67, 63, –11), (48, 224, 240, 96, –32)}.

2. Are the following subspaces of $\mathbb{R}^2$ isomorphic?

(i) $\{(x,y,z,t,w): 159x-372y+423z-9t-120w=0$ and $369x-1125y+720z+36t+279w=0$ and $261x-1134y+171z+99t+918w=0\}$;

(ii) $\{(x,y,z,t,w): x+4y+9z+16t+25w=0$ and $x+8y+27z+64t+125w=0\}$.

3. Is either subspace from Problem 1 actually equal to one of the subspaces of Problem 2?

11 · Eigenvalues and Eigenvectors

In this chapter we see how problems which are simple versions of those of interest in real life lead us, necessarily, to the introduction of the eigenvalues and eigenvectors of a (square) matrix A. (The eigenvalues of A are the roots of the characteristic polynomial of A which is the value of a certain determinant derived from A.)

Many problems lead naturally to the desire to diagonalise A. Examples (and theory) reveal that, if an $n \times n$ matrix A has an LI set of n eigenvectors or if A is symmetric, then diagonalisation is possible. Where diagonalisation is not possible the best that can be achieved is to convert A to Jordan form.

Applications of the diagonalising process abound. We end the chapter with two more: (i) to recognising conics and (ii) to finding maximum and minimum values of functions of several variables.

We now come to what many mathematicians would see as one of the most important topics in linear algebra – the notions of eigenvalue and eigenvector. There are many ways to motivate their introduction. One way is to return to the first Application at the end of Chapter 3. (For alternative motivation see Examples 3 and 9 and the Applications.)

Recall our most surprising assertion: if T is the matrix

$$\begin{bmatrix} 0.8 & 0.15 & 0.05 \\ 0.2 & 0.7 & 0.35 \\ 0 & 0.15 & 0.6 \end{bmatrix}$$

then, as k is made larger and larger, each colum of T^k approximates to

$$\begin{bmatrix} 0.380 & \dots \\ 0.451 & \dots \\ 0.169 & \dots \end{bmatrix}$$

How might we confirm this? Put another way, we are asking how we might show (for large k) the result of multiplying T^k with each of

$$\begin{bmatrix} 1 \\ 0 \\ 0 \end{bmatrix}, \quad \begin{bmatrix} 0 \\ 1 \\ 0 \end{bmatrix} \quad \text{and} \quad \begin{bmatrix} 0 \\ 0 \\ 1 \end{bmatrix}$$

is (approximately) the above vector.

These queries raise the fundamental question as to how we can possibly have the remotest idea how to calculate $T^k\mathbf{v}$ when a 3×3 matrix T, a (large) positive integer k and a vector $\mathbf{v}(\in \mathbb{R}^3)$ are given.

Of course there is one case in which $T^k\mathbf{v}$ is easy to calculate: namely if, for some number λ, $T\mathbf{v} = \lambda\mathbf{v}$. For then,

$$T^2\mathbf{v}\{= T(T(\mathbf{v}))\} = T(\lambda\mathbf{v}) = \lambda(T\mathbf{v}) = \lambda^2\mathbf{v}$$

and, more generally, $T^k\mathbf{v} = \lambda^k\mathbf{v}$ for all positive integers k – as is easily proved by mathematical induction. But surely such $\mathbf{v}$ are rare? In a sense, yes! (Certainly there can be no more than n such λ corresponding to any given $n \times n$ matrix – as we shall see below.)

However suppose, by chance, that we can find a basis $\{\mathbf{v}_1, \mathbf{v}_2, \mathbf{v}_3\}$ for $\mathbb{R}^3$ and numbers λ_1, λ_2 and λ_3 such that $T\mathbf{v}_i = \lambda_i\mathbf{v}_i$ for $i = 1, 2$ and 3. Then each $\mathbf{u} \in \mathbb{R}^3$ can be expressed in the form $\mathbf{u} = \alpha_1\mathbf{v}_1 + \alpha_2\mathbf{v}_2 + \alpha_3\mathbf{v}_3$ for suitable numbers α_1, α_2 and α_3 and so

$$T^k\mathbf{u} = \alpha_1\lambda_1^k\mathbf{v}_1 + \alpha_2\lambda_2^k\mathbf{v}_2 + \alpha_3\lambda_3^k\mathbf{v}_3$$

(using the linearity of T).

We shall return to this example after Corollary 1. But one thing is already clear. It certainly seems appropriate, given a (square) matrix T, to investigate the problem of finding (non-zero) vectors $\mathbf{v}$ (and associated numbers λ) for which $T\mathbf{v} = \lambda\mathbf{v}$. We give such numbers and vectors special names.

Definition 1

Let T be an $n \times n$ matrix. If λ is a scalar[a] and if $\mathbf{v}$ is a (non-zero) vector[a] such that $T\mathbf{v} = \lambda\mathbf{v}$, we say that $\mathbf{v}$ is an **eigenvector** for T corresponding to the **eigenvalue** λ.

[The reason for not accepting the zero vector as an eigenvector is that it would be an eigenvector corresponding to every scalar and we surely do not want every scalar to be an eigenvalue! Note also that, for each scalar α ($\neq 0$), $\alpha\mathbf{v}$ is also an eigenvector for T – since $T(\alpha\mathbf{v}) = \alpha(T\mathbf{v}) = \alpha(\lambda\mathbf{v}) = \lambda(\alpha\mathbf{v})$.]

There are many significant problems queueing up to be solved by the eigenvalue/eigenvector method but to see how, given a (square) matrix T, we can actually find λ and $\mathbf{v}$ such that $T\mathbf{v} = \lambda\mathbf{v}$ – and to see that there are associated geometrical interpretations – we begin with a simple example.

Example 1

Let T be the matrix

$$T = \begin{bmatrix} -4 & -2 \\ 3 & 3 \end{bmatrix}.$$

Suppose the vector

$$\mathbf{v} = \begin{bmatrix} x \\ y \end{bmatrix}$$

and the number λ are such that

[a]Real or complex? Wait and see!

$$\begin{bmatrix} -4 & -2 \\ 3 & 3 \end{bmatrix}\begin{bmatrix} x \\ y \end{bmatrix} = \lambda\begin{bmatrix} x \\ y \end{bmatrix}.$$

Then, since

$$\lambda\begin{bmatrix} x \\ y \end{bmatrix} = \lambda I_2\begin{bmatrix} x \\ y \end{bmatrix}$$

(I_2 being the 2×2 identity matrix), we may write

$$\begin{bmatrix} -4 & -2 \\ 3 & 3 \end{bmatrix}\begin{bmatrix} x \\ y \end{bmatrix} - \lambda I_2\begin{bmatrix} x \\ y \end{bmatrix} = \begin{bmatrix} 0 \\ 0 \end{bmatrix}.$$

This we may rewrite as

$$\left\{\begin{bmatrix} -4 & -2 \\ 3 & 3 \end{bmatrix} - \lambda\begin{bmatrix} 1 & 0 \\ 0 & 1 \end{bmatrix}\right\}\begin{bmatrix} x \\ y \end{bmatrix} = \begin{bmatrix} 0 \\ 0 \end{bmatrix}, \text{ i.e. } \begin{bmatrix} -4-\lambda & -2 \\ 3 & 3-\lambda \end{bmatrix}\begin{bmatrix} x \\ y \end{bmatrix} = \begin{bmatrix} 0 \\ 0 \end{bmatrix}.$$

But this is nothing other than the pair of simultaneous (homogeneous) linear equations

$$\begin{aligned} (-4-\lambda)x - 2y &= 0 \\ 3x + (3-\lambda)y &= 0 \end{aligned} \qquad (11.1)$$

and (by Theorem 6 of Chapter 5) these will have a non-zero solution if and only if

$$\text{Det}\begin{bmatrix} -4-\lambda & -2 \\ 3 & 3-\lambda \end{bmatrix} = 0,$$

i.e. $(-4 - \lambda)(3 - \lambda) - (-6) = 0$. Consequently $\lambda^2 + \lambda - 6 = 0$, which implies that $\lambda = 2$ or $\lambda = -3$. These are the eigenvalues of T. To find corresponding eigenvectors we now solve equations (11.1) taking, first, $\lambda = 2$ and then $\lambda = -3$.

(i) $\lambda = 2$: Here equations (11.1) are equivalent to the single equation $3x + y = 0$. Thus, for $\lambda = 2$, equations (11.1) have solution

$$\begin{bmatrix} x \\ y \end{bmatrix} = \begin{bmatrix} \alpha \\ -3\alpha \end{bmatrix} = \alpha\begin{bmatrix} 1 \\ -3 \end{bmatrix} = \alpha\mathbf{v}_2, \text{ say, where } \mathbf{v}_2 = \begin{bmatrix} 1 \\ -3 \end{bmatrix}$$

and α is any non-zero number (non-zero since we want

$$\begin{bmatrix} x \\ y \end{bmatrix} \neq 0).$$

(ii) $\lambda = -3$: Here equations (11.1) are each equivalent to the single equation $x + 2y = 0$. Thus, for $\lambda = -3$, equations (11.1) have solution

$$\begin{bmatrix} x \\ y \end{bmatrix} = \begin{bmatrix} -2\beta \\ \beta \end{bmatrix} = \beta\begin{bmatrix} -2 \\ 1 \end{bmatrix} = \beta\mathbf{v}_{-3}, \text{ say, where } \mathbf{v}_{-3} = \begin{bmatrix} -2 \\ 1 \end{bmatrix}$$

and β is any (non-zero) number.

Summarising: the given matrix T has two eigenvalues, $\lambda = 2$ and $\lambda = -3$, with associated sets of eigenvectors $\{\alpha\mathbf{v}_2: \alpha \in \mathbb{R} \text{ and } \alpha \neq 0\}$ and $\{\beta\mathbf{v}_{-3}: \beta \in \mathbb{R} \text{ and } \beta \neq 0\}$ respectively where

$$\mathbf{v}_2 = \begin{bmatrix} 1 \\ -3 \end{bmatrix} \quad \text{and} \quad \mathbf{v}_{-3} = \begin{bmatrix} -2 \\ 1 \end{bmatrix}.$$

The geometrical interpretation of the above is as follows. Regarding T as the linear transformation of $\mathbb{R}^2$ which sends

$$\begin{bmatrix} x \\ y \end{bmatrix} \text{ to } \begin{bmatrix} -4 & -2 \\ 3 & 3 \end{bmatrix}\begin{bmatrix} x \\ y \end{bmatrix}$$

we see that, since

$$\begin{bmatrix} -4 & -2 \\ 3 & 3 \end{bmatrix}\begin{bmatrix} \alpha \\ -3\alpha \end{bmatrix} = 2\begin{bmatrix} \alpha \\ -3\alpha \end{bmatrix},$$

each point $(\alpha, -3\alpha)$ on the line $3x + y = 0$ is 'stretched out', by the action of T, to a point on the same line but twice as far away from (but on the same side of) the origin. Likewise, since

$$\begin{bmatrix} -4 & -2 \\ 3 & 3 \end{bmatrix}\begin{bmatrix} -2\beta \\ \beta \end{bmatrix} = -3\begin{bmatrix} -2\beta \\ \beta \end{bmatrix}.$$

each point $(-2\beta, \beta)$ on the line $x + 2y = 0$ is 'stretched out', by the action of T, to a point on the same line but three times as far away from (and, because of the minus sign, on the opposite side of) the origin. (Likewise, the eigenvectors of an $n \times n$ matrix T correspond exactly to lines in n-dimensional space which T maps to themselves – that is, lines which are *invariant under* T.)

This interpretation helps us to 'see' the overall effect that

$$T = \begin{bmatrix} -4 & -2 \\ 3 & 3 \end{bmatrix}$$

has on the whole plane.

Expressing each vector $\mathbf{u}$ in $\mathbb{R}^2$ in the form $\mathbf{u} = a\mathbf{v}_2 + b\mathbf{v}_{-3}$, we see that $T\mathbf{u} = 2a\mathbf{v}_2 - 3b\mathbf{v}_{-3}$. That is, the point whose 'coordinates', with respect to the basis $\{\mathbf{v}_2, \mathbf{v}_{-3}\}$ of $\mathbb{R}^2$ and $<a, b>$, is mapped to the point whose 'coordinates' (with respect to the same basis $\{\mathbf{v}_2, \mathbf{v}_{-3}\}$) are $<2a, -3b>$ – see Fig. 11.1. (You might care to plot other points – and the lines joining them, for example.)

PROBLEM 11.1

Consider the (linear) transformation of $\mathbb{R}^2$ given by sending the point

$$\begin{bmatrix} x \\ y \end{bmatrix} \text{ to the point } \begin{bmatrix} 1 & 0 \\ 0 & -1 \end{bmatrix}\begin{bmatrix} x \\ y \end{bmatrix} = \begin{bmatrix} x \\ -y \end{bmatrix} = T\begin{bmatrix} x \\ y \end{bmatrix},$$

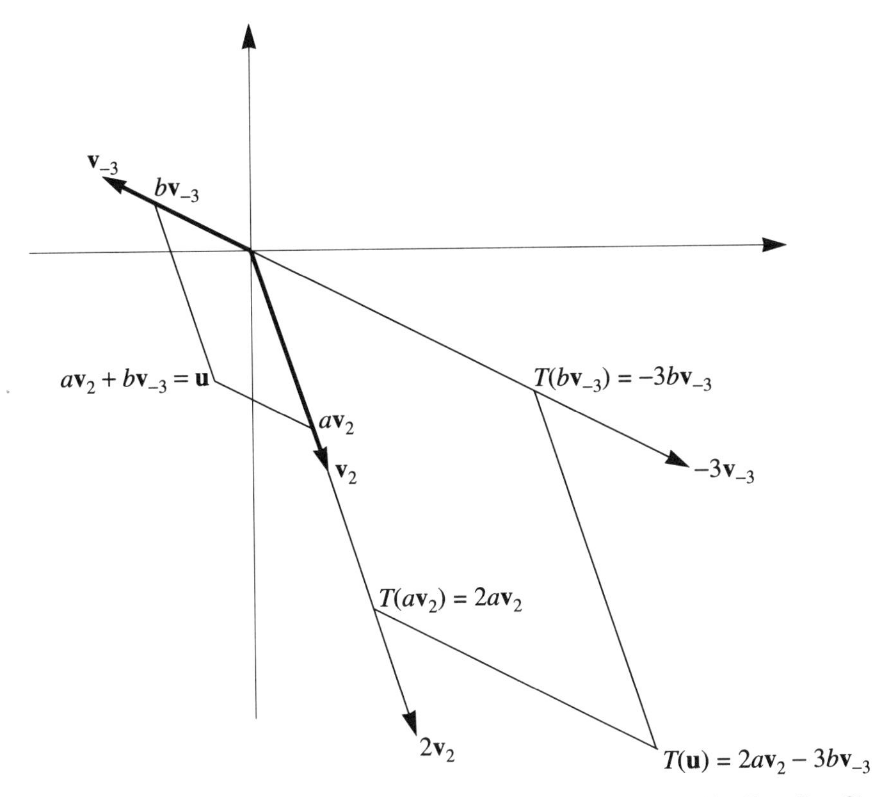

Fig 11.1 With respect to axes $\mathbf{v}_2$, $\mathbf{v}_{-3}$, the point $\mathbf{u} = \langle a, b \rangle$ is mapped by T to $\langle 2a, -3b \rangle$

say. This mapping reflects the plane $\mathbb{R}^2$ in the x-axis: in particular it maps the line $y = 0$ to itself. Can you 'see' geometrically any other line which T maps to itself? If you can then check you are right (and if you can't then find it) by obtaining the eigenvalues and corresponding eigenvectors of

$$\begin{bmatrix} 1 & 0 \\ 0 & -1 \end{bmatrix}$$

by the method of Example 1. ●

PROBLEM 11.2

Show that the map

$$T\begin{bmatrix} x \\ y \end{bmatrix} = \begin{bmatrix} 4 & 2 \\ 2 & 1 \end{bmatrix}\begin{bmatrix} x \\ y \end{bmatrix}$$

sends every point of $\mathbb{R}^2$ to the line $x = 2y$. (Consequently T certainly maps the line $x = 2y$ to itself. In other words

$$\begin{bmatrix} 2 \\ 1 \end{bmatrix} \text{ is an eigenvector of } \begin{bmatrix} 4 & 2 \\ 2 & 1 \end{bmatrix}.$$

Is any 'stretching' factor involved? That is, what is the corresponding eigenvalue?) Are there any other lines which map to themselves under the action of T? First think it out geometrically – then use the method of Example 1 to find any other eigenvalues and eigenvectors of

$$\begin{bmatrix} 4 & 2 \\ 2 & 1 \end{bmatrix}.$$

(You might be surprised!) ●

PROBLEM 11.3

The map $\mathbb{R}^2 \to \mathbb{R}^2$ given, in Problem 10.1(iii)(α), by

$$T\begin{bmatrix} x \\ y \end{bmatrix} = \begin{bmatrix} 4 & 2 \\ 2 & 4 \end{bmatrix}\begin{bmatrix} x \\ y \end{bmatrix}$$

has a shearing effect on the plane. Can you see, geometrically, any line(s) which are left unchanged by the action of T? Again check your answer (or find it) by obtaining eigenvalues and eigenvectors for

$$\begin{bmatrix} 4 & 2 \\ 2 & 4 \end{bmatrix}.$$

●

Must a linear transformation from $\mathbb{R}^2$ to $\mathbb{R}^2$ always fix at least one line? By no means. This can be seen most easily geometrically via the following example.

Example 2

Consider the (linear) transformation of the plane determined by the matrix

$$R = \begin{bmatrix} \cos\vartheta & -\sin\vartheta \\ \sin\vartheta & \cos\vartheta \end{bmatrix}.$$

This rotates everything in $\mathbb{R}^2$ anticlockwise through an angle ϑ about the origin. Consequently no non-zero vector (and hence no line) can be left fixed by R (unless ϑ is . . . what?)

Let us check the algebra. If, for certain

$$\mathbf{v} = \begin{bmatrix} x \\ y \end{bmatrix}$$

and λ we have $R\mathbf{v} = \lambda\mathbf{v}$ then

$$\begin{bmatrix} \cos\vartheta - \lambda & -\sin\vartheta \\ \sin\vartheta & \cos\vartheta - \lambda \end{bmatrix}\begin{bmatrix} x \\ y \end{bmatrix} = \begin{bmatrix} 0 \\ 0 \end{bmatrix}.$$

This requires

$$\text{Det}\begin{bmatrix} \cos\vartheta - \lambda & -\sin\vartheta \\ \sin\vartheta & \cos\vartheta - \lambda \end{bmatrix} = 0,$$

so that $\lambda = \cos\,\vartheta \pm \surd(-\sin^2\vartheta)$ [agreed?] and so λ is a real number only when $\sin\,\vartheta = 0$, that is $\vartheta = n\pi$, where n is an integer.

In particular, the matrix

$$S = \begin{bmatrix} 0 & -1 \\ 1 & 0 \end{bmatrix}$$

(corresponding to an anticlockwise rotation through $\pi/2$) leaves no line in $\mathbb{R}^2$ fixed. However, if we are prepared to allow complex numbers we can find (complex) eigenvalues and eigenvectors for S. To find these eigenvalues we (still) only have to solve the quadratic equation

$$\text{Det}\begin{bmatrix} -\lambda & -1 \\ 1 & -\lambda \end{bmatrix} = 0.$$

Thus $\lambda^2 + 1 = 0$, giving $\lambda = i$ or $-i$. Then, solving the equations

$$\begin{aligned} -\lambda x - 1y &= 0 \\ 1x - \lambda y &= 0 \end{aligned}$$

for $\lambda = i$, we find

$$\begin{bmatrix} x \\ y \end{bmatrix} = \alpha\begin{bmatrix} i \\ 1 \end{bmatrix} \quad (\alpha \neq 0)$$

whilst if $\lambda = -i$ we find

$$\begin{bmatrix} x \\ y \end{bmatrix} = \beta\begin{bmatrix} -i \\ 1 \end{bmatrix} \quad (\beta \neq 0).$$

(Geometrically the matrix maps the imaginary lines $x = iy$ and $x = -iy$ to themselves – using an imaginary scaling factor!)

Complex eigenvalues can be a bit of a nuisance. The problem is this: if we refuse to accept them then we have to admit that some matrices have no eigenvalues at all – and hence no eigenvectors. If we do accept them then, even for real matrices, we may be forced to accept eigenvectors with complex components. Now if

$$\begin{bmatrix} x \\ y \end{bmatrix} \text{ is an eigenvector, so is } c\begin{bmatrix} x \\ y \end{bmatrix}$$

for all non-zero scalars c. Should we allow c to range over all non-zero complex numbers or shall we (especially in the case of a real matrix) restrict it to being real? If the latter then, whereas I took

$$\alpha\begin{bmatrix} i \\ 1 \end{bmatrix} \text{ and } \beta\begin{bmatrix} -i \\ 1 \end{bmatrix}$$

as the eigenvectors in Example 2 you might have decided to take

$$\gamma\begin{bmatrix} -1 \\ i \end{bmatrix} \text{ and } \delta\begin{bmatrix} 1 \\ -i \end{bmatrix}$$

and, if only real scalars α, β, γ, δ are used, my set of eigenvectors would be different from yours. So it seems that we must accept c being complex throughout. But then we ask: do you really want to consider vectors such as

$$i\begin{bmatrix} -2 \\ 1 \end{bmatrix} \text{ and } (3+4i)\begin{bmatrix} -1 \\ 1 \end{bmatrix}$$

as eigenvectors in Example 1? Put another way: do we really want to accept

$$\begin{bmatrix} x \\ y \end{bmatrix} = \begin{bmatrix} -2i \\ i \end{bmatrix} \text{ and } \begin{bmatrix} -3-4i \\ 3+4i \end{bmatrix} \text{ etc.}$$

as solutions to the system of equations (11.1) above?

The resolution of this problem seems to be: either (i) always accept complex numbers under all circumstances or (ii) accept complex eigenvalues, eigenvectors and scalars only when forced to (as, for instance, in Example 8 below). We have tended to adopt this latter approach.

As a second (entirely different type of) motivating example let us look at a problem which, you might imagine, describes a chemical or physical system involving two variable quantities which are interrelated – and see how it forces on us the study of eigenvalues and eigenvectors.

Example 3

Consider the system of simultaneous differential equations

$$\begin{aligned} \frac{dx}{dt} &= \tfrac{5}{3}x(t) - \tfrac{1}{3}y(t) \\ \frac{dy}{dt} &= -\tfrac{2}{3}x(t) + \tfrac{4}{3}y(t) \end{aligned} \tag{11.2}$$

where $x(t)$ and $y(t)$ are functions of the time t. [Notice how the rate of increase of each function is lessened if the other function is able to increase in size. An intriguing question, therefore, is what will happen to $x(t)$ and $y(t)$ in the long run. For example, will either die out – that is, will $x(t)$ or $y(t)$ eventually become 0?]

If only this system had been of the form

$$\begin{aligned} \frac{dx}{dt} &= \alpha x(t) \\ \frac{dy}{dt} &= \beta y(t) \end{aligned}$$

the solution would have been easy! Indeed $\int dx/x = \int \alpha\, dt$; $\int dy/y = \int \beta\, dt$ giving $x = G\, e^{\alpha t}$, $y = H\, e^{\beta t}$ (where G and H are constants which are determined by the initial conditions of the problem). What we do with the given problem is to reduce it to one of this kind, as follows.

Let us assume that $X(t)$ and $Y(t)$ are functions of t which are such that

$$\begin{aligned} X(t) &= p_1 x(t) + p_2 y(t) \\ Y(t) &= q_1 x(t) + q_2 y(t), \end{aligned}$$

p_1, p_2, q_1 and q_2 being constants which are to be determined. We may write this in matrix form as

$$\begin{bmatrix} X(t) \\ Y(t) \end{bmatrix} = \begin{bmatrix} p_1 & p_2 \\ q_1 & q_2 \end{bmatrix} \begin{bmatrix} x(t) \\ y(t) \end{bmatrix} = M \begin{bmatrix} x(t) \\ y(t) \end{bmatrix},$$

say. If we write

$$\begin{bmatrix} dx/dt \\ dy/dt \end{bmatrix} \text{ briefly as } \begin{bmatrix} \dot{x} \\ \dot{y} \end{bmatrix}$$

and use similar notation for $X(t)$ and $Y(t)$, so that

$$\begin{bmatrix} \dot{X} \\ \dot{Y} \end{bmatrix} = M \begin{bmatrix} \dot{x} \\ \dot{y} \end{bmatrix},$$

we find we can write equations (11.2) first as

$$\begin{bmatrix} \dot{x} \\ \dot{y} \end{bmatrix} = \begin{bmatrix} \frac{5}{3} & -\frac{1}{3} \\ -\frac{2}{3} & \frac{4}{3} \end{bmatrix} \begin{bmatrix} x \\ y \end{bmatrix} = T \begin{bmatrix} x \\ y \end{bmatrix},$$

say, and then (assuming M to be invertible!) as

$$M \begin{bmatrix} \dot{x} \\ \dot{y} \end{bmatrix} = M \begin{bmatrix} \frac{5}{3} & -\frac{1}{3} \\ -\frac{2}{3} & \frac{4}{3} \end{bmatrix} M^{-1} M \begin{bmatrix} x \\ y \end{bmatrix}, \text{ that is, as } \begin{bmatrix} \dot{X} \\ \dot{Y} \end{bmatrix} = MTM^{-1} \begin{bmatrix} X \\ Y \end{bmatrix} \left(\text{since } \begin{bmatrix} X \\ Y \end{bmatrix} = M \begin{bmatrix} x \\ y \end{bmatrix} \right).$$

Now, if only we could (also) assume MTM^{-1} to be a diagonal matrix

$$D = \begin{bmatrix} \alpha & 0 \\ 0 & \beta \end{bmatrix},$$

say, then we would have

$$\begin{aligned} \dot{X}(t) &= \alpha X \\ \dot{Y}(t) &= \beta Y \end{aligned}$$

which would imply $X = U\, e^{\alpha t}$ and $Y = V\, e^{\beta t}$ (for suitable constants U and V). Then, from

$$\begin{bmatrix} X(t) \\ Y(t) \end{bmatrix} = M \begin{bmatrix} x(t) \\ y(t) \end{bmatrix}$$

we could deduce that

$$\begin{bmatrix} x(t) \\ y(t) \end{bmatrix} = M^{-1} \begin{bmatrix} X(t) \\ Y(t) \end{bmatrix}$$

and hence solve for $x(t)$ and $y(t)$.

So, the question is: given a (square) matrix T can one find an invertible matrix M such that $MTM^{-1} = D$, a diagonal matrix? If so, we say that T is **diagonalisable**. Let us investigate this possibility.

For tidyness, we set $N = M^{-1}$ and then rewrite $N^{-1}TN = D$ as $TN = ND$. Now, if

$$N = \begin{bmatrix} r_1 & r_2 \\ s_1 & s_2 \end{bmatrix} \quad \text{and} \quad D = \begin{bmatrix} \alpha & 0 \\ 0 & \beta \end{bmatrix},$$

it follows from $TN = ND$, that

$$T\begin{bmatrix} r_1 \\ s_1 \end{bmatrix} = \alpha \begin{bmatrix} r_1 \\ s_1 \end{bmatrix} \quad \text{whilst} \quad T\begin{bmatrix} r_2 \\ s_2 \end{bmatrix} = \beta \begin{bmatrix} r_2 \\ s_2 \end{bmatrix}.$$

That is, the columns of the matrix N which we hope will diagonalise T are forced to be eigenvectors of T. Furthermore, the diagonal entries of D are the corresponding eigenvalues.

We summarise the above: given the (2×2) matrix

$$T = \begin{bmatrix} a & b \\ c & d \end{bmatrix}$$

we first find its eigenvalues by solving

$$\text{Det}\begin{bmatrix} a-\lambda & b \\ c & d-\lambda \end{bmatrix} = \begin{vmatrix} a-\lambda & b \\ c & d-\lambda \end{vmatrix} = 0$$

for λ. Then, for each eigenvalue λ_1, λ_2 of T, we solve the resulting system of equations

$$\begin{bmatrix} a-\lambda & b \\ c & d-\lambda \end{bmatrix}\begin{bmatrix} x \\ y \end{bmatrix} = \begin{bmatrix} 0 \\ 0 \end{bmatrix}.$$

It follows that if

$$\begin{bmatrix} \rho \\ \sigma \end{bmatrix} \quad \text{and} \quad \begin{bmatrix} \mu \\ \nu \end{bmatrix}$$

are (non-zero) solutions for λ_1 and λ_2 respectively (and if $\lambda_1 \neq \lambda_2$)[b] then the sets

$$\left\{\alpha\begin{bmatrix} \rho \\ \sigma \end{bmatrix} : \alpha \neq 0\right\} \quad \text{and} \quad \left\{\beta\begin{bmatrix} \mu \\ \nu \end{bmatrix} : \beta \neq 0\right\}$$

are the sets of eigenvectors corresponding to λ_1 and λ_2 respectively. Furthermore we then have

[b] And if λ_1, λ_2 are equal? See Examples 5–8.

$$\begin{bmatrix} \rho & \mu \\ \sigma & \nu \end{bmatrix}^{-1} \begin{bmatrix} a & b \\ c & d \end{bmatrix} \begin{bmatrix} \rho & \mu \\ \sigma & \nu \end{bmatrix} = \begin{bmatrix} \lambda_1 & 0 \\ 0 & \lambda_2 \end{bmatrix}.$$

Let us now complete the solution of the system (11.2).

We first put

$$\text{Det}\begin{bmatrix} \frac{5}{3}-\lambda & -\frac{1}{3} \\ -\frac{2}{3} & \frac{4}{3}-\lambda \end{bmatrix} = 0.$$

We find that $\lambda_1 = 1$ and $\lambda_2 = 2$ are the eigenvalues. Corresponding to $\lambda_1 = 1$ we obtain the equations

$$\begin{matrix} \left(\frac{5}{3}-1\right)r - \frac{1}{3}s = 0 \\ -\frac{2}{3}r + \left(\frac{4}{3}-1\right)s = 0 \end{matrix} \quad \text{with solutions} \quad \begin{bmatrix} r \\ s \end{bmatrix} = \alpha \begin{bmatrix} 1 \\ 2 \end{bmatrix} \quad (\alpha \neq 0),$$

whilst, for $\lambda_2 = 2$, we have

$$\begin{matrix} \left(\frac{5}{3}-2\right)r - \frac{1}{3}s = 0 \\ -\frac{2}{3}r + \left(\frac{4}{3}-2\right)s = 0 \end{matrix} \quad \text{with solutions} \quad \begin{bmatrix} r \\ s \end{bmatrix} = \beta \begin{bmatrix} 1 \\ -1 \end{bmatrix} \quad (\beta \neq 0).$$

If we now set

$$N = \begin{bmatrix} 1 & 1 \\ 2 & -1 \end{bmatrix} \quad \left(\text{so that } N^{-1} = \frac{1}{3}\begin{bmatrix} 1 & 1 \\ 2 & -1 \end{bmatrix}\right) \quad \text{and} \quad D = \begin{bmatrix} 1 & 0 \\ 0 & 2 \end{bmatrix},$$

we find that $N^{-1}TN = D$. In particular, this working tells us to put

$$\begin{bmatrix} X \\ Y \end{bmatrix} = \frac{1}{3}\begin{bmatrix} 1 & 1 \\ 2 & -1 \end{bmatrix}\begin{bmatrix} x \\ y \end{bmatrix}.$$

That is, we should set $X = (x + y)/3$ and $Y = (2x - y)/3$ or, equivalently, $x = X + Y$ and $y = 2X - Y$. Doing this we find that

$$\begin{aligned} \frac{\mathrm{d}X}{\mathrm{d}t} &= \frac{1}{3}\left\{\frac{\mathrm{d}x}{\mathrm{d}t} + \frac{\mathrm{d}y}{\mathrm{d}t}\right\} = \frac{1}{3}\{x(t) + y(t)\} = X(t) \\ \frac{\mathrm{d}Y}{\mathrm{d}t} &= \frac{1}{3}\left\{2\frac{\mathrm{d}x}{\mathrm{d}t} - \frac{\mathrm{d}y}{\mathrm{d}t}\right\} = \frac{1}{3}\{4x(t) - 2y(t)\} = 2Y(t), \end{aligned} \tag{11.3}$$

as expected. It follows that $X(t) = K\,\mathrm{e}^t$ and $Y(t) = L\,\mathrm{e}^{2t}$ (where K and L are constants) and hence that

$$x(t) = K\,\mathrm{e}^t + L\,\mathrm{e}^{2t} \quad \text{and} \quad y(t) = 2K\,\mathrm{e}^t - L\,\mathrm{e}^{2t} \tag{11.4}$$

[The change from functions $x(t)$, $y(t)$ to new functions might be thought of as a sort of 'change of coordinates' technique.]

If, in particular, we are told that at time $t = 0$, $x = 1001$ and $y = 1999$ then, from (11.4) we obtain $1001 = K + L$ and $1999 = 2K - L$. This implies that $K = 1000$ and $L = 1$. Since, from (11.4), $y(t) = 0$ when $\mathrm{e}^t = 2K/L$, we see that there is none of y left when $\mathrm{e}^t = 2000$, i.e. when $t = \log_e 2000 = 7.601$ (years?) approximately.

PROBLEM 11.4

(a) In Example 3, what eventually happens to $x(t)$ and $y(t)$ given that $x(0) = 1000$ and $y(0) = 2000$?

(b) Given that, on 1 January 1995, $x = 1005$ and $y = 1995$ determine whether or not the 'population' y will survive to 1 January 2001.[c] Free oscillations of connected systems of particles give rise to similar problems except that the vector

$$\begin{bmatrix} dx/dt \\ dy/dt \end{bmatrix} \text{ is replaced by } \begin{bmatrix} d^2x/dt^2 \\ d^2y/dt^2 \end{bmatrix}.$$

●

Clearly, then, it is most useful to be able to diagonalise a matrix. Unfortunately not all 2×2 matrices are diagonalisable.

PROBLEM 11.5

(i) Show that the matrix

$$\begin{bmatrix} 1 & 1 \\ 0 & 1 \end{bmatrix}$$

cannot be diagonalised.

(ii) Show that the matrix

$$\begin{bmatrix} 0 & -1 \\ 1 & 0 \end{bmatrix}$$

is diagonalisable – but only if one allows the use of matrices with complex entries.

(iii) For the matrix T of Example 1 write down an invertible matrix N and a diagonal matrix D such that $N^{-1}TN = D$. ●

Of course, calculations of the above kind can be attempted for a (square) matrix T of any size. But note that, if T is $n \times n$, then $\mathrm{Det}(T - \lambda I)$ is a polynomial $p(\lambda)$, say, (called the **characteristic polynomial of T**) of degree n in λ. And, whilst there are (quite tricky) formulae for solving cubic and quartic equations in terms of the coefficients appearing in the equations (the formula in the quadratic case being well known), a very famous theorem of algebra, due to Evariste Galois (1811–1832), says that *there can exist no such general formula in cases of higher degree*. [Thus, for equations of higher degree, numerical methods (with all the attendant hazards of approximation, rounding errors and calculating computer running time) suggest themselves.] On the other hand, admitting complex roots gives the advantage that $p(\lambda)$ will factorise into a product ot n linear factors. (This assertion is the so-called **fundamental theorem of algebra** – see the Appendix.) That is, counting roots according to their multiplicities, $p(\lambda)$ will have n roots. (Cf. Example 8.)

Here, then, are some further examples. However, you should realise (see Exercise

[c] That is, the start of the next millennium!

11.13) that those chosen have been specially 'cooked' so that the **characteristic equation**[d] Det($T - \lambda I$) = 0 is fairly readily solved.[e]

Example 4

Find the eigenvalues and, to each eigenvalue, all the corresponding eigenvectors for the matrix

$$A = \begin{bmatrix} 3 & 11 & -11 \\ 1 & 3 & -2 \\ 1 & 5 & -4 \end{bmatrix}.$$

First we solve

$$\begin{vmatrix} 3-\lambda & 11 & -11 \\ 1 & 3-\lambda & -2 \\ 1 & 5 & -4-\lambda \end{vmatrix} = 0.$$

Now

$$\begin{vmatrix} 3-\lambda & 11 & -11 \\ 1 & 3-\lambda & -2 \\ 1 & 5 & -4-\lambda \end{vmatrix} = \begin{vmatrix} 3-\lambda & 11 & 0 \\ 1 & 3-\lambda & 1-\lambda \\ 1 & 5 & 1-\lambda \end{vmatrix} = (1-\lambda)\begin{vmatrix} 3-\lambda & 11 & 0 \\ 1 & 3-\lambda & 1 \\ 1 & 5 & 1 \end{vmatrix}$$

$$= (1-\lambda)\begin{vmatrix} 3-\lambda & 11 & 0 \\ 0 & 3-\lambda & 1 \\ 0 & 5 & 1 \end{vmatrix} = (1-\lambda)[(3-\lambda)(3-\lambda-5)].$$

Hence

$$\begin{vmatrix} 3-\lambda & 11 & -11 \\ 1 & 3-\lambda & -2 \\ 1 & 5 & -4-\lambda \end{vmatrix} = 0$$

if and only if $\lambda = 1, 3$ or -2.

To find corresponding eigenvectors we solve the system of equations

$$\begin{bmatrix} 3-\lambda & 11 & -11 \\ 1 & 3-\lambda & -2 \\ 1 & 5 & -4-\lambda \end{bmatrix}\begin{bmatrix} x \\ y \\ z \end{bmatrix} = 0$$

for $\lambda = 1, 3$ and -2 in turn.

For $\lambda = 1$ we obtain

[d]Eigenvalues and eigenvectors are sometimes called characteristic roots and characteristic vectors
[e]In practice the coefficients of the characteristic equation will not be integers. Even worse, because of various errors which can arise (and swiftly multiply) in solving equations, iteratively, by computer, eigenvalues are usually not found by solving the characteristic equation!

$$\begin{aligned} 2x+11y-11z&=0\\ x+2y-2z&=0\\ x+5y-5z&=0 \end{aligned}$$

from which $x = 0$ and $y = z = \alpha$, say, where α ($\neq 0$) is arbitrary. Thus the eigenvectors corresponding to $\lambda = 1$ are precisely the vectors

$$\alpha\begin{bmatrix}0\\1\\1\end{bmatrix}$$

where α ($\neq 0$) is arbitrary. In a similar manner we obtain, for $\lambda = -2$, the system of equations

$$\begin{aligned} 5x+11y-11z&=0\\ x+5y-2z&=0\\ x+5y-2z&=0 \end{aligned} \quad \text{from which} \quad \begin{bmatrix}x\\y\\z\end{bmatrix}=\beta\begin{bmatrix}33\\-1\\14\end{bmatrix},$$

β ($\neq 0$) being arbitrary. Finally, for $\lambda = 3$ we have

$$\begin{aligned} 11y-11z&=0\\ x-2z&=0\\ x+5y-7z&=0 \end{aligned} \quad \text{giving} \quad \begin{bmatrix}x\\y\\z\end{bmatrix}=\lambda\begin{bmatrix}2\\1\\1\end{bmatrix},$$

γ ($\neq 0$) being arbitrary.

Geometrically the linear transformation of $\mathbb{R}^3$ given by sending

$$\begin{bmatrix}x\\y\\z\end{bmatrix} \quad \text{to} \quad \begin{bmatrix}3&11&-11\\1&3&-2\\1&5&-4\end{bmatrix}\begin{bmatrix}x\\y\\z\end{bmatrix}$$

leaves three lines in $\mathbb{R}^3$ invariant, namely those through the origin which contain the points (0, 1, 1), (33, –1, 14) and (2, 1, 1) respectively.

PROBLEM 11.6

Identify three planes which are left invariant (i.e. mapped to themselves) by the action of A.

Following the same methods as in the 2×2 case we can now diagonalise the matrix A immediately. We have

$$\begin{bmatrix}0&33&2\\1&-1&1\\1&14&1\end{bmatrix}^{-1}\begin{bmatrix}3&11&-11\\1&3&-2\\1&5&-4\end{bmatrix}\begin{bmatrix}0&33&2\\1&-1&1\\1&14&1\end{bmatrix}=\begin{bmatrix}1&0&0\\0&-2&0\\0&0&3\end{bmatrix}.$$

Example 5

Find the eigenvalues and, to each eigenvalue, all the corresponding eigenvectors for the matrix

$$B=\begin{bmatrix}1 & -1 & -1\\ -1 & 1 & -1\\ -1 & -1 & 1\end{bmatrix}.$$

Here the eigenvalues are given by solving

$$\begin{vmatrix}1-\lambda & -1 & -1\\ -1 & 1-\lambda & -1\\ -1 & -1 & 1-\lambda\end{vmatrix}=0.$$

It is readily seen that $\lambda = 2$ is an eigenvalue (since then all rows are $-1\ -1\ -1$.) Also, adding rows 2 and 3 to the first row of the determinant shows $(-1-\lambda)$ is a factor. In fact we obtain

$$\begin{vmatrix}1-\lambda & -1 & -1\\ -1 & 1-\lambda & -1\\ -1 & -1 & 1-\lambda\end{vmatrix}=(-1-\lambda)(2-\lambda)(2-\lambda)$$

so that one eigenvalue is repeated. We find the corresponding eigenvectors as above.

For $\lambda = -1$ we solve

$$\begin{aligned}2x-y-z&=0\\ -x+2y-z&=0\\ -x-y+2z&=0\end{aligned}\quad\text{giving}\quad\begin{bmatrix}x\\ y\\ z\end{bmatrix}=\alpha\begin{bmatrix}1\\ 1\\ 1\end{bmatrix},$$

α being arbitrary (and non-zero!)

For $\lambda = 2$ we solve

$$\begin{aligned}-x-y-z&=0\\ -x-y-z&=0\\ -x-y-z&=0\end{aligned}$$

giving $x = -y - z$ so that

$$\begin{bmatrix}x\\ y\\ z\end{bmatrix}=\begin{bmatrix}-y-z\\ y\\ z\end{bmatrix}=\begin{bmatrix}-y\\ y\\ 0\end{bmatrix}+\begin{bmatrix}-z\\ 0\\ z\end{bmatrix}.$$

Taking $y = \beta$ and $z = \gamma$ arbitrarily (and, of course, not *both* zero) we have

$$\begin{bmatrix}x\\ y\\ z\end{bmatrix}=\beta\begin{bmatrix}-1\\ 1\\ 0\end{bmatrix}+\gamma\begin{bmatrix}-1\\ 0\\ 1\end{bmatrix}.$$

Thus, as distinct from previous cases, we have here a (repeated) eigenvalue with a 'double infinity' of corresponding eigenvectors – since β and γ may be chosen arbitrarily.

PROBLEM 11.7

Are there three invariant planes for this matrix B?

Can we diagonalise the matrix B? Indeed we can:

$$\begin{bmatrix}1 & -1 & -1\\1 & 1 & 0\\1 & 0 & 1\end{bmatrix}^{-1}\begin{bmatrix}1 & -1 & -1\\-1 & 1 & -1\\-1 & -1 & 1\end{bmatrix}\begin{bmatrix}1 & -1 & -1\\1 & 1 & 0\\1 & 0 & 1\end{bmatrix}=\begin{bmatrix}-1 & 0 & 0\\0 & 2 & 0\\0 & 0 & 2\end{bmatrix}.$$

Note, also, that, on taking eigenvalues and their corresponding eigenvectors in a different order, we could equally well have written

$$\begin{bmatrix}-1 & 1 & -1\\0 & 1 & 1\\1 & 1 & 0\end{bmatrix}^{-1}\begin{bmatrix}1 & -1 & -1\\-1 & 1 & -1\\-1 & -1 & 1\end{bmatrix}\begin{bmatrix}-1 & 1 & -1\\0 & 1 & 1\\1 & 1 & 0\end{bmatrix}=\begin{bmatrix}2 & 0 & 0\\0 & -1 & 0\\0 & 0 & 2\end{bmatrix}.$$

The 'double infinity' of eigenvectors referred to above can be described more precisely by reintroducing the zero vector via the following definition.

• *Definition 2*

For each eigenvalue λ of a given $n \times n$ matrix T, say, let $E_\lambda = \{\mathbf{x}: T\mathbf{x} = \lambda\mathbf{x}\}$. (Thus E_λ comprises all eigenvectors of T corresponding to λ together with the zero vector.) Then E_λ is a subspace of $\mathbb{R}^n$ called the **eigenspace of T corresponding to λ.**

We leave you to check (Exercise 30) that E_λ is a subspace of $\mathbb{R}^n$. In particular the invariant lines (and planes) which we have considered are not arbitrary lines and planes but lines (planes) through the origin. That is, they are subspaces.

In the above example the 'double infinity' of eigenvectors corresponding to the eigenvalue 2, is more neatly described by saying that E_2 has dimension 2 (whilst E_{-1} has dimension 1).

• *Example 6*

Find the eigenvalues and, to each eigenvalue, all the corresponding eigenvectors for the matrix

$$C=\begin{bmatrix}3 & 6 & 2\\0 & -3 & -8\\1 & 0 & -4\end{bmatrix}.$$

Here

$$\begin{vmatrix}3-\lambda & 6 & 2\\0 & -3-\lambda & -8\\1 & 0 & -4-\lambda\end{vmatrix}=0$$

yields the characteristic equation $-\lambda^3 - 4\lambda^2 + 11\lambda - 6 = 0$ (so that $\lambda = 1$, 1 and -6 are the eigenvalues of C). We leave you to check that the eigenvectors corresponding to $\lambda = -6$ are

$$\left\{\alpha\begin{bmatrix}-6\\8\\3\end{bmatrix} : \alpha \neq 0\right\}.$$

To find those corresponding to $\lambda = 1$ we solve

$$\begin{aligned}2x+6y+2z&=0\\-4y-8z&=0\\x-5z&=0.\end{aligned}$$

We find

$$\begin{bmatrix}x\\y\\z\end{bmatrix}=\beta\begin{bmatrix}5\\-2\\1\end{bmatrix},$$

β ($\neq 0$) being arbitrary. Here we have a new development: a repeated eigenvalue with only a 'single infinity' of corresponding eigenvectors, that is, the eigenspace E_1 has dimension 1.

PROBLEM 11.8

How many invariant planes can you find for C?

PROBLEM 11.9

Is C diagonalisable? (Certainly we cannot find scalars β, γ such that

$$N=\begin{bmatrix}-6 & 5\beta & 5\gamma\\8 & -2\beta & -2\gamma\\3 & \beta & \gamma\end{bmatrix} \quad \text{and} \quad N^{-1}CN=\begin{bmatrix}-6 & 0 & 0\\0 & 1 & 0\\0 & 0 & 1\end{bmatrix}.$$

Why not?)

Example 7

Find the eigenvalues and, to each eigenvalue, all the corresponding eigenvectors for the matrix

$$E=\begin{bmatrix}5 & -17 & 9\\2 & -6 & 3\\2 & -7 & 4\end{bmatrix}.$$

The characteristic equation is found to be $(1-\lambda)^3=0$ so that $\lambda = 1$ is a triply occurring root. The equations

$$\begin{aligned}4x-17y+9z&=0\\2x-7y+3z&=0\\2x-7y+3z&=0\end{aligned} \quad \text{have solution} \quad \begin{bmatrix}x\\y\\z\end{bmatrix}=\alpha\begin{bmatrix}2\\1\\1\end{bmatrix}.$$

Thus we have a triply repeated eigenvalue with a corresponding eigenspace of dimension 1.

Here there is clearly no invertible matrix N such that

$$N^{-1}EN = \begin{bmatrix} 1 & 0 & 0 \\ 0 & 1 & 0 \\ 0 & 0 & 1 \end{bmatrix}!$$

PROBLEM 11.10

Is there any invariant plane here?

Certainly, a (real) 3×3 matrix always gives rise to at least one invariant line in $\mathbb{R}^3$ – as Exercise 23 below indicates. (As in Problem 11.2 above we allow the case where all the points on a line are mapped to the same (one) point on it.)

PROBLEM 11.11

Does each 3×3 matrix give rise to at least one invariant plane? (Thus we require every point of such a plane to be mapped to some point in it.)

Finally we look at just one example of a 4×4 matrix.

Example 8

Let

$$A = \begin{bmatrix} 3 & 5 & -7 & 2 \\ 8 & -5 & 8 & 7 \\ 6 & -6 & 9 & 5 \\ 0 & -6 & 8 & 1 \end{bmatrix}.$$

It is not difficult to check that the characteristic polynomial for A is $(\lambda - \{2+i\})^2(\lambda - \{2-i\})^2$ and that the corresponding eigenspaces, each of dimension 2, are:

$$\text{for } \lambda = 2-i\left\{\alpha\begin{bmatrix} 1 \\ 4+4i \\ 3+3i \\ 0 \end{bmatrix} + \beta\begin{bmatrix} -1-i \\ 1 \\ 1 \\ 1+i \end{bmatrix}\right\}; \quad \text{for } \lambda = 2+i{:}\left\{\gamma\begin{bmatrix} 1 \\ 4-4i \\ 3-3i \\ 0 \end{bmatrix} + \delta\begin{bmatrix} -1+i \\ 1 \\ 1 \\ 1-i \end{bmatrix}\right\}$$

where $\alpha, \beta, \gamma, \delta \in \mathbb{C}$. (We leave you to find N and diagonal D such that $N^{-1}AN = D$.)

From all the above examples it appears that what makes the $n \times n$ matrix T diagonalisable is the existence of an LI set of n vectors all of which are eigenvectors for T. See Exercise 32(a). (Whether or not an eigenvalue is repeated is, to some extent, irrelevant – though see Theorem 1 and Corollary 1.)

There are two cases in which such an LI set of n eigenvectors is guaranteed! To obtain the first we prove the following theorem.

• *Theorem 1*

Let T be an $n \times n$ matrix and let $\lambda_1, \lambda_2, \ldots, \lambda_m$ be distinct eigenvalues of T. To each λ_i $(1 \le i \le m)$ let $\mathbf{v}_i$ be a corresponding eigenvector. Then $\{\mathbf{v}_1, \mathbf{v}_2, \ldots, \mathbf{v}_m\}$ is a linearly independent set of vectors in $\mathbb{R}^n$.

PROOF

(This is another pretty proof, so once again we invite you to sit back and enjoy it!)

For each k $(1 \le k \le m)$ we show that $\{\mathbf{v}_1, \mathbf{v}_2, \ldots, \mathbf{v}_k\}$ is an LI set. We prove this by induction on k. If $k = 1$ we have to show that the set $\{\mathbf{v}_1\}$ is a linearly independent set. But this is immediate since $\mathbf{v}_1$ is, by definition, not the zero vector. (See Exercise 18(a) of Chapter 8.)

Now suppose the desired result is known to be valid if $k = t$ where $1 \le t < m$. We show it to be true for $k = t + 1$. To do this consider the equation

$$\alpha_1\mathbf{v}_1 + \alpha_2\mathbf{v}_2 + \ldots + \alpha_t\mathbf{v}_t + \alpha_{t+1}\mathbf{v}_{t+1} = \mathbf{0}. \tag{11.5}$$

We deduce that

$$T(\alpha_1\mathbf{v}_1 + \alpha_2\mathbf{v}_2 + \ldots + \alpha_t\mathbf{v}_t + \alpha_{t+1}\mathbf{v}_{t+1}) = \mathbf{0},$$

that is

$$\alpha_1(T\mathbf{v}_1) + \alpha_2(T\mathbf{v}_2) + \ldots + \alpha_t(T\mathbf{v}_t) + \alpha_{t+1}(T\mathbf{v}_{t+1}) = \mathbf{0},$$

in other words

$$\alpha_1\lambda_1\mathbf{v}_1 + \alpha_2\lambda_2\mathbf{v}_2 + \ldots + \alpha_t\lambda_t\mathbf{v}_t + \alpha_{t+1}\lambda_{t+1}\mathbf{v}_{t+1} = \mathbf{0}. \tag{11.6}$$

Subtracting λ_{t+1} times equation (11.5) from (11.6) we obtain

$$\alpha_1(\lambda_1 - \lambda_{t+1})\mathbf{v}_1 + \alpha_2(\lambda - \lambda_{t+1})\mathbf{v}_2 + \ldots + \alpha_t(\lambda_t - \lambda_{t+1})\mathbf{v}_t = \mathbf{0}. \tag{11.7}$$

But, since $\mathbf{v}_1, \mathbf{v}_2, \ldots, \mathbf{v}_t$ are (by assumption) linearly independent, we deduce that

$$\alpha_1(\lambda_1 - \lambda_{t+1}) = \alpha_2(\lambda_2 - \lambda_{t+1}) = \ldots = \alpha_t(\lambda_t - \lambda_{t+1}) = 0.$$

However, for $1 \le i \le t$, none of the $(\lambda_i - \lambda_{t+1})$ is zero (since the λ_i are distinct). Hence each of the α_i $(1 \le i \le t)$ is zero. Thus, from (11.5) we deduce that $\alpha_{t+1}\mathbf{v}_{t+1} = \mathbf{0}$. But this implies that $\alpha_{t+1} = 0$ (since $\mathbf{v}_{t+1} \ne \mathbf{0}$). Therefore, in (11.5), all the α_i are zero. This proves that $\{\mathbf{v}_1, \mathbf{v}_2, \ldots, \mathbf{v}_t, \mathbf{v}_{t+1}\}$ is a linearly indpendent set, as claimed.

Mathematical induction then does the rest. •

TUTORIAL PROBLEM 11.1

> Surely mathematical induction can only be applied to infinite sets?

From Theorem 1 there follows immediately the corollary below.

• *Corollary 1*

Let T be an $n \times n$ matrix. If the characteristic polynomial of T has n distinct (possibly com lex) roots then T is diagonalisable. (Of course, even if T is a real matrix, the diag‿nalised version may have complex entries.) •

Examples 3 and 4 above show this corollary in action.

Whilst we could have completed, earlier, the explanation of the behaviour of the particular matrix T which introduced the chapter, it is easier to give this explanation by making use of Theorem 1. We begin by observing that the matrix

$$T = \begin{bmatrix} 0.8 & 0.15 & 0.05 \\ 0.2 & 0.7 & 0.35 \\ 0 & 0.15 & 0.6 \end{bmatrix}$$

has eigenvalues given by solving the equation

$$\text{Det}\begin{bmatrix} 0.8-\lambda & 0.15 & 0.05 \\ 0.2 & 0.7-\lambda & 0.35 \\ 0 & 0.15 & 0.6-\lambda \end{bmatrix} = 0.$$

Adding rows 2 and 3 to row 1 changes the determinant to

$$\text{Det}\begin{bmatrix} 1-\lambda & 1-\lambda & 1-\lambda \\ 0.2 & 0.7-\lambda & 0.35 \\ 0 & 0.15 & 0.6-\lambda \end{bmatrix} = (1-\lambda)\text{Det}\begin{bmatrix} 1 & 1 & 1 \\ 0.2 & 0.7-\lambda & 0.35 \\ 0 & 0.15 & 0.6-\lambda \end{bmatrix} = 0$$

giving $\lambda = 1$, $\lambda = \frac{1}{2}\{1.1 \pm\sqrt{(1.21 - 1.11)}\}$. Calling these eigenvalues $\lambda_1(= 1)$, λ_2 and λ_3 we find corresponding eigenvectors

$$\mathbf{v}_1 = \alpha\begin{bmatrix} 6.75 \\ 8.0 \\ 3.0 \end{bmatrix},$$

$\mathbf{v}_2$ and $\mathbf{v}_3$. Since λ_1, λ_2 and λ_3 are distinct Theorem 1 tells us that $\{\mathbf{v}_1, \mathbf{v}_2, \mathbf{v}_3\}$ is an LI set and, hence, a basis of $\mathbb{R}^3$. Consequently for every element $\mathbf{u} = \alpha_1\mathbf{v}_1 + \alpha_2\mathbf{v}_2 + \alpha_3\mathbf{v}_3$ of $\mathbb{R}^3$ and for every positive integer k,

$$T^k\mathbf{u} = \alpha_1\lambda_1^k\mathbf{v}_1 + \alpha_2\lambda_2^k\mathbf{v}_2 + \alpha_3\lambda_3^k\mathbf{v}_3.$$

Since $\lambda_1 = 1$, whilst $0 < \lambda_2, \lambda_3 < 1$, we see that, as k increases, $T^k\mathbf{u}$ approaches the value $\alpha_1\mathbf{v}_1$. (Are the conditions on λ_2 and λ_3 just flukes? See Exercise 34.) Since the entries in each of the columns of the matrices T,

$$\begin{bmatrix} 1 \\ 0 \\ 0 \end{bmatrix}, \quad \begin{bmatrix} 0 \\ 1 \\ 0 \end{bmatrix} \quad \text{and} \quad \begin{bmatrix} 0 \\ 0 \\ 1 \end{bmatrix}$$

add up to 1, so, for each k, do the entries in each of

$$T^k\begin{bmatrix} 1 \\ 0 \\ 0 \end{bmatrix}, \quad T^k\begin{bmatrix} 0 \\ 1 \\ 0 \end{bmatrix} \quad \text{and} \quad T^k\begin{bmatrix} 0 \\ 0 \\ 1 \end{bmatrix}$$

(by Exercise 25 of Chapter 3). Thus, for large k, each of these three vectors is (approximately) that scalar multiple of $\mathbf{v}_1$ whose entries add up to 1. That is, each is (almost) equal to

$$\frac{1}{17.75}\begin{bmatrix} 6.75 \\ 8.0 \\ 3.0 \end{bmatrix} = \frac{1}{71}\begin{bmatrix} 27 \\ 32 \\ 12 \end{bmatrix} = \begin{bmatrix} 0.3802817... \\ 0.4507042... \\ 0.1690141... \end{bmatrix}.$$

A second case of diagonalisability (not needing n distinct eigenvalues) is given by the following theorem.

Theorem 2

Let T be a real symmetric matrix. Then T is diagonalisable. Furthermore the diagonalising matrix N can be chosen so that $N^{-1} = N^T$.

(That is, N may be chosen orthogonal (see Exercise 24(c) of Chapter 3). That this can be helpful is outlined in Example 9 below. Incidentally, the diagonalising matrix for the matrix B of Example 5 above is not orthogonal. It could be replaced by an orthogonal diagonalising matrix – if we wished – by the method described in Example 10 below.)

Before indicating how to prove Theorem 2 we give the promised example.

Example 9

Consider the equation $-2x^2 - 2y^2 + 5z^2 - 12xy + 2yz - 2xz - 4 = 0$. This equation defines a (so-called *quadric*) surface in 3-dimensional space. The question is: what does this surface look like? Let us first note that, provided we are prepared to identify a 1×1 matrix with its only entry, we may write $-2x^2 - 2y^2 + 5z^2 - 12xy + 2yz - 2xz$ as

$$\begin{bmatrix} x & y & z \end{bmatrix}\begin{bmatrix} -2 & -6 & -1 \\ -6 & -2 & 1 \\ -1 & 1 & 5 \end{bmatrix}\begin{bmatrix} x \\ y \\ z \end{bmatrix}.$$

Suppose we choose new coordinates X, Y, Z by setting

$$\begin{bmatrix} X \\ Y \\ Z \end{bmatrix} = M\begin{bmatrix} x \\ y \\ z \end{bmatrix},$$

M being some (invertible) 3×3 matrix still to be determined. Then

$$\begin{bmatrix} x & y & z \end{bmatrix}\begin{bmatrix} -2 & -6 & -1 \\ -6 & -2 & 1 \\ -1 & 1 & 5 \end{bmatrix}\begin{bmatrix} x \\ y \\ z \end{bmatrix} = \begin{bmatrix} x & y & z \end{bmatrix}M^T\left\{(M^T)^{-1}\begin{bmatrix} -2 & -6 & -1 \\ -6 & -2 & 1 \\ -1 & 1 & 5 \end{bmatrix}M^{-1}\right\}M\begin{bmatrix} x \\ y \\ z \end{bmatrix}.$$

Now since

$$\begin{bmatrix} -2 & -6 & -1 \\ -6 & -2 & 1 \\ -1 & 1 & 5 \end{bmatrix}$$

(= S, say) is symmetric, Theorem 2 tells us that there is an orthogonal matrix M such that

$$(M^T)^{-1}\begin{bmatrix} -2 & -6 & -1 \\ -6 & -2 & 1 \\ -1 & 1 & 5 \end{bmatrix} M^{-1} = \begin{bmatrix} \alpha & 0 & 0 \\ 0 & \beta & 0 \\ 0 & 0 & \gamma \end{bmatrix}$$

is diagonal. Thus

$$\begin{bmatrix} x & y & z \end{bmatrix}\begin{bmatrix} -2 & -6 & -1 \\ -6 & -2 & 1 \\ -1 & 1 & 5 \end{bmatrix}\begin{bmatrix} x \\ y \\ z \end{bmatrix}$$

can be rewritten as

$$\begin{bmatrix} X & Y & Z \end{bmatrix}\begin{bmatrix} \alpha & 0 & 0 \\ 0 & \beta & 0 \\ 0 & 0 & \gamma \end{bmatrix}\begin{bmatrix} X \\ Y \\ Z \end{bmatrix} = \alpha X^2 + \beta Y^2 + \gamma Z^2, w$$

a form from which the type of the surface (hyperbolic paraboloid, ellipsoid, parabolic cylinder, etc.) can be determined from a well-documented list. (See, for example, Fraleigh J. B. and Beauregard R. A. *Linear Algebra*.)

PROBLEM 11.12

Find the eigenvalues α, β and γ above and hence determine the type of surface the given equation represents. Then find the corresponding orthogonal diagonalising matrix M. (See Example 10 below if you need help with this.)

TUTORIAL PROBLEM 11.2

Why does the change of coordinates need to be orthogonal here? After all we did not find it necessary in Example 3.

To indicate why Theorem 2 is true we first prove the following theorem.

• *Theorem 3*

All the eigenvalues of the real symmetric matrix T are themselves real.

PROOF

Suppose that $T\mathbf{v} = \lambda\mathbf{v}$. Then $\overline{T\mathbf{v}}$ (= $T\overline{\mathbf{v}}$, why?) = $\overline{\lambda}\overline{\mathbf{v}}$ (the bar signifying that all entries are replaced by their complex conjugates). From the first equality we obtain $\overline{\mathbf{v}}^T T\mathbf{v} = \lambda\overline{\mathbf{v}}^T\mathbf{v}$, from the second $\mathbf{v}^T T\overline{\mathbf{v}} = \overline{\lambda}\mathbf{v}^T\overline{\mathbf{v}}$. Taking transposes in this last equality gives $\overline{\mathbf{v}}^T T\mathbf{v} = \overline{\lambda}\overline{\mathbf{v}}^T\mathbf{v}$. We then see that $\lambda\overline{\mathbf{v}}^T\mathbf{v} = \overline{\mathbf{v}}^T T\mathbf{v} = \overline{\lambda}\overline{\mathbf{v}}^T\mathbf{v}$ so that $(\lambda - \overline{\lambda})\overline{\mathbf{v}}^T\mathbf{v} = 0$. But $\overline{\mathbf{v}}^T\mathbf{v} \neq 0$ (Exercise 12(c) of Chapter 3), hence we must deduce that $\lambda = \overline{\lambda}$. That is, λ is real.

Thus the (real) characteristic polynomial of T factors as a product of n *real* (though maybe not distinct) linear factors.

In the case of distinct linear factors we have the following theorem.

• Theorem 4

In a real symmetric matrix eigenvectors corresponding to distinct eigenvalues are orthogonal [and hence form an LI set – cf. Exercise 15 of Chapter 8].

PROOF

Suppose T is an $n \times n$ symmetric matrix with distinct eigenvalues λ and μ and corresponding eigenvectors $\mathbf{u}$ and $\mathbf{v}$. Regarding $\mathbf{u}$ and $\mathbf{v}$ as column matrices we may write $T\mathbf{u} = \lambda\mathbf{u}$ and $T\mathbf{v} = \mu\mathbf{v}$. From the first of these we deduce that $\mathbf{u}^T T^T = \lambda\mathbf{u}^T$ – and hence that $\mathbf{u}^T T\mathbf{v} = \lambda\mathbf{u}^T\mathbf{v}$ – and from the second $\mathbf{u}^T T\mathbf{v} = \mu\mathbf{u}^T\mathbf{v}$, directly. It follows that $\lambda\mathbf{u}^T\mathbf{v} = \mathbf{u}^T T\mathbf{v} = \mu\mathbf{u}^T\mathbf{v}$, so that $(\lambda - \mu)\mathbf{u}^T\mathbf{v} = 0$ (the zero 1×1 matrix!). Since $\lambda \neq \mu$, we see that $\mathbf{u}^T\mathbf{v} = 0$; that is, $\mathbf{u}$ and $\mathbf{v}$ are orthogonal. ●

We readily deduce the 'diagonalisability' part of Theorem 2 if the n real eigenvalues of T are distinct.

What can we do if the symmetric matrix T of Theorem 2 has a repeated eigenvalue? We show you by means of a particular example.

Example 10

The symmetric matrix

$$\begin{bmatrix} 1 & -4 & 2 \\ -4 & 1 & -2 \\ 2 & -2 & -2 \end{bmatrix}$$

has eigenvalues 6, –3 and –3. For $\lambda = 6$ the corresponding eigenvectors are the

$$\alpha\begin{bmatrix} 2 \\ -2 \\ 1 \end{bmatrix} \quad (\alpha \neq 0).$$

To find the eigenvectors corresponding to $\lambda = -3$ we must solve

$$\begin{bmatrix} 4 & -4 & 2 \\ -4 & 4 & -2 \\ 2 & -2 & 1 \end{bmatrix}\begin{bmatrix} x \\ y \\ z \end{bmatrix} = \begin{bmatrix} 0 \\ 0 \\ 0 \end{bmatrix}, \text{ giving eigenvectors } \begin{bmatrix} x \\ y \\ z \end{bmatrix} = \begin{bmatrix} \gamma - \beta \\ \gamma \\ 2\beta \end{bmatrix} = \beta\begin{bmatrix} -1 \\ 0 \\ 2 \end{bmatrix} + \gamma\begin{bmatrix} 1 \\ 1 \\ 0 \end{bmatrix}$$

where β, γ are not both zero. Now

$$\begin{bmatrix} -1 \\ 0 \\ 2 \end{bmatrix} \text{ and } \begin{bmatrix} 1 \\ 1 \\ 0 \end{bmatrix}$$

are not orthogonal since

$$\begin{bmatrix} -1 & 0 & 2 \end{bmatrix}\begin{bmatrix} 1 \\ 1 \\ 0 \end{bmatrix} = [-1] \, (\neq [0]).$$

However

$$\begin{bmatrix}-1\\0\\2\end{bmatrix} \text{ and } \begin{bmatrix}-1\\0\\2\end{bmatrix}+\gamma\begin{bmatrix}1\\1\\0\end{bmatrix}$$

will be orthogonal if $-1.(-1+\gamma)+2.(2+0\gamma)=0$, i.e. if $\gamma=5$. We therefore choose

$$\begin{bmatrix}-1\\0\\2\end{bmatrix} \text{ and } \begin{bmatrix}-1\\0\\2\end{bmatrix}+\begin{bmatrix}5\\5\\0\end{bmatrix}=\begin{bmatrix}4\\5\\2\end{bmatrix} \text{ rather than } \begin{bmatrix}-1\\0\\2\end{bmatrix} \text{ and } \begin{bmatrix}1\\1\\0\end{bmatrix}$$

as 'basic' vectors corresponding to $\lambda=-3$. Of course

$$\begin{bmatrix}-1\\0\\2\end{bmatrix}, \begin{bmatrix}1\\1\\0\end{bmatrix} \text{ and } \begin{bmatrix}4\\5\\2\end{bmatrix} \text{ must all be orthogonal to } \begin{bmatrix}2\\-2\\1\end{bmatrix} \quad \text{(why?)}$$

We thus have

$$\begin{bmatrix}2&-1&4\\-2&0&5\\1&2&2\end{bmatrix}^{-1}\begin{bmatrix}1&-4&2\\-4&1&-2\\2&-2&-2\end{bmatrix}\begin{bmatrix}2&-1&4\\-2&0&5\\1&2&2\end{bmatrix}=\begin{bmatrix}6&0&0\\0&-3&0\\0&0&-3\end{bmatrix}.$$

However

$$\begin{bmatrix}2&-1&4\\-2&0&5\\1&2&2\end{bmatrix}$$

is *still not orthogonal*![f] However this is easily arranged. We simply 'normalise' each column vector by dividing it by its length. That is, we replace each

$$\begin{bmatrix}2&-1&4\\-2&0&5\\1&2&2\end{bmatrix} \text{ by } \begin{bmatrix}2/3&-1/\sqrt{5}&4/3\sqrt{5}\\-2/3&0&5/3\sqrt{5}\\1/3&2/\sqrt{5}&2/3\sqrt{5}\end{bmatrix}.$$

TUTORIAL PROBLEM 11.3

Determine the lengths of the row vectors of this matrix. Do the results surprise you? Explain.

NOTE 1 The above procedure can be extended to repeated eigenvalues (no matter how often repeated) of real $n \times n$ symmetric matrices, a full set of (real) linearly independent

[f]It is unfortunate that a matrix is called orthogonal when and only when its rows (columns), when regarded as vectors, are pairwise orthogonal *and each of length 1*.

mutually orthogonal eigenvectors being obtainable via the so-called **Gram–Schmidt orthogonalisation process**. In this way we can establish Theorem 2.

We end with some comments on topics which will be dealt with fully in the follow-up volume to this but whose introduction here ties up one or two loose ends.

In Exercise 22(b) of Chapter 10 we saw that each linear transformation $T: V \to W$ from one finite dimensional vector space to another could be represented by matrix multiplication acting on appropriate coordinate vectors. Suppose that $V = W$ (so that each may, by Theorem 3 of Chapter 10, be assumed to be $\mathbb{R}^n$ for some integer n). Choosing $B_V = B_W$ as the standard basis in $\mathbb{R}^n$ we may find the corresponding matrix A, say, as in Exercise 22(a) of Chapter 10. Now let us assume that A has an LI set E, say, of n eigenvectors. Taking E as a basis for V we find (from Exercise 22(b)) that T can now be represented by a diagonal matrix. The message is that, given a linear transformation $T:V \to V$, we may, by picking a suitable basis for V, be able to obtain a very simple matrix representation of T. We give just one example.

Example 11

Let $T:\mathbb{R}^3 \to \mathbb{R}^3$ be given by

$$T\begin{bmatrix} x \\ y \\ z \end{bmatrix} = \begin{bmatrix} x-y-z \\ -x+y-z \\ -x-y+z \end{bmatrix}.$$

Then, with respect to the (standard) basis

$$\begin{bmatrix} 1 \\ 0 \\ 0 \end{bmatrix}, \begin{bmatrix} 0 \\ 1 \\ 0 \end{bmatrix}, \begin{bmatrix} 0 \\ 0 \\ 1 \end{bmatrix}$$

the matrix representing T is

$$A = \begin{bmatrix} 1 & -1 & -1 \\ -1 & 1 & -1 \\ -1 & -1 & 1 \end{bmatrix}.$$

However, with respect to the basis

$$\mathbf{u} = \begin{bmatrix} 1 \\ 1 \\ 1 \end{bmatrix}, \quad \mathbf{v} = \begin{bmatrix} -1 \\ 1 \\ 0 \end{bmatrix}, \quad \mathbf{w} = \begin{bmatrix} -1 \\ 0 \\ 1 \end{bmatrix}$$

of $\mathbb{R}^3$ comprising eigenvectors of A (see Example 15 above), the matrix representing T changes to diagonal form with diagonal entries -1, 2 and 2 (in some order).

Of course we have seen that not all $n \times n$ matrices are diagonalisable. The best that can be obtained, in general, is given by the following definition.

• Definition 3

Any (square) matrix of the form

$$J_i = \begin{bmatrix} \lambda_i & 1 & 0 & \vdots & \vdots & \vdots & 0 \\ 0 & \lambda_i & 1 & \vdots & \vdots & \vdots & 0 \\ \vdots & \vdots & \vdots & \vdots & \vdots & \vdots & 1 \\ 0 & 0 & 0 & \vdots & \vdots & \vdots & \lambda_i \end{bmatrix}$$

with a fixed scalar λ_i along the diagonal, 1s lying directly above these λs and 0s in every other position, is called a **Jordan matrix**. ●

The following can then be proved.

• Theorem 5

Let T be any $n \times n$ matrix. Then T is similar to a (so-called **block diagonal**) matrix of the form

$$D = \begin{bmatrix} J_1 & \mathbf{0} & \dots & \mathbf{0} \\ \mathbf{0} & J_2 & \dots & \mathbf{0} \\ \vdots & \vdots & \dots & \vdots \\ \mathbf{0} & \mathbf{0} & \dots & J_k \end{bmatrix}$$

where the J_i, each a Jordan matrix of the above kind, lie with their λ_i along the diagonal of D and where the $\mathbf{0}$s indicate that all other matrix entries are zero. The λs occurring in the Jordan matrices in D are clearly the eigenvalues of D (and hence of T). ●

These concepts of Gram–Schmidt orthogonalisation and Jordan form of a matrix (along with the intimately related discussion of change of basis in a vector space), you will meet in a second linear algebra course (see, for example, Hirst, A. E. *Vectors in Two and Three Dimensions*).

Applications

Above we indicated how surfaces in $\mathbb{R}^3$ can be identified by diagonalisation. Here we show how a similar approach can be used to distinguish conics in the plane.

The equation $ax^2 + 2bxy + cy^2 + dx + ey + f = 0$ defines a **conic section** in the x–y plane. [To avoid trivialities we assume that at least one of a, b, c, is not zero.] How can we tell what type of conic it is (e.g. ellipse, hyperbola, circle)? We concentrate first on the **quadratic form** $ax^2 + 2bxy + cy^2$. According to Theorem 2 we can find a real orthogonal matrix N such that

$$N^{-1}\begin{bmatrix} a & b \\ b & c \end{bmatrix}N \quad \text{is a diagonal matrix} \quad \begin{bmatrix} \lambda_1 & 0 \\ 0 & \lambda_2 \end{bmatrix},$$

say. Writing

$$\begin{bmatrix} X \\ Y \end{bmatrix} = N^{-1}\begin{bmatrix} x \\ y \end{bmatrix}$$

changes $ax^2 + 2bxy + cy^2 + dx + ey + f = 0$ to the form

$$\lambda_1 X^2 + \lambda_2 Y^2 + \delta X + \epsilon Y + \phi = 0$$

for suitable real numbers δ, ϵ, ϕ. It is then easy to 'complete the square' (see Exercise 36 below) on the parts $\lambda_1 X^2 + \delta X$ and $\lambda_2 Y^2 + \epsilon Y$ so that $\lambda_1 X^2 + \lambda_2 Y^2 + \delta X + \epsilon Y + \phi$ is rewritten in the form $\lambda_1 U^2 + \lambda_2 V^2 + \nu = 0$. [If, say, $\lambda_1 = 0$, we may have to accept the form $\lambda_2 V^2 + \delta U + \nu = 0$ where $U = X$.] Since N is orthogonal, the change of coordinates from x, y to X, Y represents a rotation of the plane (fixing the origin) or a reflection in a line through the origin; the change from X, Y to U, V represents a translation. The original figure is now placed centrally and symmetrically with respect to the new (U–V) axes and its form read off immediately from the equation $\lambda_1 U^2 + \lambda_2 V^2 + \nu = 0$ [or $\lambda_2 V^2 + \delta U + \nu = 0$]. Note, incidentally, that

$$\lambda_1 \lambda_2 = \text{Det}\begin{bmatrix} a & b \\ b & c \end{bmatrix} = ac - b^2.$$

Indeed if $\lambda_1 \lambda_2 = 0$ (with, say, $\lambda_1 = 0$, $\lambda_2 \neq 0$: why can we not have $\lambda_1 = \lambda_2 = 0$?) we find that the conic is either a parabola (e.g. $U \pm V^2 (\pm 1) = 0$), a pair of parallel straight lines (e.g. $V^2 - 1 = 0$), a single straight line (e.g. $V^2 = 0$) or the empty set (e.g. $V^2 + 1 = 0$). If $\lambda_1 \lambda_2 > 0$ we may obtain an ellipse, a single point or the empty set. [We leave you to find examples.] If $\lambda_1 \lambda_2 < 0$ we obtain either a pair of (distinct) straight lines (e.g. $U^2 - V^2 = 0$) or a hyperbola (e.g. $2U^2 - 3V^2 - 5 = 0$).

For a second application consider the function

$$f(x, y) = 10 + (5x - 7y) + (4x^2 - 6xy + 3y^2) + (x^3 - x^2 y + y^3) + (x^4 - 6y^4).$$

The question might arise as to whether or not $f(x, y)$ has a (local) maximum or minimum at the origin (0, 0). [To answer a similar question at the point (a, b) one simply 'changes the origin' by making the substitution $X = x - a$, $Y = y - b$.] Very near to the origin the dominant term in $f(x, y)$ is, of course, the constant 10, but, the term most dominant in determining how $f(x, y)$ changes when the point (x, y) is near the origin is the linear term $5x - 7y$. Now, clearly, if $x = 0$ then $f(x, y) < 10$ if y is small and positive, whilst $f(x, y) > 0$ if y is small and negative (since the terms in the higher powers of y are negligible in comparison with $-7y$). In this way one sees that $f(x, y)$ can have neither a (local) maximum nor minimum value at (0, 0). But suppose now that

$$g(x, y) = 10 + (4x^2 - 6xy + 3y^2) + (x^3 - x^2 y + y^3) + (x^4 - 6y^4).$$

Here the dominant term, near (0, 0) is $4x^2 - 6xy + 3y^2$. How does it behave for points (x, y) near (0, 0)? Since the eigenvalues λ_1, λ_2 of

$$\begin{bmatrix} 4 & -3 \\ -3 & 3 \end{bmatrix}$$

are both positive, we see that $4x^2 - 6xy + 3y^2$ ($= \lambda_1 X^2 + \lambda_2 Y^2$ for suitably chosen X, Y) is non-negative for all x, y. Thus $g(x, y)$ does indeed have a (local) minimum at (0, 0).

Summary

Given the $n \times n$ matrix T, the scalar λ is an **eigenvalue** and the (non-zero) vector $\mathbf{x}$ is an **eigenvector** of T if $T\mathbf{x} = \lambda\mathbf{x}$. All such λ can (in theory) be found by solving the **characteristic equation** $\mathrm{Det}(T - \lambda I) = 0$ of T. If we allow complex number solutions the **characteristic polynomial** $\mathrm{Det}(T - \lambda I)$ will have n roots (counting multiple roots according to their multiplicities). A root λ repeated r times will supply an LI set $\{\mathbf{v}_1, \mathbf{v}_2, \ldots, \mathbf{v}_s\}$ $(1 \leq s \leq r)$ of eigenvectors each corresponding to λ. The set $E_\lambda = \{\mathbf{x}: T\mathbf{x} = \lambda\mathbf{x}\}$ – which includes the zero vector – is a subspace of $\mathbb{R}^n$ (or $\mathbb{C}^n$) called the **eigenspace corresponding to** λ. The dimension of E_λ is less than or equal to r. If T has an LI set of n eigenvectors then T can be **diagonalised**, i.e. there exist an invertible matrix N and a diagonal matrix D, say, such that $N^{-1}TN = D$. The diagonal entries of D are the eigenvalues of T in some order and the columns of N must be corresponding eigenvectors (i.e. taken in the same order.).

Not all matrices can be diagonalised. Those which cannot may be reduced to **Jordan canonical form**. Amongst those which can are those whose eigenvalues are all distinct and the **real symmetric matrices** – these may even be diagonalised using an **orthogonal** matrix. The use of orthogonal matrices is relevant when simplifying the equations of conics and quadrics by changing from one set of (orthogonal) axes to another.

Other applications in which eigenvalues play an important role include (i) the solving of systems of simultaneous linear differential equations (where, again, a sort of 'change of basis' from $x(t)$, $y(t)$ to $X(t)$, $Y(t)$ was found helpful), (ii) determination of invariant lines in geometry and invariant vectors in population distribution problems and (iii) in calculating large powers of matrices in general.

The eigenvalue concept was introduced by Euler in his search for solutions of certain differential equations. In 1743 he solved the equation

$$\frac{d^n y}{dx^n} + a_{n-1}\frac{d^{n-1}y}{dx^{n-1}} + \ldots + a_0 y = 0$$

by assuming $y = e^{\lambda x}$ so that $\lambda^n + a_{n-1}\lambda^{n-1} + \ldots + a_1\lambda + a_0 = 0$, thus restricting the choice of λ, a technique he had also used earlier. Later Lagrange and Laplace came across what is essentially the characteristic equation of a matrix when investigating systems of linear differential equations associated with questions in mechanics. 'Knowing' by physical reasoning that the 'eigenvalues' were real, Laplace later recognised that the reality followed from the symmetry of the coefficients of the equations, but it was not until 1829 that Theorem 3 (and Theorem 2) was first proved rigorously by Cauchy whose interest lay in quadratic forms arising in geometry. Gradually quadratic forms gave way to (symmetric) matrices representing them, with Weierstrass, an important figure in this development, using a canonical form very like that introduced later by Jordan.

Leonhard Euler, the greatest mathematician of the 18th century, was born in Basel, the son of a preacher, on 15 April 1707. At 18 he produced his first research paper; aged 19, winning his first of many prizes from the French Academy of Sciences, John Bernoulli addressed him as 'a most gifted and learned man of science'.

Euler's research output was phenomenally great and wide ranging; Kline lists his mathematical interests under 11 separate categories. Even total blindness in the last 12 years of his life could not stem the flow, 1000 pages of research being produced in 1776. Perhaps, here, he was helped by his prodigious memory; it is said he knew Virgil's *Aeneid* by heart. On 18 September he suffered a brain haemorrhage, uttered the words 'I am dying', and did so the very same day.

EXERCISES ON CHAPTER 11

1. (a) Check that $\begin{bmatrix}1\\3\end{bmatrix}$ is an eigenvector for the matrix $\begin{bmatrix}8 & 9\\12 & 31\end{bmatrix}$. What is the corresponding eigenvalue?

(b) Check that $\begin{bmatrix}-2\\1\end{bmatrix}, \begin{bmatrix}10\\-5\end{bmatrix}$ and $\begin{bmatrix}7\\-4\end{bmatrix}$ are eigenvectors for $\begin{bmatrix}-9 & -28\\8 & 21\end{bmatrix}$. Is $\begin{bmatrix}-2\\1\end{bmatrix}+\begin{bmatrix}10\\-5\end{bmatrix}$ also an eigenvector for $\begin{bmatrix}-9 & -28\\8 & 21\end{bmatrix}$? What about $\begin{bmatrix}-2\\1\end{bmatrix}+\begin{bmatrix}7\\-4\end{bmatrix}$?

2. Find a matrix A whose eigenvalues are 1 and 3 and whose corresponding eigenvectors are $\begin{bmatrix}5\\7\end{bmatrix}$ and $\begin{bmatrix}11\\13\end{bmatrix}$. Is A unique?

3. To each eigenvalue correspond infinitely many eigenvectors. Show that, to each eigenvector corresponds only one eigenvalue. (That is, each eigenvector is determined by a unique eigenvalue.)

4. (i) Let A and B be $n \times n$ matrices. Give examples to show that if λ, μ are eigenvalues of A and B respectively then $\lambda + \mu$ need not be an eigenvalue of $A + B$ and $\lambda\mu$ need not be an eigenvalue of AB.

(ii) Show that, if **x** is an eigenvector of both A and B, then **x** is an eigenvector of AB.

(iii) If A and B are row equivalent do A, B necessarily have the same eigenvalues or eigenvectors?

5. Find eigenvalues and all corresponding eigenvectors for the matrices:

$$\text{(a)} \begin{bmatrix} 5 & 39 \\ 3 & 9 \end{bmatrix}; \quad \text{(b)} \begin{bmatrix} 1 & 1 \\ 1 & 1 \end{bmatrix}; \quad \text{(c)} \begin{bmatrix} 0 & 0 \\ 0 & 0 \end{bmatrix}; \quad \text{(d)} \begin{bmatrix} -27 & 98 \\ -8 & 29 \end{bmatrix}.$$

Where possible exhibit an invertible matrix N and a diagonal matrix D such that the given matrix is expressible as NDN^{-1}.

6. Let A be a diagonalisable matrix so that $C^{-1}AC = D$ for suitable C and diagonal D. Show that for each such D there are infinitely many such C.

7. Under what conditions on a, b, c, d is $\begin{bmatrix} a & b \\ c & d \end{bmatrix}$ diagonalisable?

8. Draw the rectangle with corners at (–1, 0), (1, 0), (1, 2), (–1, 2) and the 'squashed then sheared rectangle' with vertices at $\left(-\frac{3}{2},0\right), \left(\frac{3}{2},0\right), \left(\frac{5}{2},\frac{3}{2}\right), \left(-\frac{1}{2},\frac{3}{2}\right)$. Can you identify, visually, an eigenvector (other than (1, 0)) of the linear transformation producing this change? In any case find, algebraically, the eigenvectors of the matrix representing the transformation.

9. Given the simultaneous equations

$$\frac{dx}{dt} = -x + 2y$$
$$\frac{dy}{dt} = \frac{1}{2}x - y$$

and given that $x = 600$ and $y = 1300$ when $t = 0$ determine what happens to the sizes of $x(t)$ and $y(t)$ as t increases. (This is a so-called *symbiotic* (cooperative) system. If x and y represent the populations of two species at time t it will be found that the species can live side by side with neither species dying out.)

10. At the end of year n the number of rats and cats in a certain area is given, in terms of the numbers in the year n–1, by the matrix formula

$$\begin{bmatrix} r_n \\ c_n \end{bmatrix} = \begin{bmatrix} 5 & -2 \\ 1 & 2 \end{bmatrix} \begin{bmatrix} r_{n-1} \\ c_{n-1} \end{bmatrix}.$$

By diagonalising $\begin{bmatrix} 5 & -2 \\ 1 & 2 \end{bmatrix}$ determine a formula for r_n and c_n in terms of r_0 and c_0. Given that $r_0 = 1000$ and $c_0 = 300$ determine whether or not the rat population is killed off. (This kind of equation is a difference equation of the *predator–prey* type.)

11. Find eigenvalues and all corresponding eigenvectors for the matrices:

$$\text{(a)} \begin{bmatrix} 3 & 0 & 0 \\ 0 & -1 & 0 \\ 0 & 0 & 10 \end{bmatrix}; \quad \text{(b)} \begin{bmatrix} 1 & 2 & 3 \\ 0 & 4 & 5 \\ 0 & 0 & 6 \end{bmatrix}; \quad \text{(c)} \begin{bmatrix} 1 & 2 & 3 \\ 4 & 5 & 0 \\ 6 & 0 & 0 \end{bmatrix};$$

(d) $\begin{bmatrix} 9 & 20 & 42 \\ 10 & 19 & 42 \\ -6 & -14 & -29 \end{bmatrix}$; (e) $\begin{bmatrix} 13 & -6 & -6 \\ 16 & -7 & -8 \\ 12 & -6 & -5 \end{bmatrix}$.

Where possible exhibit an invertible matrix N and a diagonal matrix D such that the given matrix is expressible as NDN^{-1}. (Note: a fairly common error in an exercise like (b) is to deduce that the eigenvector corresponding to the eigenvalue $\lambda = 1$, is $\begin{bmatrix} 0 \\ 0 \\ 0 \end{bmatrix}$. You should, of course never obtain the zero vector as an eigenvalue. Why not? The proper answer is not 'Because $\begin{bmatrix} 0 \\ 0 \\ 0 \end{bmatrix}$ cannot be an eigenvector – by definition.' It's a deeper reason than that.)

12. Find eigenvalues and to each eigenvalue a corresponding eigenvector of length 1 for the matrices:

(a) $\begin{bmatrix} 8 & -6 & 0 \\ 5 & -5 & 2 \\ 4 & -4 & 2 \end{bmatrix}$; (b) $\begin{bmatrix} 15 & -8 & -2 \\ 26 & -14 & -4 \\ 0 & 0 & 2 \end{bmatrix}$; (c) $\begin{bmatrix} 13 & -7 & -1 \\ 24 & -13 & -3 \\ -2 & 1 & 3 \end{bmatrix}$; (d) $\begin{bmatrix} 1 & 0 & -2 \\ -2 & 2 & -2 \\ 6 & -3 & -3 \end{bmatrix}$.

Show that (a) and (b) have infinitely many eigenvectors of unit length. What about (d)?

13. Find, if you can, the eigenvalues of the matrix

$$\begin{bmatrix} 0 & 5 & 6 \\ 3 & 0 & 4 \\ 1 & 2 & 0 \end{bmatrix}.$$

(See Exercise 14(ii) of Chapter 5.)

14. Let A be the matrix

$$\begin{bmatrix} -\frac{9}{2} & 0 & 5 \\ 3 & \frac{1}{2} & -3 \\ -1 & 0 & \frac{3}{2} \end{bmatrix}.$$

Is it true that as n increases all the entries in A^n eventually tend towards zero? [Hint: diagonalise.] Do the same for the matrix in Computer Package Problem 4 in Chapter 3.

15. Write down two 3×3 matrices A, B whose entries differ in only one place and yet A is diagonalisable whilst B is not. (Cf. Problem 11.5(i).)

16. Find eigenvalues and corresponding eigenvectors for the matrices:

(a) $\begin{bmatrix} 1 & 1 & 0 \\ 0 & 1 & 0 \\ 0 & 0 & 1 \end{bmatrix}$; (b) $\begin{bmatrix} 1 & 1 & 1 \\ 0 & 1 & 0 \\ 0 & 0 & 1 \end{bmatrix}$; (c) $\begin{bmatrix} 1 & 1 & 0 \\ 0 & 1 & 1 \\ 0 & 0 & 1 \end{bmatrix}$; (d) $\begin{bmatrix} 1 & 1 & 1 \\ 0 & 1 & 1 \\ 0 & 0 & 1 \end{bmatrix}$;

(e) $\begin{bmatrix} 1 & 1 & 1 & 1 \\ 1 & 1 & 1 & 1 \\ 1 & 1 & 1 & 1 \\ 1 & 1 & 1 & 1 \end{bmatrix}$; (f) $\begin{bmatrix} 0 & 1 & 1 & 1 \\ 1 & 0 & 1 & 1 \\ 1 & 1 & 0 & 1 \\ 1 & 1 & 1 & 0 \end{bmatrix}$.

17. Let A, B and C be diagonalisable 3×3 matrices.
(i) Let A have triply repeated eigenvalue λ. Describe A (as best you can).
(ii) Let α be any scalar. Is αB diagonalisable?
(iii) Is C^T diagonalisable? (Please give proofs – or counterexamples – for (ii) and (iii).)

18. (i) Show that $\lambda = 0$ is an eigenvalue for the $(n \times n)$ matrix A if and only if A is singular. [Hint: use the $A - \lambda I$ definition.]
(ii) Show that if $\text{Det}(A - \lambda I) \neq 0$ then λ is definitely not an eigenvalue of A.

19. Show (i) that if $A\mathbf{x} = \lambda\mathbf{x}$ then, for all positive integers s, $A^S\mathbf{x} = \lambda^S\mathbf{x}$ and (ii) that, if A is invertible then $\lambda \neq 0$ and λ^{-1} is an eigenvalue for A^{-1} with corresponding eigenvector $\mathbf{x}$. Deduce from (i) that, if $A^t = 0$, the zero matrix, for some integer t, then $\lambda = 0$ is the only eigenvalue of A. Deduce from Exercise 18(i) that, if $\lambda_1, \lambda_2, \ldots, \lambda_n$ are (all) the eigenvalues (possibly complex) of the $n \times n$ matrix A, then A^{-1} exists if and only if $\lambda_1\lambda_2 \ldots \lambda_n \neq 0$.

20. The eigenvalues and corresponding eigenvectors of the matrix A are

$$1, \begin{bmatrix} 1 \\ 1 \\ 1 \end{bmatrix}; \quad -1, \begin{bmatrix} 1 \\ 1 \\ 0 \end{bmatrix}; \quad 2, \begin{bmatrix} 1 \\ 0 \\ 0 \end{bmatrix}.$$

Find A and A^{10}.

21. Show that if A has λ as an eigenvalue then $A + \mu I$ has $\lambda + \mu$ as an eigenvalue.

22. Show that each polynomial is the characteristic polynomial of some suitable matrix. [Hint: what is the characteristic polynomial of

$$\begin{bmatrix} 0 & 1 & 0 \\ 0 & 0 & 1 \\ c & b & a \end{bmatrix}?]$$

23. Use the fact (deducible from the fundamental theorem of algebra) that the complex roots of a real polynomial occur in pairs of complex conjugates, to prove that each 3×3 matrix (indeed each $n \times n$ matrix for odd n) with real entries has at least one real eigenvalue and hence at least one eigenvector with all entries real.

24. Let A be a real 3×3 matrix with just one real eigenvalue. Show that:
(i) the two other eigenvalues must be complex conjugate numbers;
(ii) A can always be diagonalised;
(iii) the eigenvectors corresponding to the complex eigenvalues (a) cannot be real; (b) occur in conjugate pairs.

25. Find the eigenvalues and corresponding eigenvectors of the matrix

$$\begin{bmatrix} 0 & 1 & -3 & 0 \\ -7 & 0 & 0 & -3 \\ -2 & 0 & 0 & -1 \\ 0 & -2 & 7 & 0 \end{bmatrix}.$$

(This is fairly easy even though it is 4×4.)

26. (a) Let λ_i $(1 \leq i \leq 3)$ be the eigenvalues of the 3×3 matrix A. Show that $\sum_{i=1}^{3} \lambda_i$ is the trace of A and that $\prod_{i=1}^{3} \lambda_i$ is Det A. [Hint: write Det$(A - \lambda I)$ out as a cubic and compare its coefficients with those of $-(\lambda - \lambda_1)(\lambda - \lambda_2)(\lambda - \lambda_3)$.] (These equalities hold for all square matrices – of course some of the λ_i may be complex numbers.)

(b) Given that

$$\begin{bmatrix} 2 & 12 & -1 & 2 \\ 1 & 1 & -1 & 0 \\ 0 & 6 & 1 & 2 \\ -3 & 9 & 0 & 4 \end{bmatrix}$$

has eigenvalues -1 and $\frac{1}{2}(5 + i\sqrt{15})$, find the others.

27. Let A and B be $(n \times n)$ matrices. Show that if either (I) $B = A^T$ or (II) B is similar to A, then A and B have the same characteristic polynomial and hence (α) the same eigenvalues, (β) the same determinant and (γ) the same trace. [Hint: see Exercise 22(iii) of Chapter 5.]

28. (a) Let A be the matrix

$$\begin{bmatrix} -2 & -12 & -2 \\ 0 & -2 & 2 \\ 0 & 0 & 5 \end{bmatrix}.$$

Show

$$\begin{bmatrix} x & y & z \end{bmatrix} \begin{bmatrix} -2 & -12 & -2 \\ 0 & -2 & 2 \\ 0 & 0 & 5 \end{bmatrix} \begin{bmatrix} x \\ y \\ z \end{bmatrix}$$

is the equation in Example 9. Is A similar to a diagonal matrix D, say? [Hint: the eigenvalues of A and D would be the same.] Now explain why, in Example 9, the symmetric matrix M is a better choice than A.

(b) Let A and B be $n \times n$ matrices with exactly the same set of eigenvalues. Must A and B necessarily be similar? (Proof or counterexample please!)

29. Write down the 3×3 matrices A and B corresponding to rotations of $\mathbb{R}^3$ about the x and y-axes each through an (anticlockwise) angle of $\pi/2$ (when looking along the positive axes to the origin). By finding the eigenvectors of their product find the resulting axis of rotation when A and B are applied in succession.

30. Let λ be an eigenvalue of the $n \times n$ matrix A. Show that the eigenspace $E_\lambda = \{\mathbf{x}:A\mathbf{x} = \lambda\mathbf{x}\}$ is a subspace of $\mathbb{R}^n$.

31. Show that the eigenvalues of a skew symmetric matrix are pure imaginary. (That is, are of the form $a + ib$ with $a = 0$.)

32. (a) Show that the $n \times n$ matrix A is diagonalisable if – and only if – it has an LI set of n eigenvectors.

(b) Show that the (real) matrix A is orthogonally diagonalisable if – and only if – A is symmetric. [Hint: Corollary 1 does one half of (a); for (b) see Note 1.]

33. Show that the column vectors and the row vectors of an $n \times n$ orthogonal matrix have unit length. Replace the matrix

$$\begin{bmatrix} 1 & -1 & -1 \\ 1 & 1 & 0 \\ 1 & 0 & 1 \end{bmatrix}$$

in Example 5 above by an orthogonal matrix.

34. A **real stochastic** (or **Markov**) **matrix** is a square matrix in which (i) all the entries are non-negative and (ii) the entries in each column add up to 1. (See the Applications in Chapter 3.) Prove that (I) the product of two such matrices is again stochastic, that (II) each stochastic matrix has 1 as an eigenvalue and that (III) each of its eigenvalues λ satisfies $|\lambda| \leq 1$. [Hint: use $A^s\mathbf{x} = \lambda^s\mathbf{x}$.]

35. When the Laboratory Party came to power 42% of the population were 'poor', 36% were 'middle income' and 22% classified as 'rich'. The party claimed the number of poor would decrease if all future governments followed its policies based on the 5-year transition matrix

p	m	r	
0.8	0.2	0.1	p
0.1	0.7	0.2	m
0.1	0.1	0.7	r

which indicates, for example, of the p(oor), $\frac{1}{10}$ will, in 5 years, become m(iddle income) and $\frac{1}{10}$ become r(ich). Is this transition matrix compatible with the party's claims?

36. In Example 9 we saw that $Q = -2x^2 - 2y^2 + 5z^2 - 12xy + 2yz - 2xz$ can be expressed in the form $3X^2 + 6Y^2 - 8Z^2$. We might have proceeded more directly by writing Q as $-2(x + 3y + (z/2))^2 + \{\text{terms in } y, z \text{ only}\}$. Repeat this procedure (called *completing the square*) on the y, z-only terms to obtain Q in the form $-2U^2 + \alpha V^2 + \beta W^2$. [This is Lagrange's reduction.] This seems to give a different sum- and difference-of-squares expression for Q, and hence a different quadric surface for Example 9. Am I right? [You should find that both α and β are positive. The fact that $3X^2 + 6Y^2 - 8Z^2$ also has two positive coefficients and one negative is an instance of *Sylvester's law of inertia* which says that such a result always holds.]

37. (a) Is $17x^2 + 50y^2 + z^2 + 60xy + 14yz + 8zx \geq 0$ for all (real) x, y, z?

(b) Express xy and $xy + yz + zt + tx$ as sums and differences of squares.

38. Let $T:\mathbb{R}^3 \to \mathbb{R}^3$ be given by

$$T\begin{bmatrix} x \\ y \\ z \end{bmatrix} = \begin{bmatrix} -17x-6y+12z \\ -35x-8y+22z \\ -40x-12y+27z \end{bmatrix}.$$

Find the matrix A, say, corresponding to T with respect to the standard basis for $\mathbb{R}^3$ and with respect to a basis of $\mathbb{R}^3$ comprising three eigenvectors of A.

COMPUTER PROBLEMS (THEORY)

1. (a) Let A be the matrix

$$\begin{bmatrix} 0 & 0 & 0 & h \\ 1 & 0 & 0 & 0 \\ 0 & 1 & 0 & 0 \\ 0 & 0 & 1 & 0 \end{bmatrix}$$

where $h = \frac{1}{16}$. Show that

$$A^4 = \begin{bmatrix} h & 0 & 0 & 0 \\ 0 & h & 0 & 0 \\ 0 & 0 & h & 0 \\ 0 & 0 & 0 & h \end{bmatrix}.$$

Deduce that $\frac{1}{2}$ is an eigenvalue for A.

(b) Find all the eigenvalues of the 100×100 matrices $[a_{ij}]$ and $[b_{ij}]$ where $a_{ii} = 1$ if $2 \le i \le 100$ and $a_{11} = 1 + (1/2)^{100}$; $b_{i+1,i} = 1$ and $b_{1,100} = (1/2)^{100}$ and all other entries are zero. [The point of this example is to show that the eigenvalues of a matrix may – or may not – change quite substantially even when one entry in a matrix is altered by an amount which the computer would interpret as 0.]

2. Let $\{\mathbf{v}_1, \mathbf{v}_2, \ldots, \mathbf{v}_n\}$ be a basis of eigenvectors for the $n \times n$ matrix A. For any vector $\mathbf{v} \in \mathbb{R}^n$ express $\mathbf{v} = \alpha_1\mathbf{v}_1 + \alpha_2\mathbf{v}_2 + \ldots + \alpha_n\mathbf{v}_n$. Deduce that, for all positive integers s,

$$A^s\mathbf{v} = \alpha_1\lambda_1^s\mathbf{v}_1 + \alpha_2\lambda_2^s\mathbf{v}_2 + \ldots + \alpha_n\lambda_n^s\mathbf{v}_n.$$

[If $|\lambda_1| > |\lambda_i|$ for $2 \le i \le n$, one sees that, for large s, $A^s\mathbf{v}$ approximates to $\alpha_1\lambda_1^s\mathbf{v}_1$ Since one knows A and $\mathbf{v}$ one can hope to compute $\mathbf{v}_1$ – and then λ_1 – since $\mathbf{v}_1$ and $A^s\mathbf{v}$ will be almost parallel. Variations permit the evaluation of the other eigenvalues.

COMPUTER PACKAGE PROBLEMS

1. Find the characteristic polynomial of the matrix

$$A = \begin{bmatrix} 1 & 2 & 3 \\ 8 & 9 & 4 \\ 7 & 6 & 5 \end{bmatrix}.$$

Hence find the eigenvalues of A by solving the characteristic equation. Alternatively, find these eigenvalues directly by using (with MAPLE) 'eigenvals(A)'. [The point of this is that (in MAPLE) the eigenvalues are given here in two completely different formats.]

2. Find the characteristic polynomial of the matrix

$$A = \begin{bmatrix} 1 & 0 & 0 & 0 & 2 \\ 0 & 3 & 0 & 4 & 0 \\ 0 & 0 & 5 & 0 & 0 \\ 0 & 6 & 0 & 7 & 0 \\ 8 & 0 & 0 & 0 & 9 \end{bmatrix}.$$

See if you can spot its roots. [You should be able to see one!] If not, use your computer package to solve the characteristic equation. [I got answers involving square roots rather than in decimal form.]

3. Find the eigenvalues and the eigenvectors of the following matrices:

$$A = \begin{bmatrix} 1 & 2 & 3 & 4 & 5 \\ 2 & 1 & 2 & 3 & 4 \\ 3 & 2 & 1 & 2 & 3 \\ 4 & 3 & 2 & 1 & 2 \\ 5 & 4 & 3 & 2 & 1 \end{bmatrix} \text{ and 'Pascal's matrix' } B = \begin{bmatrix} 1 & 1 & 1 & 1 & 1 \\ 1 & 2 & 3 & 4 & 5 \\ 1 & 3 & 6 & 10 & 15 \\ 1 & 4 & 10 & 20 & 35 \\ 1 & 5 & 15 & 35 & 70 \end{bmatrix}.$$

4. Find the eigenvalues and corresponding eigenvectors of the matrix

$$A = \begin{bmatrix} -121 & 384 & 92 & -32 & -100 \\ 156 & -1085 & -508 & 236 & 180 \\ -48 & -840 & -613 & 312 & 48 \\ 234 & -3900 & -2286 & 1115 & 462 \\ 738 & -4212 & -1918 & 868 & 745 \end{bmatrix}.$$

Is A diagonalisable?

5. Check that the eigenvalues and corresponding eigenvectors of the matrix

$$\begin{bmatrix} 3 & 5 & -7 & 2 \\ 8 & -5 & 8 & 7 \\ 6 & -6 & 9 & 5 \\ 0 & -6 & 8 & 1 \end{bmatrix}$$

of Example 8 are as claimed.

6. Find the eigenvalues and eigenvectors of the symmetric matrix

$$\begin{bmatrix} 1 & 0 & 0 & 0 & 1 \\ 0 & 1 & 0 & 1 & 0 \\ 0 & 0 & 1 & 0 & 0 \\ 0 & 1 & 0 & 1 & 0 \\ 1 & 0 & 0 & 0 & 1 \end{bmatrix}.$$

Hence find a matrix P which diagonalises the given matrix orthogonally.

7. (a) Show that it is possible to change the x–y–z axes to new rectangular axes X–Y–Z so that the equation

$$859x^2 + 769y^2 + 1163/2z^2 - 656xy + 454yz - 202zx = 1$$

can be expressed in the form $X^2/\alpha^2 + Y^2/\beta^2 + Z^2/\gamma^2 = c$. (The equation therefore represents a . . . what?)

(b) Do the same for the equation

$$0.521x^2 + 0.271y^2 + 0.568z^2 + 0.458xy + 0.272yz + 0.728zx = 50.$$

8. Find the Jordan forms of the matrices

$$A = \begin{bmatrix} -7 & -16 & 13 & -5 \\ 2 & -5 & -2 & 0 \\ -13 & 7 & 18 & -4 \\ -25 & 67 & 26 & 0 \end{bmatrix} \quad \text{and} \quad B = \begin{bmatrix} -10 & 33 & -15 & -13 \\ 11 & -43 & 19 & 17 \\ 40 & -121 & 56 & 48 \\ -12 & 7 & -6 & -3 \end{bmatrix}.$$

9. Write down some 3×3 matrix (A, say) – with integer coefficients for convenience. Find Det$(A - \lambda I)$ in the form $c_3\lambda^3 + c_2\lambda^2 + c_1\lambda + c_0$. Now evaluate the (3×3) matrix $c_3A^3 + c_2A^2 + c_1A + c_0I_3$. Do the same for several more 3×3 – and 4×4 (and 5×5 . . . etc.) matrices. The results you should get should allow you to conjecture the result known as the Cayley–Hamilton theorem. (Cf. Exercise 11 of Chapter 3.)

10. (a) Find the eigenvalues of the 'nearly equal' matrices

$$A = \begin{bmatrix} -5.1 & 21.6 & 15.2 & -6.4 \\ 0.5 & 2.2 & 1.5 & -0.5 \\ 17.9 & -73.3 & -51.4 & 21.7 \\ 48.9 & -185.3 & -130.1 & 55.4 \end{bmatrix}$$

and

$$B = \begin{bmatrix} -5.0 & 21.7 & 15.3 & -6.3 \\ 0.6 & 2.3 & 1.6 & -0.4 \\ 18.0 & -73.2 & -51.3 & 21.8 \\ 49.0 & -185.2 & -130.0 & 55.5 \end{bmatrix} = A + (0.1)F \quad \text{where } F = \begin{bmatrix} 1 & 1 & 1 & 1 \\ 1 & 1 & 1 & 1 \\ 1 & 1 & 1 & 1 \\ 1 & 1 & 1 & 1 \end{bmatrix}.$$

(b) Now find, for each t, $1 < t < 10$, the eigenvalues of $A + (0.01)tF$ in order to see how the eigenvalues change in passing from A to B.

[The eigenvalues of a matrix change continuously with the elements of the matrix.]

11. (a) Write the quadratic form

$$f(x, y, z, t) = 211x^2 + 172y^2 + 90z^2 + 175t^2 + 4xy + 204xz - 224xt + 48yz - 148yt + 228zt$$

in the form XAX^T where $X = [x, y, z, t]$ and A is a symmetric 4×4 matrix. Find the eigenvalues of A and write $f(x, y, z, t)$ in the form $\alpha X^2 + \beta Y^2 + \gamma Z^2 + \delta T^2$ for suitable functions X, Y, Z, T of x, y, z, t. Determine whether or not $f(x, y, z, t)$ can take negative values for suitable values of x, y, z, t.

(b) Do the same for the quadratic form

$$4.728x^2 + 2.665y^2 + 2.590z^2 + 4.177t^2 + 0.796xy + 3.748xz - 3.546xt + 0.326yz - 3.240yt + 3.586zt.$$

Brief Appendix – on sets, functions and proofs

One of the most important concepts in mathematics is that of *function*. Because of this importance and because, two centuries ago there was some disagreement as to exactly what constituted a function, it was subsequently shown how the concept could be defined entirely in terms of the basic mathematical concept of 'set'. There is no need for us to be quite so pedantic here and we shall adopt a sort of 'half way house' definition which still employs the 'set' notion and is essentially that which was good enough for Dirichlet in the 1820s. So we begin with a paraphrase of Cantor's definition (1895) of set.

• Sets

By a **set** we understand any collection M (say) of definite distinct objects (the objects being called the **elements** or **members** of the set) gathered together into a single entity. ●

Thus examples of sets include the sets of whole numbers (also called integers), rational numbers, real numbers and complex numbers. These sets are denoted, respectively, by the symbols $\mathbb{Z}$, $\mathbb{Q}$, $\mathbb{R}$ and $\mathbb{C}$.

To indicate that the object m is (is not) an element of the set M we write $m \in M$ ($m \notin M$). Thus $\frac{1}{3} \notin \mathbb{Z}$, $\frac{3}{7} \in \mathbb{Q}$, $\sqrt{2} \notin \mathbb{Q}$, $\sqrt{6} \in \mathbb{R}$, $-2.8\pi + 3.17i \in \mathbb{C}$.

TUTORIAL PROBLEM I

> If a set is defined as a collection – what is a collection?

Some sets can be described by listing all their elements between brackets {also called braces}. For example $S = \{1, 9, 25\}$ is the set of all squares, which are less than 41, of odd integers. Of course $\mathbb{Z}$, $\mathbb{Q}$, $\mathbb{R}$ and $\mathbb{C}$ cannot be so listed. However some infinite sets can be defined by the properties their elements possess. For example $T = \{x : x \in \mathbb{Z}\ \&\ x^3 > 4441\}$ – which we read as 'the set of (all those) x such that x belongs to $\mathbb{Z}$ and x cubed is greater than 4441' – is just a silly way of describing the set of all integers greater than 16. Hence $T = \{x : x \in \mathbb{Z}\ \&\ x > 16\} = \{x : x \in \mathbb{Z}\ \&\ x \geq 17\}$ etc. This example suggests we define two sets A and B to be **equal** when and only when they contain exactly the same elements. Note that S, above, may also be described by $S = \{x : x \in \mathbb{Z}\ \&\ x = 1 \text{ or } x = 9 \text{ or } x = 25\}$. In this book we are much concerned with the set $\mathbb{R}^3 = \{(x, y, z) : x, y, z \in \mathbb{R}\}$ of all triples of real numbers and similar sets of n-tuples where $n = 2, 4, 5, \ldots$ etc. (Note that, in $\mathbb{R}^3$ for example, the six triples $(1, -1, 5)$, $(5, 1, -1) \ldots (1, 5, -1)$ are distinct since, as with coordinates of a point, the ordering of the numbers $1, -1, 5$ is relevant.) We shall also come across the set with no elements – for example: the set of all (living) human beings who are more than 1000 years old or the set of all negatives integers which are squares of real numbers. We denote this set, the **empty set**, by the symbol $\varnothing$.

We shall occasionally need to operate with pairs of sets. Let sets A and B be given. We say A is **contained in** B (written $A \subseteq B$ or $B \supseteq A$) if each element of A is also an element of B. If, in addition, there are elements in B which are not in A we may write, instead, $A \subset B$ or $B \supset A$, if we wish to draw attention to the fact. We then describe A as a **subset** (**proper subset**) of A (respectively). The set of all elements of A which are not members of B is denoted by $A \setminus B$. The **intersection** $A \cap B$ of A and B is (by definition) the set $\{x{:}x \in A \;\&\; x \in B\}$ of all elements belonging to both A and B, their **union**, $A \cup B$, is the set $\{x{:}x \in A \text{ or } x \in B\}$ comprising the elements which are in A or in B (or, perhaps, in both).[a]

As examples: (i) $\mathbb{Z} \subseteq \mathbb{Q}$, $\mathbb{Q} \subseteq \mathbb{R}$, $\mathbb{R} \subseteq \mathbb{C}$ (or, more pedantically, $\mathbb{Z} \subset \mathbb{Q} \subset \mathbb{R} \subset \mathbb{C}$); (ii) $\mathbb{Z} \cap \{x{:}x \in \mathbb{R} \;\&\; -2.1 < x < 3.07\} = \{-2, -1, 0, 1, 2, 3\}$ $(= \{x{:}x \in \mathbb{Z} \;\&\; -2 \le x \le 3\})$; (iii) $\{x{:}x = a^2 + b^2 + c^2 \;\&\; a, b, c \in \mathbb{Z}\} \cup \{x{:}x = 4^m(8n + 7) \;\&\; m, n \in \mathbb{Z} \;\&\; m, n, \ge 0\} = \{x{:}x \in \mathbb{Z} \;\&\; x \ge 0\}$ (since each integer $x \ge 0$ is either a sum of three integer squares or of the form $4^m(8n + 7)$); (iv) For all sets A, B; $A \setminus B = A \setminus (A \cap B)$.

One may equally well consider the intersection and union of more than two sets – even of infinitely many.

• Functions

A definition of function can be given solely in terms of the set concept. We give a somewhat less precise definition.

Given sets A and B a **function** (also called **map(ping)** or **transformation**) f, say, from A to B is any rule which associates, with each element a of A some (uniquely determined) element, denoted by $f(a)$, of B. To express this association we may write $f{:}A \to B$. ●

Some examples are:

(I) $A = \mathbb{R}$, $B = \mathbb{R}$: $f(a) = |a|$ (the *modulus* or *absolute value* of a). [Thus, for example, $f(-3.21) = f(3.21)$ but there is no $a \in \mathbb{R}$ for which $f(a) = -1$.]

(II) $A = \mathbb{Q}$, $B = \mathbb{R}$: $f(a) = a^3$ [Here, if $f(a) = f(b)$ then $a = b$. Note that there is no a (in $\mathbb{Q}$) for which $f(a) = 2$.]

(III) [b]$A = \{\text{so, la}\}$, $B = \{365, 375\}$: $f(\text{so}) = 365$, $f(\text{la}) = 375$.

(IV) $A = \mathbb{R}$, $B = \mathbb{Z}$: $f(x) = [x]$ (meaning the greatest integer not exceeding x). Then $f(\pi) = 3$ whilst $f(-\pi) = -4$.

TUTORIAL PROBLEM 2

If a function is defined as a rule – what is a rule?

Given $f{:}A \to B$, we shall call the subset $\{f(a){:}a \in A\}$ of B the **image** or **range** of f, denoting it by $\text{im}(f)$. If the equality $f(a) = f(b)$ always implies that $a = b$ we say that f is

[a]In mathematics 'or' includes the possibility of both conditions being satisfied. (Host: 'Would you like white wine or red?' Mathematician: 'Yes, please' [Thinks: 'If the host is a mathematician I might just get a glass of each!']

[b]I presume no cricket buffs need any explanation of the set A??!

one to one (1 – 1). If im(f) is the whole of B we say f is **onto** B. Thus, above, (I) is neither 1–1 nor onto, (II) is 1–1 but not onto, (III) is both 1–1 and onto, (IV) is onto but not 1–1. {Note that an onto function f:$A \to B$ may cease to remain onto if B is changed.}

If f:$A \to B$ is 1–1 and onto it is possible to define an **inverse function** g of f which 'undoes' f. In fact g:$B \to A$ is given by: $g(b) = a$ if and only if $f(a) = b$.

Let S be any set and suppose that T is a set whose elements can (i) be added together and (ii) be multiplied by real (or complex) numbers. (For example T might be $\mathbb{R}$ or $\mathbb{C}$ or, maybe, a real (or complex) vector space.) Then for any two functions f:$S \to T$. and g:$S \to T$. and for any real (or complex) number α we can define the **sum** function $f \oplus g$ and the **scalar multiple** function αf by: for all $s \in S$, $(f \oplus g)(s) = f(s) + g(s)$ and $(\alpha f)(s) = \alpha.f(s)$. Thus the **set of all functions from S to T** is now a set whose elements can be added in pairs and multiplied by scalars.

Finally, in this section, a remark about polynomials. Suppose that $p(x) = a_0 + a_1x + a_2x^2 + \ldots + a_n x^n$ is a polynomial of degree n with real coefficients a_i. The **factor theorem** states that if t is a number such that $p(t) = 0$ then $(x - t)$ is a factor of $p(x)$, i.e. $p(x) = (x - t)q(x)$ where $q(x)$ is a polynomial of degree $n - 1$. Such a t is called a **root** (or **zero**) of $p(x)$. Note that the real polynomial $x^2 + 1$ {$= (x - i)(x + i)$ where $i = \sqrt{-1}$} needs the complex numbers if it is to be factorised. So it is remarkable that, if we replace the real a_i in $p(x)$ by complex numbers, we still only need complex numbers (and not any kind of super-complex numbers) to factorise it. This is the famous **Fundamental theorem of Algebra** (Gauss). Let $p(x)$ be as above but with complex coefficients a_i. Then $p(x) = a_n(x - c_1)(x - c_2) \ldots (x - c_n)$ factorises completely into a product of n linear factors where the $c_j \in \mathbb{C}$.

One immediate consequence of this is: If, in $p(x)$, the $a_i \in \mathbb{R}$ then the complex roots of $p(x)$ occur in pairs $u + iv$, $u - iv$ of complex conjugates.

PROOFS

Consider the following statements concerning numbers:

(i) Every positive integer is expressible as a sum of no more than 8 cubes of positive integers;

(ii) for every positive integer n,

$$1.2 + 2.3 + \ldots + n.(n+1) = \frac{n(n+1)(n+2)}{3};$$

(iii) $\sqrt{2}$ is not a rational number.

Do you believe any of these statements to be true? In fact (i) is false! To show that a general statement is false one only needs to give *a single instance* of its failure. Thus I can show (i) is false by saying '(i) is false because 23 is a counterexample' (i.e. 23 is not expressible as a sum of only 8 cubes of positive integers – as you may check quickly by hand. Curiously enough only one other integer fails to satisfy (i). You could get your computer to help you search – if I tell you it is less than 1000.) So, infinitely many integers do satisfy statement (i). This is not easy to prove. But some statements concerning infinitely many integers can be proved fairly readily (and not by examining each of the infinitely many integers in turn!!!!) Of course checking the validity of a statement for the first 50 million integers is also of no use here. The first integer for which the statement fails may be lurking just round the corner!! (See Exercise 2.) We use instead the method of mathematical induction which is stated as follows.

The principle of mathematic induction Suppose $S(n)$ is, for each positive integer n, a statement involving the number n. (For example we might have $S(n)$: $1 + \frac{1}{2} + \frac{1}{3} + \frac{1}{4} + \ldots + \frac{1}{n} < 10$.) *IF* (i) $S(1)$ is a true statement and *IF* (ii) for each integer k, the truth of $S(k + 1)$ can be deduced from the assumption of the truth of $S(k)$ *THEN* $S(n)$ is true for all integers n.

We won't explain here why this works. Rather we'll prove (ii) above is true by this means. So, we claim: For each positive integer n, the statement $S(n)$:

$$1.2+2.3+\ldots+n.(n+1)=\frac{n(n+1)(n+2)}{3},$$

is true.

Proof of claim. For $n = 1$ the claim would read

$$1.2=\frac{1(1+1)(1+2)}{3}.$$

Since this is indeed true we see that the statement $S(1)$ is true.

Now suppose the claim has been verified for $n = k$. That is, we assume $S(k)$ is true. In other words we are supposing

$$1.2+2.3+\ldots+k.(k+1)=\frac{k(k+1)(k+2)}{3}.$$

To prove $S(k + 1)$ is true we must look at $1.2 + 2.3 + \ldots + k.(k + 1) + (k + 1)(k + 2)$. By our assumption about $S(k)$ this sum is

$$\frac{k(k+1)(k+2)}{3}+(k+1)(k+2)$$

which is equal to

$$(k+1)(k+2)\left\{\frac{k}{3}+1\right\}=\frac{(k+1)(k+2)(k+3)}{3}=\frac{\{k+1\}(\{k+1\}+1)(\{k+1\}+2)}{3},$$

which is the right hand side of $S(k + 1)$. We have, then, by assuming $S(k)$ is true shown that $S(k + 1)$ is true. The principle of mathematical induction tells us that the statement $S(n)$ is true for every integer n. [If you feel that a proof by induction can only succeed if you 'know the answer in advance' or that 'any formula can be proved by induction' see Exercise 2 below and Exercise 11 of Chapter 1.]

Now consider statement (iii) above. It is true. How do I convince you? One ploy is this. We assume $\sqrt{2}$ is a rational number. If, then, logical reasoning throws up a blatant contradiction we shall be forced to conclude that our assumption is wrong and hence accept that $\sqrt{2}$ is *not* a rational number.

[The following proof is not to be taken as a model of argument. It is extra verbose in an attempt to help you through it. Perhaps, as an exercise, you can rewrite it retaining the mathematics and omitting the extraneous waffle.]

Claim: $\sqrt{2} \notin \mathbb{Q}$. Proof. Assume $\sqrt{2} \in \mathbb{Q}$. This means that there exist integers a, b such that $\sqrt{2} = a/b$. Now if a, b have any common divisor greater than 1 let us cancel it now and write a/b in 'lowest terms' r/s $(= \sqrt{2})$. Squaring up gives $2 = r^2/s^2$, that is $2s^2 = r^2$. This equality implies that r^2 is even – and hence that r itself is even (since, if r were odd so would r^2 be). Consequently r is of the form $2m$ for some integer m. It follows that $2s^2 = r^2 = (2m)^2$ and, on cancelling a factor 2, that $s^2 = 2m^2$. But this means that s^2, and then s itself, is even. But this is absurd – we (quite legitimately) took r and s to have no common divisor greater than 1 and we've arrived at the conclusion that each is even!!! This blatant contradiction tells us something is wrong! If you examine the logic of our argument you should see that the only possible problem is with our assumption, namely that $\sqrt{2} \in \mathbb{Q}$. Hence this assumption must be wrong and it must be the case that $\sqrt{2} \notin \mathbb{Q}$ as claimed. This method of proof is called **reductio ad absurdum** (I wonder why?!!)

Finally we comment on the expression 'if and only if' which you will find occasionally in the book. Suppose that from a statement A you can deduce a statement B (Example[c]: A is: 'Tom is a university student'; B is: 'Tom is more than two years old') We write this briefly as $A \Rightarrow B$ (or, even, $B \Leftarrow A$) and we say 'If A [holds] then B [holds]' or 'B [holds] if A [holds]'. We also use the curious expression 'A [holds] only if B [holds]'. Thus 'C if and only if D' means that $C \Leftarrow D$ and $C \Rightarrow D$ – in brief $C \Leftrightarrow D$ – meaning that each of C and D implies the other. In other words C and D are equivalent statements. Being longwinded 'if and only if' is abbreviated to **iff**. (In the days when secretaries did all a mathematician's typing 'iff' would often be returned corrected to 'if' with, presumably, some doubt in the secretary's mind concerning the general intelligence of unworldly mathematicians' . . .)

EXERCISES

1. Prove, by mathematical induction:

$$\text{(a)} \sum_{k=1}^{n} k = \frac{k(k+1)}{2}; \qquad \text{(b)} \sum_{k=1}^{n} k^2 = \frac{k(k+1)(2k+1)}{6}.$$

2. Evaluate $\dfrac{n(n-1)(n^2-5n+18)}{24}$ for $n = 1, 2, 3, 4, 5$. Use intuition to make a conjecture involving 2^{n-1}. Try to prove the conjecture by induction.

3. Follow the proof that $\sqrt{2} \notin \mathbb{Q}$ as closely as you can to try to prove $\sqrt{3} \notin \mathbb{Q}$. Identify where the proof breaks down if you try to prove, similarly, that $\sqrt{4} \notin \mathbb{Q}$.

[c] As things are going in mathematics, the 'two' in statement B might soon need reducing!!

Bibliography

For the subject of Linear Algebra it would be possible to compile a useful bibliography of very great length. However, the few books listed below should be sufficient to enable the reader to find more details of those results, theories and ideas mentioned, but not expanded on, in the text.

Allenby, R.B.J.T. *Rings, Fields and Groups*, 2nd ed., Edward Arnold, London, 1991.

Allenby, R.B.J.T. and Redfern E.J. *Introduction to Number Theory with Computing*, Edward Arnold, London, 1989.

Bell, E.T. *Men of Mathematics*, Simon and Schuster, New York, 1962.

Ciarlet, P.G. *Introduction to Numerical Linear Algebra and Optimization*, Cambridge University Press, 1989.

Finkbeiner, D.T. *Matrices and Linear Transformations*, Freeman, San Francisco, 1966.

Fraleigh, J.B. and Beauregard, R.A. *Li near Algebra*, 2nd ed., Addison Wesley, Reading, Massachusetts, 1990.

Gass, S.I. *Linear Programming; Methods and Applications*, McGraw-Hill, New York, 1975.

Grossman S.I. *Elementary Linear Algebra*, 4th ed., Saunders College Publishing, Philadelphia, 1991.

Hager, W.W. *Applied Numerical Linear Algebra*, Prentice Hall, Englewood Cliffs, New Jersey, 1988.

Hawkins, T. *The Theory of Matrices in the 19th Century*, Proceedings of the International Conference of Mathematicians, Vancouver, 1970.

Hirst, A.E. *Vectors in Two and Three Dimensions*, Edward Arnold, London, (in this series) 1995.

Jeffrey, A. *Linear Algebra and Ordinary Differential Equations*, Blackwell Scientific Publications, Boston, 1990.

Jordan, C. and Jordan D. *Groups*, Edward Arnold, London, (in this series) 1994.

Kline, M. *Mathematical Thought from Ancient to Modern Times*, Oxford University Press, New York, 1972.

Roberts, F.S. *Applied Combinatorics*, Prentice Hall, Englewood Cliffs, New Jersey, 1984.

Strang, G. *Linear Algebra and its Applications*, 2nd ed. Academic Press, New York, 1980.

For help with the computer package problems you might use:

Johnson, E.W. *Linear Algebra with Maple V*, Brooks/Cole Publishing Company, Pacific Grove, California, 1993.

Index